Ion and Enzyme Electrodes
in Biology and Medicine

Ion and Enzyme Electrodes in Biology and Medicine

International Workshop at Schloß Reisensburg near Ulm.
Sponsored by the Deutsche Forschungsgemeinschaft
and the Max-Planck-Gesellschaft
zur Förderung der Wissenschaften

Edited by
Manfred Kessler, Dortmund, Germany
Leland C. Clark Jr., Cincinnati, USA
Dietrich W. Lübbers, Dortmund, Germany
Ian A. Silver, Bristol, England
Wilhelm Simon, Zürich, Switzerland

251 Figures

University Park Press
Baltimore · London · Tokyo

This workshop was supported by

Bayer A.G., Wuppertal-Elberfeld

C.F. Boehringer & Söhne, Mannheim

Colora Meßtechnik, Lorch

Siemens A.G., Dortmund

Deutsche Forschungsgemeinschaft, Bonn-Bad Godesberg

Max-Planck-Gesellschaft zur Förderung der Wissenschaften e.V., München

Published in USA by University Park Press Inc, International Publishers in Science and Medicine, Chamber of Commerce Building, Baltimore, Maryland, USA

Copyright © Urban & Schwarzenberg, München-Berlin-Wien 1976 (ISBN 3-541-07461-2)

All Rights reserved. This book, or parts thereof, must not be reproduced without permission.

Library of Congress Cataloging in Publication Data

Main entry under title:

Ion and enzyme electrodes in biology and medicine.

Sponsored by the Deutsche Forschungsgemeinschaft and the Max-Planck-Gesellschaft zur Förderung der Wissenschaften.
1. Electrodes, Ion selective--Congresses. 2. Electrodes, Enzyme--Congresses. 3. Biological transport--Congresses. I. Kessler, Manfred. II. Deutsche Forschungsgemeinschaft (Founded 1949). III. Max-Planck-Gesellschaft zur Förderung der Wissenschaften.

QP519.9.E43I66   574.1'92'028   76-5470
ISBN 0-8391-0847-8

Printed in Germany.

Preface

At present we are witnessing a fascinating period of research work in biology and in medicine which was made possible by the development of new types of ion-selective and enzyme electrodes. These new electrodes which created the opportunity to monitor directly and continuously a variety of biological signals of cells and of tissues, opened a huge field of investigations, the ultimate limits of which are still unknown. In 1972 the authors decided that the development had reached a point where an interdisciplinary meeting in the field of ion and enzyme measurement would be meaningful. Accordingly, a workshop was initiated and organized with the aim to bring together the researchers of the many different fields related to the problem of monitoring biological signals with the new sensors. Thanks to Dr. *Degkwitz*, Urban & Schwarzenberg agreed to publish the proceedings of the International Workshop.

The preliminary editorial work for this book was done already at the time of the meetings by Prof. *Ian Silver* who took over the difficult and time-consuming task of making first corrections of the discussions and, quite frequently, of sentences with more or less mystic content. Dr. *David Briggs* has spent much time fine tuning terminology and idioms. His linguistic skills and experience in reading scientific texts greatly assisted in editing this book. Mrs. *Gerlinde Blümel* proved to be a true „good spirit" of our workshop and of this book. For her dedicated involvement as official secretary as well as for her invaluable assistance during the preparation of this book, we owe her our appreciation and thanks. We also wish to thank Miss *Valerie Jeal*, Miss *Thelma Norman* and Mrs. *Elke Runte* for their quick and able transcription of the discussion and for the final typing. We also acknowledge the assistance of Dr. *Bärbel Krumme*, Dr. *David Briggs*, Mr. *Friedrich Keinemann*, Mr. *Reinhard Strehlau* and Mr. *Walter Thutewohl* during the meetings. Mr. *Krüger* and his staff created an atmosphere of hospitality at Schloss Reisensburg that made us feeling like members of the „Schloss Family".

The sponsorship of the Deutsche Forschungsgemeinschaft and of the Max-Planck-Gesellschaft made it possible to organize the „International Workshop at Schloss Reisensburg". In addition we received financial support from several firms. We sincerely hope that the fruitful scientific exchange between the participants as well as the effort of the authors in preparing, presenting, and writing their papers justify the generous help of our sponsors.

A special note of gratitude goes to Prof. *Müller*, Mr. *Mees* and the staff of Urban & Schwarzenberg for their excellent collaboration in preparing this book.

Dortmund, April 1976

Manfred Kessler

Contents

SESSION I: Theoretical and Practical Aspects of Ion-Selective Electrodes. Chairman: *W. Simon* 1

Further Studies on Ion Selectivity. *G. Eisenman, S. Krasne, S. Ciani* 3

Synthetic Neutral Carriers for Cations. *D. Amman, R. Bissig, Z. Cimerman, U. Fiedler, M. Güggi, W. E. Morf, M. Oehme, H. Osswald, E. Pretsch, W. Simon* 22

Application of Solid State Principles to Response Interpretation and Design of Solid Membrane Electrodes. *R. Buck* 38

Theoretical Treatment of the Dynamic Response of Ion-Sensitive Membrane Electrodes. *W. E. Morf, W. Simon* 51

Compensation of Measuring Errors Produced by Finite Response Time with Ion-Selective Electrodes. *R. Wodick* 54

Some Aspects of the Application of Ion-Selective Electrodes in Flowing Systems. *E. Pungor, K. Tóth, G. Nagy* 56

Investigation of pH Electrode Membrane Glasses by Means of an Ion Sputtering Method — Discussion of Microelectrode Membranes. *F. Baucke* 77

Discussion of Session I 85

SESSION II: Microelectrodes. Chairman: *I. A. Silver* 101

The RF Sputtering Technique as a Method for Manufacturing Needle-Shaped pH Microelectrodes. *Y. Saito, H. Baumgärtl, D. W. Lübbers* 103

Insulation Properties and Ion-Sensitivity of Thin Layers of Dielectrics Deposited by RF Sputtering on Glass Surfaces. *D. W. Lübbers, H. Baumgärtel, Y. Saito* 110

Ion-Selective Liquid Ion Exchanger Microelectrodes. *J. L. Walker* 116

Multi-Parameter Electrodes. *L. A. Silver* 119

Microelectrodes Utilizing Glass and Liquid Ion-Exchanger Sensors. *R. N. Khuri* 123

The Art of Making Small Glass Electrodes. *N. C. Hebert, R. R. Deleault* 131

Four-Barrelled Microelectrode for the Measurement of Potassium-, Sodium- and Calcium-Ion Activity. *M. Kessler, K. Hájek, W. Simon* 136

Construction and Properties of Recessed-Tip Micro-Electrodes for Sodium and Chloride Ions and pH. *R. C. Thomas* 141

Discussion of Session II 149

SESSION III: Enzyme Electrodes. Chairman: *L. C. Clark Jr.* 159

The Measurement of Cholesterol and its Esters Using a Polarographic Anodic Enzyme Electrode. *L. C. Clark Jr., C. R. Emory* 161

Amperometric Methods of Enzyme Assay. *G. D. Christian* 173

Immobilized Enzymes and Ligands in Biological Research. *W. W. C. Chan, P. Davies, K. Mosbach* 182

An Ultra Micro Glucose Electrode. *I. A. Silver* 189

An Automated Kinetic Analysis of Cholinesterase Activity by a Substrate-Selective Ion-Exchange Electrode. *G. Baum* 193

An Enzyme Electrode Based on Immobilized Glucose Oxidase. *S. C. Martiny, O. J. Jensen* 198

Reference Electrodes for Measurements with Ion-Sensitive Electrodes. The Importance of the Liquid Junction Potential. *F. Baucke* 200

Cytological Aspects of Membrane Regeneration After Experimental Injury of Amebae and Acellular Slime Molds. *R. Olfers-Weber, W. Stockem, K. E. Wohlfarth-Bottermann* 205

Alterations in Rat Liver Cells and Tissue Caused by Needle Electrodes. *D. Schäfer, J. Höper* 217

Discussion of Session III 223

SESSION IV A: Basic Effects of Ion Activity in Metabolic Reaction and in Membrane Functions Including Tissue Measurements. Chairman: *A. Kovách.* 235

The Role of Ions in the Control of Intermediary Metabolism. *T. Clausen* 237

Physical and Biological Properties of Ionophores. *B. Pressman* 251

Metallochromic Indicator for Kinetic Measurements of Magnesium Ions and Determination of Cytosolic-Free Magnesium Ions by Microspectrophotometry. *A. Scarpa* 252

Ion-Selective Electrodes in the Study of Metabolic Steady States of Potassium and Ammonium Ions in Isolated Perfused Rat Liver. *H. Sies* 261

A Mathematical Analysis of Simultaneous Transport Phenomena in the Microcirculation. *D. D. Reneau* 268

Discussion of Session IV A 274

SESSION IV B: Basic Effects of Ion Activity in Metabolic Reactions and in Membrane Function Including Tissue Measurements. Chairman: *D. W. Lübbers* 281

Resting Ionic Permeabilities in a Neuron. *A. M. Brown, J. M. Russel, D. C. Eaton* 283

The Effects of CO_2 and Bicarbonate on the Intracellular pH of Snail Neurones. *R. C. Thomas* 288

Brain Metabolism and Ion Movements in the Brain Cortex of the Rat During Anoxia. *E. Dora, T. Zeuthen* 294

Measurements of Extracellular Potassium, D. C. Potential and ECoG in the Cortex of the Conscious Rat During Cortical Spreading Depression. *W. Crowe, A. Mayevsky, L. Mela, I. A. Silver* 299

Simultaneous Measurements of Extracellular Potassium-Ion Activity and Membrane Currents in Snail Neurones. *H. D. Lux* 302

Cortical Spreading Depression in Relation to Potassium Activity, Oxygen Tension, Local Flow and Carbon Dioxide Tension. *A. Lehmenkühler, E.-J. Speckmann, H. Caspers* 311

Pial Vascular Smooth Muscle Reactions During Perivascular Microperfusion with Artificial CSF of Varying Ionic Composition. *E. Betz* 316

Measurement of Potassium Ion and Hydrogen Ion Activities by Means of Microelectrodes in Brain Vascular Smooth Muscle. *D. Heuser, E. Betz* 320

Measurements with Ion-Selective Surface Electrodes (pK, pNa, pCa, pH) During No-Flow Anoxia. *J. Höper, M. Kessler, W. Simon* 331

Contents

Disturbance and Compensation of Cellular Cation Activity During Anoxia and in Shock. M. Kessler, J. Höper, B. Krumme, H. Starlinger 335

Mitochondrial Calcium Transport Activity Following Acute and Chronic Changes of Tissue Oxygenation. L. Mela, C. W. Goodwin, L. D. Miller 341

Electrical Potentials and Ion Activities in the Epithelial Cell Layer of the Rabbit Ileum in vivo. T. Zeuthen, C. Monge 345

Intracellular Chloride Activity in Heart Muscle. J. L. Walker, R. O. Ladle 351

Use of the Sodium Microelectrode to Define Sodium Efflux and the Behavior of the Sodium Pump in the Frog Sartorius. J. F. White, J. A. M. Hinke 355

Intracellular Potassium in Single Cells of Renal Tubules. R. N. Khuri 364

The Interstitial pH of the Isolated Skeletal Muscle of the Dog at Rest and During Exercise. Hj. Hirche, C. Steinhagen, I. Hosselmann, J. Manthey, U. Bovenkamp 372

Tissue Electrolytes and Blood Flow in Skeletal Muscle. G. Gebert 376

Recent Advances in Areas of Clinical Utility and Instrumentation for the Determination of Ionized Calcium. L. Coleman 381

Ion-Selective Electrodes in Clinical Medicine. C. Fuchs 389

Measurements with Surface Electrodes of Interstitial Ion Activities and Local Tissue PO_2 on the Human Kidney (Preliminary Report). E. Sinagowitz, R. Hagist, H. Sommerkamp . . . 398

Discussion of Session IV B 400

Index 420

List of Participants

Ammann, Daniel, Organisch-chemisches Laboratorium der ETH Zürich, CH-8006 Zürich Universitätsstr. 6/8
Baucke, Friedrich, Dr., Jenaer Glaswerk Schott & Ben., Elektrochemisches Labor FS 2, 6500 Mainz, POB 2480 / GFR
Baum, George, Dr., Corning Glass Works, Sullivan Park, Corning N.Y. 14830/USA
Baumgärtl, Horst, Chem.-Ing., Max-Planck-Institut für Systemphysiologie, 4600 Dortmund, Rheinlanddamm 201 / GFR
Betz, Eberhard, Prof. Dr., Physiologisches Institut der Universität Tübingen, 7400 Tübingen Gmelinstr. 5/GFR
Blümel, Gerlinde, Max-Planck-Institut für Systemphysiologie, 4600 Dortmund, Rheinlanddamm 201/GFR
Briggs, David, Dr., Max-Planck-Institut für Systemphysiologie, 4600 Dortmund, Rheinlanddamm 201/GFR
Brown, Arthur, Prof. Dr., University of Texas Medical Branch, Department of Physiology and Biophysics, Galveston, Texas 77550/USA
Buck, Richard P., Prof., Laboratory of Chemistry University of North Carolina, Chapel Hill, N.C. 27514/USA
Chan, William W.-C., Prof. Dr., The Lund Institute of Technology, Chemical Centre, Biochemistry 2, POB 740, S-22007 Lund 7/Sweden
Christian, Gary D., Dr., The University of Washington, Dept. of Chemistry BG-10, Seattle, Wash. 98195/USA
Clark, Leland C. Jr., Prof. Dr., Children's Hospital Research Foundation, Division of Neurophysiology, University of Cincinnati, Elland & Bethesda Ave., Cincinnati, Ohio 45229/USA
Clausen, Torben, Dr., University of Aarhus, Dept. of Physiology, DK-8000 Aarhus/Denmark
Coleman, Robert, L., Dr., Orion Research Inc., Biomedical Systems Division, 11, Blackstone St., Cambridge, Mass. 02139 / USA
Crowe, Wayne, Dr., The University of Pennsylvania, School of Medicine, Harrison Dept. of Surgical Research, Philadelphia, Pa. 19174/USA
Dora, Eors, Dr., Experimentelles Forschungslaboratorium der Medizinischen Universität Budapest, Budapast VIII, Üllöi út 78a, Hungary
Eisenmann, George, Prof. Dr., The University of California, School of Medicine, Dept. of Physiology, The Center of the Health Science, Los Angeles, California 90024/USA
Fleckenstein, Wolfgang, cand. med., Physiologisches Institut der Universität Kiel, 2400 Kiel, Ohlhausenstraße 40—60/GRF
Fuchs, Christian, Dr., Physiologisches Institut der Universität Göttingen, Lehrstuhl Physiologie I, 3400 Göttingen, Humboldtallee 7/GFR
Gebert, Gerfried, Prof. Dr., Abteilung Physiologie der Universität Ulm, 7900 Ulm, Oberer Eselsberg/GRF
Grote, Jürgen, Prof. Dr. Dr., Physiologisches Institut der Universität Mainz, 6500 Mainz, Saarstr. 21
Hebert, Normand C.,ZDr., Microelectrodes Inc., Grenier Industrial Village, Londonderry, New Hamsphire 03053/USA
Heisler, Norbert, Dr., Max-Planck-Institut für experimentelle Medizin, Abteilung Physiologie, 3400 Göttingen, Hermann-Rein-Str. 3/GFR
Heuser, Dieter, Dr., Physiologisches Institut der Universität Tübingen, Lehrstuhl I, 7400 Tübingen, Gmelinstraße 5/GFR
Hinke, John A., Prof. Dr., The University of British Columbia, Faculty of Medicine, Dept. of Anatomy, Vancouver, B.C. V6T 1W5/Canada
Hirsche, Hansjürgen, Prof Dr., Lehrstuhl für angewandte Physiologie der Universität Köln, 5000 Köln-Lindenthal, Robert-Koch-Str. 39/GFR
Höper, Jens, Max-Planck-Institut für Systemphysiologie, 4600 Dortmund, Rheinlanddamm 201/GFR
Jeal, Valerie, E. University of Bristol, Dept. of Pathology The Medical School, University Walk, Bristol BS8 1TD/England

List of Participants

Jensen, Ole Jørgen, Radiometer AS, Dk 2400 Copenhagen, Emdrupvej/Denmark
Keinemann, Friedrich-Karl, Max-Planck-Institut für Systemphysiologie, 4600 Dortmund, Rheinlanddamm 201/GFR
Kessler, Manfred, Prof. Dr., Max-Planck-Institut für Systemphysiologie, 4600 Dortmund, Rheinlanddamm 201/GFR
Keup, Uwe, Dr., Farbenfabriken Bayer AG, Institut für Pharmakologie, 5600 Wuppertal 1, Postfach 130105
Khuri, Raja N., Dr., American University of Beirut, Dept. of Physiology, Beirut/Lebanon
Klingenberg, Martin, Prof. Dr., Institut für Physiologie Chemie und Physikalische Biochemie, 8000 München 15, Goethestraße 33 / GFR
Kovách, Arisztid, G. B., Prof. Dr., Experimentelles Forschungslaboratorium der Medizinischen Universität Budapest, Budapest VIII, Üllöi út 87a/Hungary
Krumme, Bärbel, Dr., Max-Planck-Institut für Systemphysiologie, 4600 Dortmund, Rheinlanddamm 201/GFR
Lehmenkühler, A., Dr., Physiologisches Institut der Universität Münster, 4400 Münster, Westring 6/GFR
Lübbers, Dietrich W., Prof. Dr., Max-Planck-Institut für Systemphysiologie, 4600 Dortmund, Rheinlanddamm 201/GFR
Lux, Hans Dieter, Prof Dr., Max-Planck-Institut für Psychiatrie, Theoretisches Institut, 8000 München 40, Kraepelinstr. 2 und 10/GFR
Martiny, S. C., Chr. Hansen's Laboratorium A/S. Fanøgade 17, DK-2100 Copenhagen/Denmark
Mela, Leena, Dr., The University of Pennsylvania School of Medicine, Harrison Dept. of Surgical Research, Philadelphia, Pa. 1974/USA
Messmer, Konrad, PD Dr., Chirurgische Klinik der Universität München, 8000 München 15, Nussbaumstr. 20/GFR
Morf, W. E., Dr., Eidgenöss. Technische Hochschule, Laboratorium f. Organische Chemie, Ch-8006 Zürich, Universitätsstraße 6—8/Switzerland
Müller, Hans, Prof. Dr., 8000 München 19, Flüggenstraße 10
Norman, Thelma, University of Bristol Medical School, Dept. of Pathology, University Walk, Bristol BS8 1TD/England
Olfers-Weber, Rosemarie, Hygiene Institut der Universitäts-Klinik Bonn, 5300 Bonn, Venusberg
Pressman, Berton C., Dr., The University of Miami, Dept. of Pharmacology, Miami, Florida 33152/USA
Pungor, Ernö, Prof. Dr., Institute for General and Analytical Chemistry, Technical University, Gellert térr 4, 1111 Budapest XI/Hungary
Reneau, Daniel D., Prof. Dr., Louisiana Technical University, Dept. of Biomedical Engineering, POB 4875, Ruston, La. 71270/USA
Runte, Elke, Max-Planck-Institut für Systemphysiologie, 4600 Dortmund, Rheinlanddamm 201/GFR
Saito, S., Dr., Nr. 3—810, Hinode-cho 27-ban, Adachi-ku, Tokyo 120/Japan
Scarpa, Toni, Dr., The University of Pennsylvania Medical School, Johnson Research Foundation, Dept. of Biophysics and Physical Biochemistry, Philadelphia, Pa. 19174/USA
Schiemann, Alexander, cand. med., Physiologisches Institut der Universität Mainz, 6500 Mainz, Saarstraße 21/GFR
Schuler, Peter, Dr., Boehringer Mannheim GmbH, Biochemica Werk Tutzing, 8132 Tutzing/Obb., Postfach 120/GFR
Sies, Helmut, PD Dr., Institut für Physiologische Chemie und Physikalische Biochemie der Universität München, 8000 München 2, Goethestraße 33/GFR
Silver, Ian A., Prof. Dr., The University of Bristol Medical School, Dept. of Pathology, University Walk, Bristol BS8 1TD/England
Simon, Wilhelm, Prof. Dr., Eidgenöss. Technische Hochschule, Laboratorium für Organische Chemie, CH-8006 Zürich, Universitätsstraße 6—8/Switzerland
Sinagowitz, Ekkehard, Dr., Urologische Abteilung der Chirurgischen Universitätsklinik, 7800 Freiburg, Hugstetter Str. 55/GFR
Steinhagen, Christian, Dr., Institut für angewandte Physiologie der Universität Köln, 5000 Köln-Lindenthal, Robert-Koch-Str. 39/GFR
Stockem, Wilhelm, Prof. Dr., Institut für Cytologie und Mikromorphologie der Universität Bonn, 5300 Bonn, Ulrich-Haberland-Straße 61a/GFR

List of Participants

Strehlau, Reinhard, Max-Planck-Institut für Systemphysiologie, 4600 Dortmund, Rheinlanddamm 201/GFR

Thomas, Roger, C., Dr., The University of Bristol Medical School, Dept. of Physiology, University Walk, Bristol BS 8 1 TD/England

Thutewohl, Walter, Max-Planck-Institut für Systemphysiologie, 4600 Dortmund, Rheinlanddamm 201/GFR

Walker, John, L., Prof. Dr., The University of Utah Medical Center, Dept. of Physiology, Salt Lake City, Utah 84112/USA

White, John, F., Dr., Emory University, Dept. of Physiology, Atlanta, Georgia 30322/USA

Wodick, Reinhard, PD Dr. Dr., Max-Planck-Institut für Systemphysiologie, 4600 Dortmund, Rheinlanddamm 201/GFR

Woidtke, Hans, stud. med., 4630 Bochum-Querenburg, Hustadtring 57/GFR

Zeuthen, Thomas, Dr., The University of Bristol Medical School. Dept. of Pathology, University Walk, Bristol BS 8 1 TD/England

Session I

**Theoretical and Practical Aspects
of Ion-Selective Electrodes**

Chairmen: *W. Simon*

G. Eisenman, S. Krasne and S. Ciani

Further Studies on Ion Selectivity*)

(11 Figures)

I. Introduction

An understanding of ion permeation and ion selectivity in ultra-thin lipid bilayers is of at least indirect relevance to ion-selective electrodes, which for practical reasons are usually based on a thicker bulk-phase membrane. Therefore, we will present the findings of our recent theoretical and experimental studies of carrier- and channel-forming molecules in lipid bilayer membranes in the hope that the information gained in these relatively well-understood systems will prove useful in advancing the further development of ion-selective electrode of more conventional construction. Not only will the group Ia cations be considered but also the non-"noble-gas" cations Tl+ and Ag+ and molecular monovalent cations such as ammonium, hydrazinium, various guanidiniums, and alkylated ammoniums such as choline and acetylcholine.

II. Electrode selectivity as defined by permeability ratios

The selectivity of electrodes among monovalent cations manifests itself, quite universally (cf. *Eisenman* 1965, 1969; *Buck* 1972; *Morf, Ammann, Pretsch* and *Simon* 1973) in the observed potential E of the familiar electrode equation

$$E = \text{const} + \frac{RT}{F} \ln \left(a_i + \left(\frac{P_j}{P_i} \right) a_j \right) \tag{1}$$

through the permeability ratio $\left(\frac{P_j}{P_i}\right)$, which is a characteristic constant**) defining the relative selectivity between ionic species J+ and I+. Equation (1) has been found to apply to a very wide variety of electrodes (including glasses, liquid ion exchangers, and neutral carriers) and is apparently quite useful in describing the data for practical electrodes (*Morf* et al. 1973; *Eisenman* 1969). However, with permeability ratios taken as constants, this equation cannot hold invariably, as we will demonstrate here, since there are certain kinetic situations, recently encountered with carriers in bilayers (*Ciani* et al. 1973), for which this equation is no longer adequate unless permeability ratios are permitted to be variables.

A. *Bilayer-neutral carrier systems in the "equilibrium domain"*

We will first consider situations where Equation 1 is completely adequate with permeability ratios constant. Typical of these is the lipid bilayer-neutral carrier system, whose electrode

---

*) Supported by NSF Grant GB 30835 and USPHS Grant NS 09931.
**) This "permeability ratio", as it is called in biophysical parlance (*Goldman* 1943; *Hodgkin* and *Katz* 1949) is usually referred to as the "potential selectivity constant" $\left(K_{ij}^{pot}\right)$ in the analytical chemical literature (*Eisenman* 1969; *Morf* et al. 1973).

selectivity has so far (for valinomycin and nonactin) been found to be identical to that of bulk-phase electrodes (cf. *Eisenman* et al. 1973, Table IX; *Morf* et al. 1973).

Figure 1 illustrates the satisfactory way in which Equation 1 accounts for the membrane potential of the neutral carriers nonactin, monactin, dinactin and trinactin in lipid bilayers made from asolectin (*Szabo, Eisenman* and *Ciani* 1969), the curves being drawn according to the indicated permeability ratios, while the points represent the experimentally observed values. The agreement between the points and the curves indicates that the permeability ratios are indeed constants, independent of experimental conditions, and in particular of the transmembrane potential (which exceeded 100 mV in some of these experiments). This constancy of permeability ratios was also observed in this system over a wide range of antibiotic and salt concentrations (*Szabo* et al. 1969) and is expected (*Ciani* et al. 1973) whenever such a system is in the "equilibrium domain" that is, when the rate limiting process is the translocation of the complexes across the membrane interior so that the interfacial reactions of loading and unloading the carriers are at equilibrium. This "equilibrium domain" situation presumably also holds for practical bulk electrodes based upon valinomycin and nonactin (*Stefanac* and *Simon* 1966; *Lev* et al. 1968, 1973; *Simon* 1971; *Simon* and *Morf* 1973; *Morf* et al. 1973) since, as noted above, their selectivities are identical to those of these molecules in the "equilibrium domain" in bilayers.

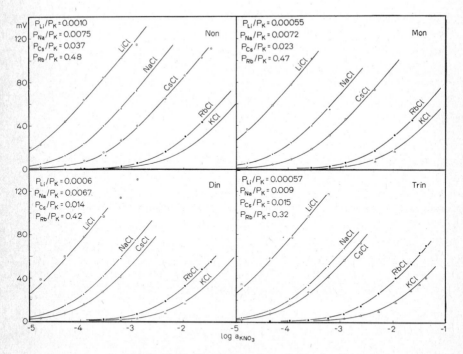

Fig. 1. Membrane potentials in alkali chloride-potassium nitrate mixtures in the presence of nonactin, monactin, dinactin or trinactin. Bilayer membranes were formed from asolectin in 10^{-2} M alkali metal chloride solution at 10^{-7} M concentration of the indicated macrotetralide actin (non-, mon-, din-, trin-). The curves are drawn according to Equation 1 for the indicated values of the permeability ratios. (After *Szabo* et al. 1969, Fig. 4.)

The selectivity of neutral carrier electrodes in the "equilibrium domain" is of the simplest type observable in practical electrodes in that there is no significant contribution due to differences in the mobilities of the permeant species, as there is in glass electrodes or liquid ion exchangers (cf. *Eisenman* 1969, Table I). Indeed, the electrode selectivity has been shown to reflect directly the equilibrium constant for solubilizing the ion in the carrier in the membrane (*Ciani* et al. 1969; *Eisenman* et al. 1969; *Szabo* et al. 1969)*).

B. *Complexities introduced by kinetic effects in bilayers*

(1) *Voltage-dependent permeability ratios, "sub-Nernst" and "supra-Nernst" potential behavior, and kinetic contributions to apparent selectivity*

In contrast to the above simple situation, recent studies on glyceryl oleate bilayers with enhanced permeabilities to cations using the more strongly complexing carriers indicate that it is possible to encounter situations for which the rate-determining step is no longer diffusion across the membrane interior. Here, the rates of loading or unloading the complexes at the interface become comparable to the rates of transport of the complexes across the membrane interior (*Läuger* and *Stark* 1970; *Stark* and *Benz* 1971; *Ciani* et al. 1973; *Laprade* et al. 1975). Under these circumstances Equation 1 is no longer expected to be adequate theoretically (*Ciani* et al. 1973, cf. Eq. 212) unless the permeability ratios are allowed to be voltage-dependent parameters. An example of this situation is illustrated in Figure 2 which presents the membrane potentials observed for trinactin in bilayers made of glyceryl dioleate**) (*Laprade* et al. 1974; *Ciani* et al. 1973) instead of the asolectin used in Figure 1. Here the experimental points are seen to fall not on the dashed curves expected from Equation 1 (which have Nernstian limiting slopes); rather, they fall on the solid curves which can have markedly non-Nernstian slopes (being either less than 58.5 mV per decade, as seen in the lower portion of the figure, or as high as 80 mV per decade as seen on the insert).

The origin of this behavior has been inferred to be a consequence of rate limitations on the kinetics of unloading the carriers at the membrane-solution interfaces (*Laprade* et al. 1975; *Ciani* et al. 1973)***); and a theoretical analysis of the expected potential behavior for an Eyring single-barrier model (*Ciani* et al. 1973) leads to the solid curves of Figure 2, which are seen to describe the data satisfactorily. This analysis indicates that the membrane potential

---

*) Incidentally, the identical selectivities of bulk electrodes and bilayers referred to above probably occur because the number of complexes in the bulk-phase electrode is a constant which balances the charge of "trapped anions" (either from the solvent or deliberately introduced). Evidence for this is the linear dependence of bulk membrane resistance on membrane thickness (*Tosteson* 1968, cf. Figs. 6, 7; *Lev* et al. 1973). The selectivity of the bulk electrode in this situation merely reflects the equilibrium selectivity for solubilizing the ion in the carrier in the solvent of the membrane.
**) Such bilayers appear to have about 180 mV greater negativity of the membrane interior due to dipole effects (*Szabo* et al. 1973).
***) While at first sight this might not be expected to occur in bulk phase electrodes where the interfacial reactions would not be expected to be rate limiting, an analogous complexity could occur if a "carrier relay" (*Markin* et al. 1969; *Simon* and *Morf* 1973) mechanism of transport occurred in the interior; for the rates of loading and unloading of the carriers in the interior could then contribute to the observed overall selectivity.

can still be described by Equation 1 provided the permeability ratios are taken as voltage-dependent functions, defined by

$$\frac{P_j}{P_i} = \left(\frac{P_j}{P_i}\right)_{eq.} \left[\frac{1 + 2 w_i \cosh \frac{FV}{2RT}}{1 + 2 w_j \cosh \frac{FV}{2RT}}\right] \quad (2)$$

where V is the transmembrane potential and w_i and w_j are kinetic parameters dependent on the ratio of the rate constants for the translocation of the complex across the bilayer to those for unloading the complex at the membrane-solution interface. $(P_j/P_i)_{eq.}$ is the voltage-

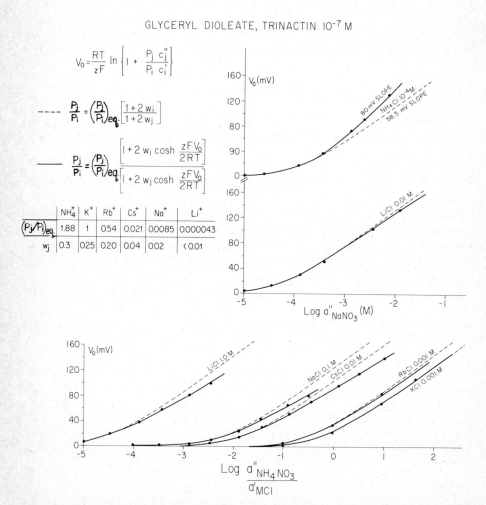

Fig. 2. Membrane potential data for trinactin measured on glyceryl dioleate bilayers in ion mixtures. The filled circles represent experimental points. The solid curves are drawn according to Equation 2 with the values of the parameters $(P_j/P_i)_{eq.}$ and w_j given in the figure. The dashed curves, which give limiting slopes of 58.5 mV/decade, correspond to the case in which the permeability ratios are constant and equal to those measured at vanishingly small voltages [see the equation in the figure]. (After *Ciani* et al. 1973, Fig. 19.)

independent (constant) "true" permeability ratio which would be seen in the "equilibrium domain".

Notice that for small transmembrane potentials (i.e. $V \to 0$) Equation 2 reduces to the simple form:

$$\left(\frac{P_j}{P_i}\right)_{V \to 0} = \left(\frac{P_j}{P_i}\right)_{eq.} \left[\frac{1 + 2w_i}{1 + 2w_j}\right]. \tag{3}$$

from which it can be seen that, although the "apparent" permeability ratio, $(P_j/P_i)_{V \to 0}$, measured in this limit does indeed become a constant, it is no longer identical to the "true" permeability ratio measured for a membrane in the "equilibrium domain", differing from it by the ratio of the "kinetic" parameters defined in the bracketed term. Thus, there can be a kinetic contribution to the apparent permeability ratio of a system which will be dependent on the composition of the lipid (in a bilayer) or solvent (in a bulkphase electrode) of which a membrane is made. Two examples are considered below.

(2) *Example, the macrotetralide actins*

Even though kinetic effects for the macrotetralide actins produce substantial departures from Nernstian electrode behavior, as illustrated in Figure 2 for trinactin in glyceryl dioleate bilayers, these kinetic complexities lead to no changes in apparent selectivity sequence. This can be seen from Table 1 where the values of "apparent" permeability ratios $[(P_j/P_i)_{V \to 0}]$ and "true" permeability ratios $[(P_j/P_i)_{eq.}]$ corresponding to the data of Figure 2 can be compared. Note that the biggest correction (corresponding to $2w_j = 0.6$ for NH_4^+ and $2w_j = 0$ for Li^+) is merely a factor of 1.6. It is of some interest to note that the correction for kinetic effects makes the "true" permeability ratios $(P_j/P_i)_{eq.}$ for trinactin in glyceryl dioleate bilayers more consistent with the values for this molecule previously estimated in Figure 1 in asolectin bilayers (compare the last two columns of Table 1).

Table 1. Kinetic contributions to permeability ratios of trinactin in GDO-decane bilayers (Eyring single barrier model)

	w_i	$(P_i/P_K)_{V \to 0}$	$(P_i/P_K)_{eq}$	P_i/P_K (asolectin)
K	.25	1.0	1.0	1.0
Rb	.20	.58	.54	.32
Cs	.04	.029	.021	.015
Na	.02	.0123	.0085	.009
Li	< .01	.000065	.000043	< .00057
NH4	.30	1.76	1.88	–

Because the above analysis of deviations from eq. (1) is probably not convincing by itself that such kinetic effects are indeed present, we will digress briefly here to illustrate the kinds of conductance-voltage behavior which are the counterparts of the data of Figure 2 and which provide independent verification that the non-Nernstian behavior is attributable to the kinetic effects mentioned above. Indeed, a detailed analysis has been carried out for this system with internally consistent results for potential, conductance-voltage, and time-dependent (voltage-jump) relaxation behavior (*Laprade et al.* 1975). Figure 3 illustrates the shape of the steady-state conductance-voltage characteristic for the various ions whose electrode behavior was characterized in Figure 2. For convenience we plot the normalized conductance (G/G_0), which is the conductance at voltage V divided by that at low voltage,

G. Eisenman, S. Krasne and S. Ciani

as a function of voltage as suggested by *Stark* and *Benz* (1971). The theoretical expectation for the Eyring single-barrier model in the limit of low ionic concentration is (*Stark* and *Benz*, 1971, eq. 12; *Laprade* et al. 1975, eq. 52):

$$\frac{G}{G_0} = \frac{2\,RT}{FV} \frac{(1 + 2\,w_i)\sinh\dfrac{FV}{2\,RT}}{1 + 2\,w_i\cosh\dfrac{FV}{2\,RT}} \tag{4}$$

For Na^+ and Cs^+ (for which $w_i \ll 1$) the data on the left-hand side illustrate that the current-voltage curve is the same "hyperbolic" function of voltage (and also the same as that seen for nonactin); that

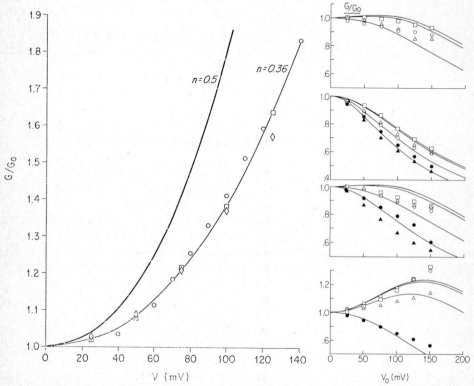

Fig. 3. Comparison of experimental and theoretical conductance-voltage relationships. Left, the relative conductance G/G_0 is plotted against the applied membrane potential for GDO bilayers in the presence of 10^{-1} M trinactin and 10^{-1} M NaCl (diamonds), 10^{-8} M trinactin and 10^{-2} N CsCl (squares), and 10^{-6} M nonactin and 1M NaCl (open circles). The solid line shows the expectation of eq. (4) in the equilibrium domain (i.e., for $w_i \ll 1$). Also shown is a curve for an empirically defined "barrier shape", labelled $n = 0.36$, as defined in footnote (19) of *Laprade* et al. (1975). Right, the relative conductance vs. voltage is plotted for GDO bilayers in the presence of 10^{-8} M trinactin and Tl, NH_4, K and Rb ions. The concentrations of the permeant ions are 10^{-3} M (open squares), 10^{-2} M (open circles), 10^{-1} M (open triangles), 1M (closed circles) or 3M (closed triangles). Up to 1M, the ionic strength is maintained constant by the presence of LiCl. The theoretical curves are drawn according to eq. (4) using best fit w_i values quite comparable to those of Figure 2 (w_i for NH_4^+ is .354, for K^+ is .259, for Rb^+ is .112, for Cs^+ is $\leq .02$, and for Tl^+ is .227). The uppermost curves are for 10^{-3} M, except for NH_4 where the 10^{-3} M and 10^{-2} M curves coincide. Lower curves are in the sequence of increasing concentrations, 10^{-2} M, 10^{-1} M 1M and 3M. The values for the parameter v_i, which define the concentration dependence, will be found in *Laprade* et al., 1975, Table 2. (After Figs. 15 and 10 of *Laprade* et al. 1975.)

Further Studies on Ion Selectivity

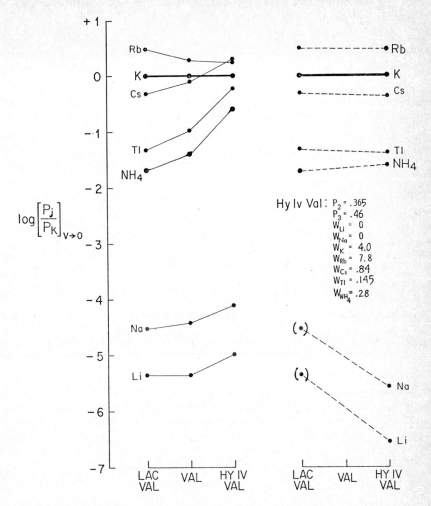

Fig. 4. Left, "apparent" permeability ratios $(P_j/P_K)_{V\to 0}$ are plotted for valinomycin and its two homologues lac-val and HyIvVal as a function of increasing degree of methylation. The permeability ratios were measured in glyceryl dioleate-decane bilayers in the case of lac-val and HyIvVal and in glyceryl monooleate-decane bilayers in the case of valinomycin. the concentrations of the carrier were: lac-val 10^{-6} M, val 10^{-7} M, and HyIvVal 3.6×10^{-8} M except for the measurements of Na and Li for which the HyIvVal concentration was 3.6×10^{-7} M.

Right, calculated "true" permeability ratios $(P_j/P_K)_{eq.}$ are plotted for lac-val and HyIvVal. The kinetic contributions are totally negligible for lac-val in this lipid so that the "true" permeability ratios for lac-val are the same as the "apparent" permeability ratios. For HyIvVal the kinetic parameters tabulated in the figure were measured from the conductance-voltage relationship using eq. (6) (unpublished results). Note that the apparent decrease of permeability ratios for Na$^+$ and Li$^+$ relative to K$^+$ seen on the righthand side probably corresponds to the fact that the lac-val measurements reflect trace contaminants of K$^+$ in the Na and Li salts, and the correction for kinetics in HyIvVal allows one to project the permeability ratio below the K$^+$-contaminant level. For this reason the data points for Na$^+$ and Li$^+$ in lac-val have been parenthesized; and it is of some interest that the kinetic limitations on K$^+$ permeation actually enable one to see further into the "mud" level for HyIvVal with Li$^+$ and Na$^+$.

is G/G_0 is an increasing function of voltage. This is expected for the "equilibrium domain". In contrast, kinetic effects are seen at the right for the species Tl^+, NH_4^+, K^+ and Rb^+, for which various degrees of "saturation" of the I–V characteristic are implied by the decreasing value of G/G_0 with increasing voltage. (N.B. The concentration dependence of these effects seen in Figure 3 is expected and is examined in detail by *Laprade* et al. (1975)). The "kinetic" effects demanded by the values of w_i from the membrane potential behavior of Figure 2 are clearly demonstrable in the conductance behavior of Figure 3 and the magnitudes of the kinetic parameters, w_i, measured by the I–V behavior agree with those assessed from the membrane potential behavior.

Incidentally, since conductometric measurements on membranes provide an alternative, and independent, way of sensing ions selectively, these conductance characteristics may not be entirely devoid of practical significance.

(3) Example, valinomycin and its methylated homologues

Similar kinetic effects are seen for valinomycin and homologues of valinomycin having different degrees of methylation; but for these molecules the effects are sufficiently severe that they can significantly alter the apparent sequence of selectivity observed in glyceryl oleate bilayers. This is illustrated by the data at the left of Figure 4, which plots the permeability ratios observed for valinomycin and two synthetic homologues, lac-val (cyclo(D-Val-L-Lac-LVal-D-Lac)$_3$ and HyIvVal (cyclo(D-Val-L-HyIv-LVal-D-HyIv)$_3$) (unpublished results). These molecules differ from Val in having three fewer and three more methyl groups, respectively; and the series of LacVal, Val, and HyIvVal are ranked from left to right as a function of the increasing methylation of the molecule. A striking change in "apparent" selectivity sequence is seen between Val and HyIvVal, where the sequence shifts to $Cs > Rb > K > Na > Li$ from that ($Rb > K > Cs > Na > Li$) characteristic of LacVal and Val. Work now in progress indicates that these effects are largely attributable to kinetic contributions to $(P_j/P_i)_{V \to 0}$, since the values, plotted at the right, for $(P_j/P_i)_{eq.}$, calculated taking into account our preliminary estimates of the kinetic parameters obtained from analysis of the conductance-voltage behavior, show no such inversions.*)

---

*) Unfortunately, our results with this system indicate that a more elaborate model than the Eyring single barrier is necessary to describe all the data adequately. We have used an extended model similar to that proposed by *Hladky* (1974), see Figure 5, to obtain the estimates of the kinetic parameters for these molecules. (The values for HyIvVal are given on the figure). Here, not only is a trapezoidal "barrier shape" (cf. *Hall, Mead* and *Szabo* 1973) allowed for ($P_2 = .365$) but also a "reaction plane" lying somewhat internal to the membrane-solution interface ($P_3 = .46$). For this model the apparent permeability ratio is defined by

$$\left(\frac{P_j}{P_i}\right)_{V \to 0} = \left(\frac{P_j}{P_i}\right)_{eq.} \left[\frac{1 + \frac{2 w_i}{P_2}}{1 + \frac{2 w_j}{P_2}}\right] \tag{5}$$

The preliminary values of the parameters w_i, P_2, and P_3 given in Figure 4 have been evaluated from conductance-voltage curves through a least squares computer fit to the expression for the theoretically expected dependence of the conductance ratio (G/G_0) on applied voltage $\left(\Phi = \frac{FV}{RT}\right)$ for this model at low permeant ion concentrations, which is

$$\frac{G}{G_0} = \frac{2(P_2 + P_1) \sinh \frac{\Phi}{2}}{\sinh P_2 \Phi + P_1 \Phi \cosh P_3 \Phi} \tag{6}$$

where $P_1 = 2 w_i$. The details of this model will be presented elsewhere.

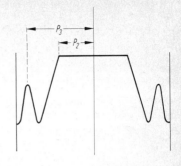

Fig. 5. Schematic representation of the molar free energy profile within the membrane in the absence of an externally applied electric field for the extended model mentioned in the footnote to p. 8. The height of the peaks of the two lateral humps with respect to the adjacent wells indicates the activation free energy barrier of the complexation and dissociation reaction, while their x-coordinates are identified with those of the plane at which the reaction occurs. P_3 is the distance of this plane from the the center of the membrane, measured in units of "membrane thickness". The trapezoidal barrier characterized by a half-width of P_2 (in "membrane thickness" units) represents the free energy profile of the ion-carrier complex, which is used for the integration of a generalized Nernst-Planck equation to deduce eqs. (5) and (6) (the height of the barrier is taken into account in the parameter P_1). When a voltage $V = \psi(0) - \psi(d)$ is applied externally and the field is assumed linear, the distortion of the profile is obtained by superimposing the function $zF\left[V - \dfrac{Vx}{d}\right]$ on the profile of the figure.

III. Selectivity among polyatomic cations

Hille (1971, 1972) clearly demonstrated that the "Na Channel" of nerve was substantially permeable to a variety of polyatomic cations; and we have recently found that typical neutral carrier molecules such as hexadecavalinomycin (*Eisenman* and *Krasne* 1974, Table 2.1) and, to

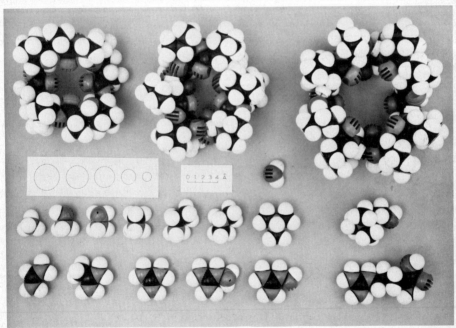

Fig. 6. Photographs of space-filling (Cory-Pauling-Koltun) models of nonactin, valinomycin, and hexadecavalinomycin and of typical cations which they carry. Cs^+, Rb^+, K^+, Na^+, Li^+ are schematized by the circles; and a water molecule is presented to the right of the angstrom scale. The polyatomic cations in the two bottom rows are: NH_4^+, $HONH_3^+$, $H_2NNH_3^+$, $CH_3NH_3^+$, $C_2H_5NH_3^+$, $(CH_3)_2NH_2^+$, $(CH_3)_3NH^+$, choline$^+$; and formamidinium, acetamidinium, guanidinium, aminoguanidinium, hydroxyguanidinium, and arginine, respectively.

Table 2. Permeability ratios for valinomycin and hexadecavalinomycin (glyceryl oleate-decane bilayers)

	Valinomycin	Hexadecavalinomycin
Small cations	$(P_i/P_K)_{V \to 0}$	$(P_i/P_K)_{V \to 0}$
Li^+	.0000042	.00011
Na^+	.000036	.001
K^+	1.0	1.0
Rb^+	1.8	3.0
Cs^+	.76	.74
Ag^+	.0000056	.0016
Tl^+	.105	.11
NH_4^+	.039	.045
Substituted NH_4's		
$CH_3NH_3^+$	.00039	.21
$(CH_3)_3NH^+$	$\leq$ .0000046	.32
$C_2H_5NH_3^+$	$\leq$ .00005	.0054
$H_2NNH_3^+$	$\leq$ .000082	.0016
$HONH_3^+$	$\leq$ .000024	.0011
Choline$^+$	$\leq$ .000031	1.3
Acetylcholine$^+$	$\leq$ .000016	4.5
Amidiniums and guanidiniums		
Formamidinium$^+$	$\leq$ .0059	.052
Guanidinium$^+$	$\leq$ .00072	.09
$NH_2 \cdot$ guan$^+$	$\leq$ .000025	.21
$OH \cdot$ guan$^+$	$\leq$ .01	.045
Arginine$^+$	–	.00066

a lesser extent, valinomycin and nonactin (*Eisenman* and *Krasne* 1973) also render bilayer membranes significantly and selectively permeable to such cations. Figure 6 presents space filling (CPK) models of these carriers and of typical polyatomic cations which they carry. Although the selectivity of the usual carriers such as nonactin and valinomycin for these larger species is too low compared to NH_4^+ and K^+ to be useful (traces of NH_4^+ and K^+ even make the characterization difficult), for molecules with only slightly larger interiors, like hexadecavalinomycin*) (*Ovchinnikov* 1971), we found the permeability to become even larger for some of the polyatomic cations than for NH_4^+ and the group Ia cations (*Eisenman* and *Krasne* 1974, cf. Table 2.) Such neutral carriers clearly have the potentiality for use in practical bulk phase electrodes for polyatomic cations, for which selectivities comparable to those reported here for bilayers are to be expected. Therefore, we will examine here the selectivities among such species in some detail to encourage development of even more appropriately tailored "host" molecules for such interesting "guests" as acetylcholine (the pioneering studies of Cram and his colleagues on "host-guest" chemistry are particularly noteworthy in this regard (cf. *Cram* and *Cram* 1974).

---

*) Hexadecavalinomycin (abbreviated here as HDVal) is a synthetic valinomycin analog based upon four, instead of the usual three repeating (D-Val-L-Lac-L-Val-D-HyIv) subunits. It was synthesized by Ovchinnikov, Ivanov and their colleagues at the Shemyakin Institute of the Soviet Academy of Sciences.

A. *Example, the neutral carriers: hexadecavalinomycin and valinomycin*

Hexadecavalinomycin (cyclo-(D-Val-L-Lac-L-Val-DHyIv)$_4$ and valinomycin (cyclo-(D-Val-L-Lac-L-Val-DHyIv)$_3$) are of particular interest because the ligands are chemically identical in these two molecules and their hydrogen-bonding "bracelet" structure in the complex is the same (*Ovchinnikov* 1971). Indeed, their only difference is that the interior cavity is one-third larger in radius (and circumference) in hexadecavalinomycin than in valinomycin.

The selectivity among the smaller cations is very similar for these two molecules (*Eisenman* and *Krasne* 1974) both selecting in the same sequence Rb > K > Cs > Tl > NH$_4$ > Na > Li (see "small cations" in Table 2) (from which observation *Eisenman* and *Krasne* have argued that the "field strength" of the ligands was more crucial than were steric restrictions in determining the selectivity among these ions). For larger cations (see "substituted NH$_4'$s" and "amidiniums and guanidiniums" in Table 2 and Fig. 7) valinomycin (val) barely renders bilayers detectably permeable (only "less than" values for permeability ratios being certain for most polyatomic cations because of the effects of possible traces of NH$_4^+$ and K$^+$). That this is due to a limitation on what can fit into the cavity, and not to an inability to replace water of hydration by the carbonyl ligands of the carrier, is clear from the substantial

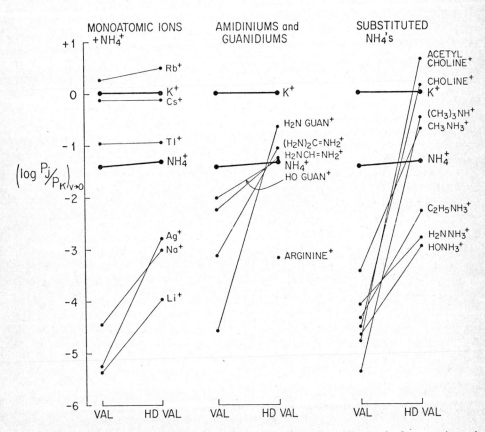

Fig. 7. Observed permeability ratios of valinomycin and hexadecavalinomycin for a variety of monovalent cations (unpublished results).

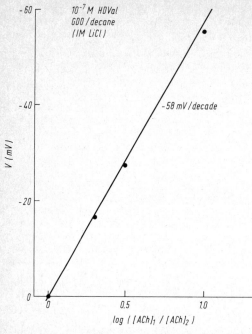

Fig. 8. Nernstian responses of hexadecavalinomycin to acetylcholine concentrations (unpublished results). The data points represent the transmembrane potentials observed as a function of the logarithm of the acetylcholine ratio in the solutions on two sides of a glyceryl dioleate-decane bilayer separating solutions containing 10^{-7} M hexadecavalinomycin and varying acetylcholine concentrations. The concentration of acetylcholine on side 2 was 10^{-3} M and that on side 1 was varied from 10^{-3} to 10^{-2} M. Ionic strength was maintained at 1.0 M with lithium chloride.

permeabilities seen for hexadecavalinomycin (HDVal), for which it is of noteworthy practical interest that acetylcholine is the most permeant species.

Indeed, the potentiometric response to acetylcholine is Nernstian, as illustrated in Figure 8. It is also apparent that hexadecavalinomycin carries acetylcholine in exactly the same way that

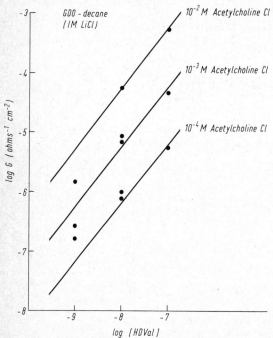

Fig. 9. Demonstration of the first order dependence of membrane conductance on concentration of hexadecavalinomycin and acetylcholine as seen in glyceryl dioleate-decane bilayers (unpublished results). (Ionic strength maintained constant with 1M lithium chloride.)

valinomycin carries a small cation such as potassium. This is illustrated in Figure 9 by the strict proportionality between membrane conductance and acetylcholine (and hexadecavalinomycin) concentration seen in the log-log plot of conductance vs. hexadecavalinomycin concentrations, which is typical of the behavior of the usual carriers and small cations in the "equilibrium domain" (cf. *Szabo* et al. 1969, 1973). A model of HDVal containing choline$^+$ is presented in Figure 10. Molecules of comparable dimensions to HDVal but more specifically "tailored" to form specific interactions with acetylcholine would appear to offer great potentiality for the design of electrodes specific to this important biological cation, as well as for other particular cations which may be of interest.

Although steric factors appear to be implicated in permitting hexadecavalinomycin (in contrast to valinomycin) to function as a cation carrier for the polyatomic cations, the observed selectivity does not correlate strictly with "size", as can be seen from the different selectivities observed among cations having similar projected dimensions (cf. hydrazinium

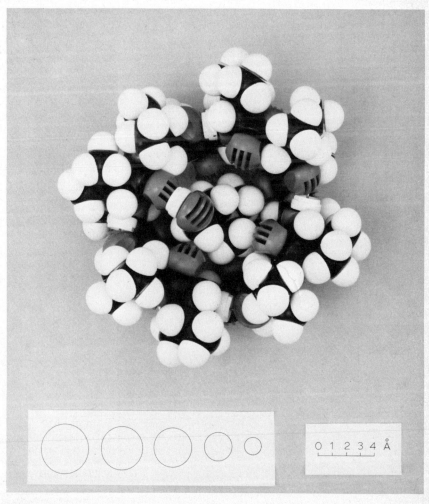

Fig. 10. Space-filling model of hexadecavalinomycin containing choline$^+$. The circles beneath the figure indicate the relative sizes of Cs$^+$, Rb$^+$, K$^+$, Na$^+$ and Li$^+$.

and methylammonium, the 3rd and 4th species in the next to last row of Fig. 5). This is not surprising since such a situation was first demonstrated by *Hille* (1971, 1972) for the permeation of the sodium channel of nerve and has been found more recently by *Moreno* and *Diamond* (1974a, 1974b) for permeation of the tight junctions of gallbladder epithelia. In the last part of this section we will demonstrate that some of the observed selectivities can be rationalized using the concept of "field strength" and reasoning entirely analogous to that previously used successfully in developing a theory for the pattern of selectivity seen among group Ia cations (*Eisenman* 1961, 1962). Before turning to this subject, we will present data observed for the selectivity among these cations manifested by the gramicidin channel, which is another system for which we have recently found substantial selective permeation by polyatomic cations.

B. *Example, the channel formed by gramicidin A*

Gramicidin A dimerizes to form cation-semipermeable channels across bilayer membranes (*Goodall* 1970; *Hladky* and *Haydon* 1972; *Myers* and *Haydon* 1972). The interior of the

Table 3. Permeability ratios for gramicidin A channel (glyceryl oleate-decane bilayers)

	P_i/P_K
Small cations	
Li^+	.083*)
Na^+	.29*) (.33)
K^+	1.0
Rb^+	1.03*)
Cs^+	1.31*)
Tl^+	50.0
NH_4^+	1.8*)
H^+	16.0*)
Substituted NH_4's	
$CH_3NH_3^+$	.24
$C_2H_5NH_3^+$	.014
$(CH_3)_2NH_2^+$	.0076
$(CH_3)_4N^+$	< .00011*)
$H_2NNH_3^+$	.96
$HONH_3^+$	1.7
*Amidiniums and guanidiniums***)	
Formamidinium$^+$	1.55
Guanidinium$^+$	≤ .027 (pH 7.9)
$NH_2 \cdot$ guan$^+$	.02 (pH 7.2)
$OH \cdot$ guan$^+$	≤ .24

*) Data of *Myers* and *Haydon* (1972).
**) Only "less than" values can be given for guanidinium$^+$ and hydroxyguanidinium$^+$ since we cannot exclude an NH_4^+ contaminant in these reagents as high as 1.3% and 10%, respectively using our measurements with nonactin as an index of NH_4^+ contaminant. Although fresh aqueous formamidinium$^+$ solutions have less than 10% NH_4^+ as a contaminant, these solutions decompose with time giving rise to an ammonium$^+$ contaminant level in the present experiment of approximately 30%. This was still sufficiently low that one can exclude NH_4^+ as contributing significantly to the apparent permeability of formamidinium$^+$.

Table 4. *Permeability sequences observed for "Main Group" polyatomic cations*

Na-channel (Hille, 1972):	Hydraz$^+$ > NH$_4^+$	> Form$^+$	> Guan$^+$	> NH$_2$·Guan$^+$	
Hexadecavalinomycin:	NH$_2$·Guan > Guan$^+$	> Form$^+$	> NH$_4^+$	> Hydraz$^+$	
Gramicidin A Channel:	NH$_4^+$	> Form$^+$	> Hydraz$^+$	> Guan$^+$	, NH$_2$·Guan$^+$
Tight Junctions (Moreno & Diamond, 1974b):	NH$_4^+$	> Form$^+$	> Hydraz$^+$	> Guan$^+$	> NH$_2$·Guan$^+$
Valinomycin, Nonactin:	NH$_4^+$	> Form$^+$	> Guan$^+$	, Hydraz$^+$	, NH$_2$·Guan$^+$
Activation of Diol Dehydrase (Toraya and Fukui, 1973):	NH$_4^+$	> Form	> Guan$^+$		

channel is lined by carbonyl oxygens from the polypeptide backbone (*Urry* 1971; *Urry* et al. 1971), which is analogous to the situation in the interior of a carrier. Among the monoatomic cations the selectivity has been reported as small (*Myers* and *Haydon* 1972); indeed it is not greatly different from that characteristic of ionic mobilities in free solution. *Myers* and *Haydon*'s permeability ratios are given in the upper portion of Table 3, and we have confirmed their values for Na$^+$ and K$^+$ permeability, as indicated in the table.*) We have also found (unpublished results), however, that the gramicidin channel is unusually selective for Tl$^+$, which is 50 times as permeant as K$^+$. This clearly illustrates that, even for small cations, permeation involves more than simple diffusion in the water within the channel.

Myers and *Haydon* reported that tetramethylammonium$^+$ ion was completely impermeant, which they attributed to its being too large to enter the pore. However, we find (unpublished results) that larger species (e.g., aminoguanidinium$^+$) are significantly permeable and that hydroxylamminium$^+$ and formamidinium$^+$ are even more permeant than the most permeant alkali metal cation (see Table 3).**) What is of present interest about the gramicidin-mediated permeabilities in Table 3 is that the selectivity sequence is quite different from that manifested by the hexadecavalinomycin carrier or by the sodium channel. This is illustrated in Table 4 for the five "main group" polyatomic cations, where it can also be seen that the "main group" permeability sequence is the same as that found by *Moreno* and *Diamond* (1974) for the gallbladder tight junction. This "main group" series of ions consists of H$_2$NNH$_3^+$ (hydrazinium$^+$), NH$_4^+$ (ammonium$^+$), H$_2$NCH = NH$_2^+$ (formamidinium$^+$), (H$_2$N)$_2$C = NH$_2^+$ (guanidinium$^+$), and H$_2$N(NH)NH$_2$C = NH$_2^+$ (aminoguanidinium$^+$), and is chosen for reasons to be discussed in the next section.

The permeability sequences among these species observed to date are summarized in Table 4, where partial sequences are also given for nonactin and valinomycin from previous studies (*Eisenman* and *Krasne* 1973)***) as well as for the activation of the enzyme dioldehydrase

---

*) Table 3 presents permeability ratios deduced in our usual way (cf. Fig. 1) from membrane potential measurements in 2-cation mixtures of K$^+$ with the listed cations (usually at .01M concentration as the chloride salts) in the presence of gramicidin A, usually at 10^{-9} M.

**) The reason we cannot comment on the permeability to the other gaunidiniums is because of the level of NH$_4^+$ contaminant in these reagents described in the footnote to Table 3.

***) Owing to NH$_4^+$ effects we cannot at present rank hydraz$^+$, guan$^+$, NH$_2^+$·guan$^+$ for nonactin and valinomycin.

(*Toraya* and *Fukui* 1974).*) Note that three different sequences are seen in Table 4, which bear no obvious relationship to ionic size, shape, or chemical formula. This situation is reminiscent of the "anomalous sequences" for the group Ia cations which were shown to be theoretically expected (*Eisenman* 1961) in that they reflected a selectivity pattern which could be deduced from considerations of "field strength" vs. hydration. We will now show that a similar simplification may be possible for the present polyatomic cations.

C. *A possible selectivity pattern for certain polyatomic cations analogous to that for the alkali metal cations*

Although it will be a long time before the factors underlying the selectivity of the polyatomic cations in a given system can be understood even at the crude level at which we now "understand" group Ia cations (one serious limitation being that hydration energies are not known for most polyatomic cations), certain regularities in the above described data suggest that a selectivity pattern may exist which is entirely analogous to that found for the group Ia cations. To show this, we begin by choosing a series of "main species" polyatomic cations of increasing "field strength" analogous to the group Ia cations whose field strength increases in the series Cs^+, Rb^+, K^+, Na^+, Li^+. These cations are: aminoguanidinium$^+$, guanidinium$^+$, formamidinium$^+$, ammonium$^+$, hydrazinium$^+$. We argue, following *Hille* (1972), that the "field strength" increases for the first four species since the (delocalized) positive charge is distributed over fewer and fewer nitrogens (i.e., 4 in aminoguanidinium, 3 in guanidinium, 2 in formamidinium, and 1 in ammonium). We also infer that the "field strength" increases from NH_4^+ to hydrazinium$^+$, again following *Hille* (1972), implying that the electron withdrawing effect of the NH_2 substituent on ammonium is more important than charge delocalization. Note that our ranking of these species is consistent with the observed trends in the proton affinities of these species. (The pK_a's are known to be 8.1 for $H_2NNH_3^+$, 9.3 for NH_4^+ and 13.6 for guanidinium$^+$ [*Sober* 1968]; and one can estimate a value of 11.2 for formamidinium$^+$ from the known value of 12.5 for acetamidinium$^+$ by subtracting 1.3 for the approximate effect of the removal of the methyl group [*Hall* and *Sprinkle* 1932 cf. p. 3478].)

We thus suggest that $NH_2 \cdot$ guanidinium$^+$ is the analogue of Cs^+, guanidinium$^+$ is the analogue of Rb^+, formamidinium$^+$ is the analogue of K^+ and so forth. Having done this, it becomes possible formally to generate a *pattern* of eleven selectivity rank orders for the polyatomic cations without necessity for quantitative calculations. How this is done is illustrated in Figure 11. First consider the group Ia cations in the upper portion of this figure. In the top row we rank them in their order of increasing positive field strength Cs^+, Rb^+, K^+, Na^+, Li^+. This is the sequence (*Eisenman* 1961, sequence I) of increasing energy of hydration and is the one in which they are expected to be selected by a site of "low anionic field strength".

By analogy, in the lower portion of Figure 11, we write the first sequence for the polyatomic ions as NH_2guan$^+$, guan$^+$, form$^+$, NH_4^+, hydraz$^+$, and call this sequence (I'). It may be thought of as representing the sequence of increasing strength of interaction with water molecules (for which there are unfortunately no direct data).

For the group Ia cations one can represent the expected changes in selectivity sequence by a site with increasing "field strength" through a series of reversals which occur with Rb^+ being the first species to cross to the left, followed by K^+, then Na^+, and lastly Li^+, as indicated on

---

*) A similar sequence has been seen for pyruvate kinase by *Wolcott* and *Boyer* (personal communication).

Monoatomic Ions

I	Cs	Rb	K	Na	Li
II	Rb	Cs	K	Na	Li
III	Rb	K	Cs	Na	Li
IV	K	Rb	Cs	Na	Li
V	K	Rb	Na	Cs	Li
VI	K	Na	Rb	Cs	Li
VII	Na	K	Rb	Cs	Li
VIII	Na	K	Rb	Li	Cs
IX	Na	K	Li	Rb	Cs
X	Na	Li	K	Rb	Cs
XI	Li	Na	K	Rb	Cs

Polyatomic Ions

I'	NH_2Guan	Guan	Form	NH_4	Hydraz	← HD Valinomycin
II'	Guan	NH_2 Guan	Form	NH_4	Hydraz	
III'	Guan	Form	NH_2 Guan	NH_4	Hydraz	
IV'	Form	Guan	NH_2 Guan	NH_4	Hydraz	
V'	Form	Guan	NH_4	NH_2 Guan	Hydraz	
VI'	Form	NH_4	Guan	NH_2 Guan	Hydraz	
VII'	NH_4	Form	Guan	NH_2 Guan	Hydraz	
VIII'	NH_4	Form	Guan	Hydraz	NH_2 Guan	
IX'	NH_4	Form	Hydraz	Guan	NH_2 Guan	← Gramicidin; Gall-bladder
X'	NH_4	Hydraz	Form	Guan	NH_2 Guan	
XI'	Hydraz	NH_4	Form	Guan	NH_2 Guan	← Na-channel

Fig. 11. Selectivity patterns for the group Ia cations and for the analogous series of "main species" polyatomic cations.

the figure, each ion waiting in turn until the previous one has crossed all the species it can. This simple procedure generates formally the pattern in which theoretically calculated selectivity isotherms cross each other (cf. *Eisenman* 1961, Fig. 1) and therefore leads to the same set of eleven sequences (indicated by the roman numerals I–XI)*).

Proceeding similarly for the polyatomic cations in the lower portion of the figure, we generate the eleven analogous sequences I'–XI' and find, *hocus-pocus*, that all the observed sequences of Table 4 are thereby generated! Thus, the sequence for hexadecavalinomycin is I', that for gramicidin (and the gallbladder) is IX', and that for the Na-channel is XI'. In addition the sequence seen for valinomycin and nonactin is consistent with any sequence between VII' and XI', as is the sequence seen for the activation of the enzyme dioldehydrase.

---

*) These sequences are rather independent of the particulars of the model chosen (*Eisenman* 1961, 1962) and were, indeed, first generated by *Eisenman, Rudin* and *Casby* (1957) by a qualitative line of reasoning similar to the present, and which can be found summarized on pp. 130–133 of *Mattock's* (1961) text. The variable in the present examples may indeed be not the "field strength" of the site, but rather its steric accessability to the ion. In consequence, the present pattern might come about because it represents the ease with which such ions shed water.

This is perhaps too good to be true, but it is important at least as an illustration that the "field strength" of a polyatomic cation as well as its "size" and "shape", should contribute importantly to its selection; and it offers some hope that one will not have to be totally empirical in developing electrodes specific to polyatomic cations of interest.

Acknowledgement

We are indebted to *V. I. Ivanov* and *Yu. Ovchinnikov*, of the Shemyakin Institute of the USSR Academy of Sciences, for their gift of the valinomycin analogs: hexadecavalinomycin, hydroxyisovalerate valinomycin, and lactate valinomycin. We also thank Dr. *Hans Bickel* and Miss *Barbara Stearns* for the samples of nonactin, monactin, dinactin and trinactin.

References

Buck, R. P.: Ion Selective Electrodes, Potentiometry, and Potentiometric Titrations. Anal. Chem. 44 (1972) 270R.

Ciani, S. M., Eisenman, G., Laprade, R., Szabo, G.: Theoretical Analysis of Carrier-mediated Electrical Properties of Bilayer Membranes. In: Membranes – A Series of Advances, Vol. 2, ed by *G. Eisenman.* Dekker, New York 1973

Ciani, S. M., Eisenman, G., Szabo, G.: A Theory for the Effects of Neutral Carriers such as the Macrotetralide Actin Antibiotics on the Electric Properties of Bilayer Membranes. J. Memb. Biol. 1 (1969) 1–36

Cram, D. J., Cram, J. M.: Host-Guest Chemistry. Complexes between Organic Compounds simulate the Substrate Selectivity of Enzymes. Science 183 (1974) 803

Eisenman, G.: On the Elementary Atomic Origin of Equilibrium Ionic Specificity. In: Symposium on Membrane Transport and Metabolism, ed. by *Kleinzeller* and *A. Kotyk*. Academic Press, New York 1961

Eisenman, G.: Cation Selective Glass Electrodes and their Mode of Operation. Biophys. J. 2 (1962) 259–323

Eisenman, G.: The Electrochemistry of Cation Sensitive Glass Electrodes. In: Advances in Analytical Chemistry and Instrumentation, IV, 215–369, ed. by *C. N. Reilley*. Interscience, New York 1965.

Eisenman, G.: The Origin of the Glass Electrode Potential. In: Glass Electrodes for Hydrogen and Other Cations: Principles and Practice, pp. 133–173. Dekker, New York 1967

Eisenman, G.: Theory of Membrane Electrode Potentials. In: Ion-selective Electrodes, ed. by *R. A. Durst*. National Bureau of Standards Special Publication 314 (1969) 1–56

Eisenman, G.: Bioelectrodes. In: Modern Techniques in Physiological Science, ed. by *J. F. Gross, R. Kaufmann,* and *E. Wetterer.* Academic Press, London 1973

Eisenman, G., Krasne, S.: The Selectivity of Carrier Antibiotics for Substituted Ammonium Ions. Biophys. Soc. Abstr. 244a (1973)

Eisenman, G., Krasne, S.: The Ion Selectivity of Carrier Molecules, Membranes and Enzymes. In: MTP International Review of Science. Biochemistry Series, pp. 1–33, ed. by *C. F. Fox.* Butterworths, London 1974

Eisenman, G., Ciani, S., Szabo, G.: The Effects of the Macrotetralide Actin Antibiotics on the Equilibrium Extraction of Alkali Metal Salts into Organic Phases. J. Memb. Biol. 1 (1969) 294–345

Eisenman, G., Rudin, D. O., Casby, J. U.: Principles of Specific Ion Interactions. 10th Ann. Conf. on Elect. Techniques in Medicine and Biology of the AIEE. I.S.A. and I.R.E. (1957)

Eisenman, G., Szabo, G., Ciani, S., McLaughlin, S. G. A., Krasne, S.: Ion Binding and Ion Transport Produced by Lipid-soluble Molecules. Prog. Surf. Memb. Sci. 6 (1973) 139–241

Goldman, D. E.: Potential, Impedance, and Rectification in Membranes. J. Gen. Physiol. 27 (1943) 37

Goodall, M. C.: Structural Effects in the Action of Antibiotics on the Ion Permeability of Lipid Bilayers. III. Gramicidins "A" and "S", and Lipid Specificity. Biochim. Biophys. Acta 219 (1970) 471

Hall, J. E., Mead, C. A., Szabo, G.: A Barrier Model for Current Flow in Lipid Bilayer Membranes. J. Memb. Biol. 11 (1973) 75

Hall, N. F., Sprinkle, M. R.: Relations between the Structure and Strength of Certain Organic Bases in Aqueous Solution. J. Amer. Chem. Soc. 54 (1932) 3469

Hillle, B.: The Permeability of the Sodium Channel to Organic Cations in Myelinated Nerve. J. gen. Physiol. 58 (1971) 599–619

Hille, B.: The Permeability of the Sodium Channel to Metal Cations in Myelinated Nerve. J. gen. Physiol. 58 (1972) 637–658

Hladky, S. B.: The Energy Barriers to Ion Transport by Nonactin across Thin Lipid Membranes. Biochim. Biophys. Acta 352 (1974) 7185

Hladky, S. R., Haydon, D. A.: Ion Transfer across Lipid Membranes in the Presence of Gramicidin A. Biohcim. Biophys. Acta 273 (1972) 294–312

Hodgkin, A. L., Katz, B.: The Effect of Sodium Ions on the Electrical Activity of the Giant Axon of the Squid. J. Physiol. (London) 108 (1949) 37–77

Laprade, R., Ciani, S. M., Eisenman, G., Szabo, G.: The Kinetics of Carrier-mediated Ion Permeation in Lipid Bilayers and its Theoretical Interpretation. In: Membranes – A Series of Advances, Vol. 3, ed. by G. Eisenman. Dekker, New York 1975, pp. 127–214

Läuger, P., Stark G.: Kinetics of Carrier-mediated Ion Transport across Lipid Bilayer Membranes. Biochim. Biophys. Acta 211 (1970) 458–466

Lev, A. A., Buzhinsky, E. P., Grenfeldt, A. E.: Electrochemistry of Bimolecular Phospholipid Membranes with Valinomycin at Nonzero Current. Proc. XXIV Internat. Congr. Physiol. Sci. p. 39. Excerpta Medica, Amsterdam 1968

Lev, A. A., Malev, V. V., Osipov, V. V.: Electrochemical Properties of Thick Membranes with Macrocyclic Antibiotics. In: Membranes-A Series of Advances, Vol. 2, ed by G. Eisenman. Dekker, New York 1973

Markin, V. S., Kristalik, L. I., Liberman, E. A., Topaly, V. P.: Mechanism of Conductivity of Artificial Phospholipid Membranes in Presence of Ion Carriers. Biofizika 14 (1969) 256

Mattock, G.: pH Measurements and Titration. Heywood, London 1961

Moreno, J. H., Diamond, J. M.: Role of Hydrogen Bonding in Organic Cation Discrimination by Gallbladder Epithelium. Nature 247 (1974a) 368

Moreno, J. H., Diamond, J. M.: Nitrogenous Cations as Probes of Permeation Channels. J. Memb. Biol. (1974b) 197–259

Morf, W. E., Ammann, D., Pretsch, E., Simon, W.: Carrier Antibiotics and Model Compounds as Components of Selective Ion-sensitive Electrodes. Pure and applied Chem. 36 (1973) 421

Myers, V. B., Haydon, D. A.: Ion Transfer across Lipid Membranes in the Presence of Gramicidin A. Biochim., Biophys. Acta 274 (1972) 313–322

Ovchinnikov, Yu. A.: Rational Synthetic Routes to Depsipeptides from the Standpoint of their Structure-function Relations. XXIIIrd Internat. Congr. Pure and Appl. Chemistry. 2 (1971) 121

Simon, W.: Alkali Cation Specificity of Antibiotics, their Behavior in Bulk Membranes and Useful Sensors Based on these. In: Symposium on Molecular Mechanisms of Antibiotic Action on Protein Biosynthesis and Membranes, ed. by D. Vasquez. Springer, Berlin 1971

Simon, W., Morf, W.: Alkali Cation Specificity of Carrier Antibiotics and their Behavior in Bulk Membranes. In: Membranes-A Series of Advances, Vol. 2, ed. by G. Eisenman. Dekker, New York 1973

Sober, H. A.: Handbook of Biochemistry, J-184. Chemical Rubber Co., Cleveland/Ohio 1968

Stark, G., Benz, R.: The Transport of Potassium through Lipid Bilayer Membranes by the Neutral Carriers Valinomycin and Monactin. J. Memb. Biol. 5 (1971) 133–153

Stefanac, Z., Simon, W.: In-vitro-Verhalten von Makrotetroliden in Membranen als Grundlage für hochselektive kationenspezifische Elektrodensysteme*. Chimia (Switzerland) 20 (1966) 436

Szabo, G., Eisenman, G., Ciani, S.: The Effects of the Macrotetralide Actin Antibiotics on the Electrical Properties of Phospholipid Bilayer Membranes. J. Memb. Biol. 3 (1969) 346–382

Szabo, G., Eisenman, G., Ciani, S. M., Laprade, R., Krasne, S.: Experimentally Observed Effects of Carriers on the Electrical Properties of Membranes. The Equilibrium Domain. In: Membranes-A Series of Advances, Vol. 2, ed. by G. Eisenman, Dekker, New York 1973

Toraya, T., Fukui, S.: Activation of Dioldehydrase by Formamidinium or Guanidinium Ion, Polyatomic Monovalent Cations Having sp^2 Nitrogen. Biochem. Biophys. Res. Comm. (1974) 54 (1973) 862–866

Tosteson, D. C.: Effect of Macrocyclic Compounds on the Ions Permeability of Artificial and Natural Membranes. Fed. Proc. 27 (1968) 1269

Urry, D. W.: The gramicidin A transmembrance channel: A proposed $\pi_{(L, D)}$ helix. Proc. Nat. Acad. Sci. 68 (1971) 672

Urry, D. W., Goodall, M. C., Glickson, J. D., Mayers, D. F.: The Gramicidin A Transmembrane Channel: Characteristics of Head-to-head Dimerized $\pi_{(L, D)}$ Helices. Proc. Nat. Acad. Sci. 68 (1971) 1907–1911

D. Ammann, R. Bissig, Z. Cimerman, U. Fiedler, M. Güggi, W. E. Morf, M. Oehme, H. Osswald, E. Pretsch and W. Simon

Synthetic Neutral Carriers for Cations

(11 Figures)

Abstract

On the basis of model calculations, a series of electrically neutral lipophilic molecules was designed and synthesized. These carrier ligands may be tailored to ion selectivities suitable for practical application as components in liquid-membrane electrodes selective for Ca^{2+}, Ba^{2+}, Li^+ and Na^+ respectively. Some of the selectivities observed are far superior to systems known to date. By incorporating these ligands in PVC membranes, life times of more than one year may be obtained. Through a special treatment with lipophilic anions (e.g., tetraphenylborate), silver surfaces may be coated with thin layers of ligand-impregnated PVC to obtain metal-contacted membrane electrodes of high EMF stability. They are perfectly suited for use as components in flow-through and in miniaturized electrode systems.

Introduction

The recent efforts in the field of ion-selective electrodes were concentrated on the fundamental understanding of the membrane processes involved [1–7], the development of newer applications of available electrode systems [3], [5–8] as well as the design of new ion sensors [3], [5–7].

The different types of ion-selective membrane electrodes known so far may be classified as follows (see however [6]):

a) *Solid Membranes* (fixed ion-exchange sites)
 – Homogeneous: Glass membrane
  Crystal membrane
 – Heterogeneous: Crystalline substance
  in inert matrix

b) *Liquid Membranes* (mobile ion-exchange sites)
 – Charged ligand
 – Neutral ligand

c) *Special Electrodes*
 – Gas-sensing electrodes
 – Enzyme-substrate electrodes

In the last few years, the development of new ion-selective electrodes was mainly directed toward special electrodes (gas-sensing electrodes [9, 10], enzyme-substrate electrodes [11–13]) as well as liquid-membrane electrodes [14–21]. Liquid-membrane sensors offer a wide range of accessible ion selectivities. In these electrodes, mobile ion-selective sites (e.g., ion-selective

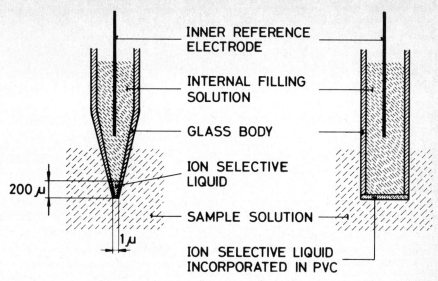

Fig. 1. Diagrammatic representation of liquid-membrane micro- [22] and PVC-electrodes [21] (The abbreviations used are summarized in the experimental part).

ligand dissolved in an appropriate solvent) are interposed between the sample solution and a reference system as shown diagrammatically in Figure 1 for a microelectrode [22] and a more conventional ion electrode. By incorporating the mobile sites into solvent-impregnated PVC, systems of high electromotive and mechanical stability with electrode life-times of more than one year may be obtained [21] (Fig. 1). Since a miniaturization of liquid-membrane electrodes is easily realized and since such tools are attractive for biomedical applications [23], efforts in our laboratory were directed toward the design of ion-selective ligands for alkali and alkaline earth metal cations.

In a first approximation, the EMF of an electrochemical cell containing a membrane electrode can be described by an extended *Nicolsky* equation:

$$E = E_0 + \frac{RT}{z_i F} \ln \left[a_i + \sum_{i \neq j} K_{ij}^{Pot} (a_j)^{z_i/z_j} \right] \quad (1)$$

where:
E: cell potential (EMF)
E_0: constant reference potential
a_i: activity of a primary ion I^{z_i} in the sample solution
a_j: activity of an interfering ion J^{z_j} in the sample solution
K_{ij}^{Pot}: selectivity factor, characteristic of a given membrane
RT/F: *Nernst* factor

If the liquid membrane does not contain a complexing agent for the monovalent cations I^+ and J^+ the selectivity factor measuring the preference of J^+ relative to I^+ by the sensor may be approximated by the ratio of the partition coefficients k_j and k_i of the respective cations between the sample solution and the membrane [14, 24, 25]:

$$K_{ij}^{Pot} = \frac{k_j}{k_i} \qquad (2)$$

Outstanding selectivities between cations may be obtained if a selective complexing agent for the ion to be measured is incorporated into the membrane phase. For electrically neutral ligands S the selectivity between cations of the same charge becomes [14, 24, 25]:

$$K_{ij}^{Pot} = \frac{k_j}{k_i} \cdot \frac{K_{js}}{K_{is}} \qquad (3)$$

where K_{js}, K_{is} are the complex formation constants between the ligand S and the cations within the membrane. The selectivity factor K_{ij}^{Pot} given in Equation 3 corresponds to the equilibrium constant for the exchange reaction ($z = 1$):

$$IS_n^{z+} \text{ (membrane)} + J^{z+} \text{ (solution)} \rightleftharpoons JS_n^{z+} \text{ (membrane)} + I^{z+} \text{ (solution)} \qquad (4)$$

For electrically charged ligands S⁻ selectivity factors are observed that lie somewhere between the limiting values given by Equations 2 and 3 [14]; thus, the potential for selective behavior as suggested by the quotient K_{js}/K_{is} of the complex formation by S⁻ often cannot be fully exploited for controlling the selectivity of the corresponding sensors. Equation 3, however, suggests that extremely high selectivities can be achieved by using neutral ion-specific ligands (ion carriers, ionophores) as membrane components. This is the reason why the design and synthesis of ligands was initated in this direction.

In order for such ligands to behave as carriers for metal cations in a lipophilic membrane the most important requirements are the following:

1. *Lipophilicity:* The ligand and the complex have to be sufficiently soluble in the membrane phase.
2. *Mobility:* An adequate mobility of both ligand and complex are guaranteed only as long as the over-all dimensions of the carrier remain within limits, but are still compatible with high lipid-solubility.
3. *Complex-formation constant* K_{is}: The electrode response becomes especially selective for the ion I if K_{ij}^{Pot} is small (Eq. 1). This is true for $K_{is} k_i \gg K_{js} k_j$. Furthermore, it can be shown theoretically and experimentally [26] that a cation response is obtained only if an excess of uncomplexed ligand is present within the membrane, i.e., K_{is} has an acceptable upper limit. The limiting value for $K_{is} \cdot k_i$ is of the order of unity if a cationic response up to 1 M sample solutions is demanded.
4. *Kinetics:* The ion-exchange kinetics (see Eq. 4) have to be compatible with the demanded response time of the membrane electrode.

High-selectivity complexing agents for hard cations are multidentate ligands which lock the cation in question into a rather rigid arrangement of coordinating sites [25, 27]. The most important molecular parameters for such a complexing agent that fulfills the requirements mentioned above are:

a) *Coordination number, cavity:* A carrier molecule should be a multidentate ligand which is able to assume a stable conformation that provides a cavity; the cavity formed by a given number of polar coordinating groups is suited for the uptake of a cation, while the nonpolar groups form a lipophilic shell around the coordination sphere. A cavity that snugly fits the cation in question is desirable [25].

b) *Ligand atoms:* For A-cations, the polar coordinating groups preferably contain oxygen as ligand atoms. Principally, amine-nitrogens conform to the specifications given [28] but were avoided here, however, in order to eliminate interference by protonation reactions.

Synthetic Neutral Carriers for Cations

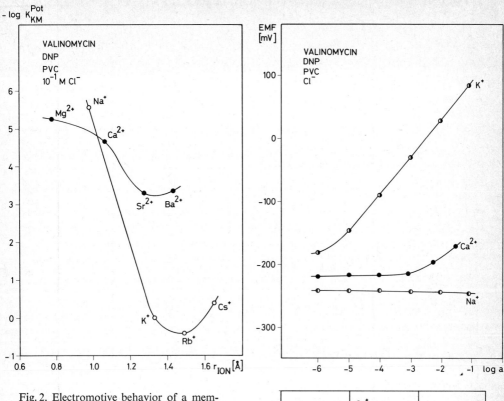

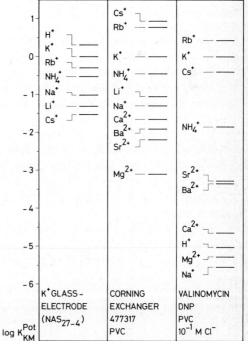

Fig. 2. Electromotive behavior of a membrane electrode assembly using valinomycin as membrane component and comparison of its selectivities with those of a glass [34] and p-chloro-tetraphenylborate liquid membrane electrode [35] showing response to K^+.

Membrane composition:
 2.7% by weight valinomycin
 67.2%   DNP
 30.1%   PVC

c) *Arrangement of the coordinating centers:* These centers should preferably be arranged so as to form five-membered chelate rings with the non-solvated cation I [29].
d) *Size of the ligand:* A small thickness of the ligand layer around the central atom leads to a preference of divalent relative to monovalent cations of the same size; this is of special importance when using polar membrane solvents [30].
e) *Dipole moment of the coordinating sites:* Increasing dipole moments increase the stability of the complex and increase the preference of divalent relative to monovalent cations of the same size, other parameters (e.g., orientation of dipole) remaining constant [25].

The antibiotics valinomycin and the macrotetrolides [25] ideally meet the requirements mentioned above and are used in a variety of commercially available electrodes for the measurement of K^+ (valinomycin) and NH_4^+ (nonactin, monactin) [31]. By incorporating these ion carriers into PVC, membrane electrodes with the electromotive behavior shown in Figures 2 and 3 are obtained (see also [20, 32, 33]). Unfortunately, other highly selective natural products similarly predestined as components for ion-selective electrodes have not yet been found.

Certain representatives of the synthetic crown compounds [37] show selectivities of K^+ over Na^+ and can be used as components in liquid-membrane electrodes [38]. The highest selectivities of K^+ over Na^+ found [19, 20] however, are still an order of magnitude lower than those obtained when using valinomycin [31, 39]. Due to low lipophilicity (see 1.) and especially to slow exchange kinetics (see 4.), the synthetic macroheterobicyclic ligands [28], which show very high selectivities for A-cations, are unfortunately unsuitable as components for liquid-membrane electrodes. In perfect agreement with requirements b), c), and d) a number of polyethyleneglycols show selectivities for divalent relative to monovalent cations; they have been used as components in sensors for Ba^{2+} [40].

Ion-Selective Electrodes Based on Synthetic Neutral Ligands

According to requirements (a) through (e), above, a series of carrier molecules suitable for liquid-membrane electrodes responsive to alkali and alkaline earth metal cations has been synthesized. Out of 154 molecules prepared, the four shown in Figure 4 are, so far, the most attractive ones.

The selectivity of such ligands can be drastically influenced by the choice of the membrane solvent. An increase in the dielectric constant of a typical membrane solvent (water-immiscible liquid of low vapor-pressure, compatible with PVC, no functional groups which can undergo protonation reactions) increases the selectivity of divalent over monovalent cations of the same size and vice-versa [25]. The power residing within this parameter is illustrated in Figure 5. To determine the selectivity factors presented in Figures 2, 3 and 6 to 9 membrane solvents have been chosen correspondingly.

Figure 6 clearly demonstrates that the selectivity of the Ca^{2+} sensor based on the neutral carrier shown in Figure 4 is especially in respect to Mg^{2+}, H^+ and Zn^{2+} far superior to the values for the Orion liquid-ion-exchange electrode [21, 31, 42]. The selectivity in respect to Na^+ for a PVC electrode incorporating a modified ion-selective component is much higher in a buffered than in an unbuffered system [16]; the same is true for the neutral carrier electrode, with the difference that the performance is even better (see Fig. 10). Since the discrimination of Na^+ and K^+ is adequate for blood serum studies and the discrimination of protons, Zn^{2+} and Mg^{2+} is exceptionally high, it appears that the electrode described here is unsurpassed as

Synthetic Neutral Carriers for Cations

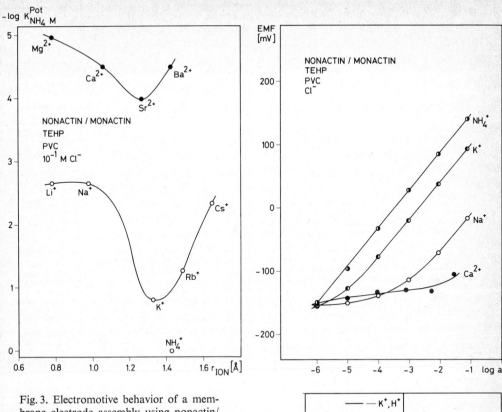

Fig. 3. Electromotive behavior of a membrane electrode assembly using nonactin/monactin as membrane component and comparison of its selectivities with those of a glass electrode [36] showing response to NH_4^+.

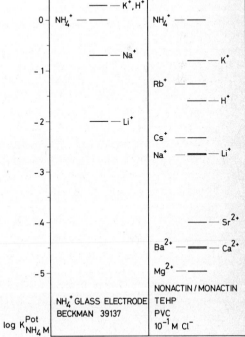

Membrane composition:
4.6% by weight nonactin/monactin
68.9%   TEHP
26.5%   PVC

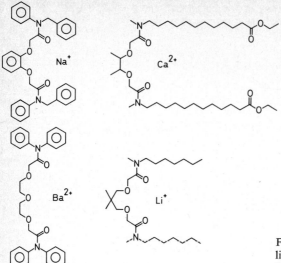

Fig. 4. Structure of synthetic ion-selective ligands showing useful selectivities for Ca^{2+}, Ba^{2+}, Na^+ and Li^+ repectively.

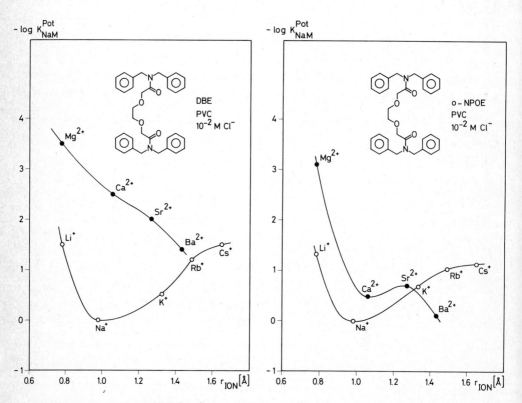

Fig. 5. Influence of the membrane solvent on the selectivity of the corresponding neutral carrier liquid membrane electrodes (see [41]).
o-NPOE: dielectric constant ≈ 24; DBE: dielectric constant ≈ 4.

Synthetic Neutral Carriers for Cations

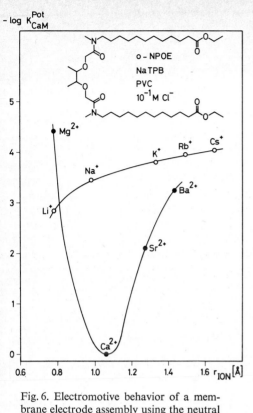

Fig. 6. Electromotive behavior of a membrane electrode assembly using the neutral Ca^{2+} carrier (Fig. 4) as membrane component and comparison of its selectivities with those of the Orion 92–20 Ca^{2+} electrode [21, 31].

Membrane composition:
0.9% by weight Ca^{2+} carrier
0.4%   Na TPB
64.3%   o-NPOE
34.4%   PVC

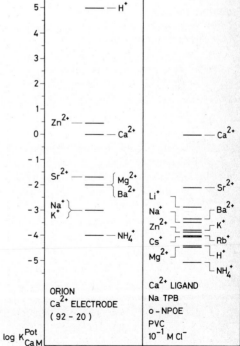

D. Ammann et al.

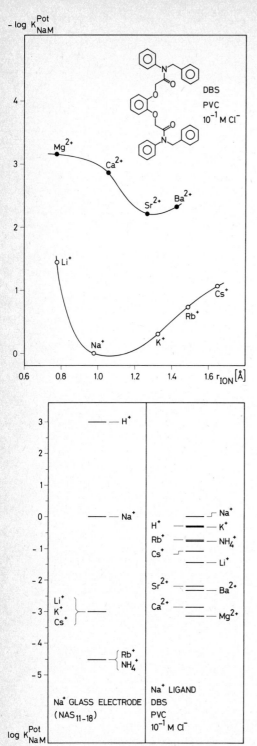

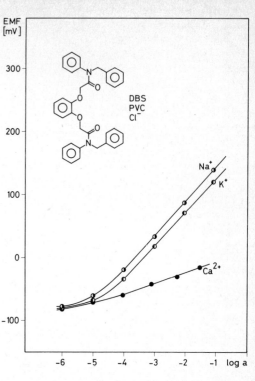

Fig. 7. Electromotive behavior of a membrane electrode assembly using the neutral Na$^+$ carrier [46] (Fig. 4) as membrane component and comparison of its selectivities with those of a glass electrode [31] showing response to Na$^+$.

Membrane composition:
0.6% by weight Na$^+$-carrier
0.1%   NaSCN
66.3% DBS
33.0% PVC

Synthetic Neutral Carriers for Cations

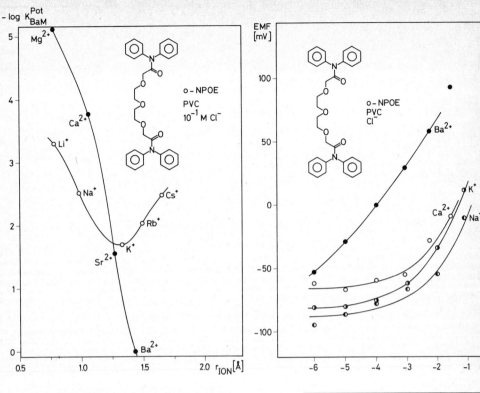

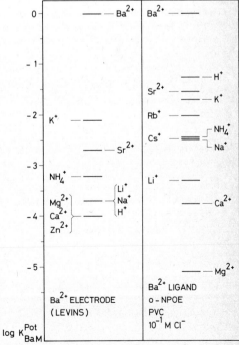

Fig. 8. Electromotive behavior of a membrane electrode assembly using the neutral Ba²⁺ carrier (Fig. 4) as membrane component and comparison of its selectivities with those of a neutral carrier Ba²⁺ electrode described by *Levins* [40].

Membrane composition:
0.6% by weight Ba^{2+}-carrier
66.3%   o-NPOE
33.1%   PVC

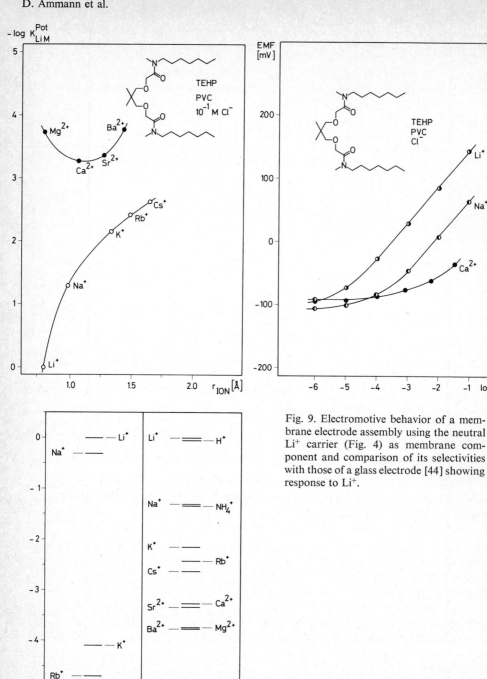

Fig. 9. Electromotive behavior of a membrane electrode assembly using the neutral Li$^+$ carrier (Fig. 4) as membrane component and comparison of its selectivities with those of a glass electrode [44] showing response to Li$^+$.

Membrane composition:
5.8% by weight Li$^+$-carrier
62.8%   TEHP
31.4%   PVC

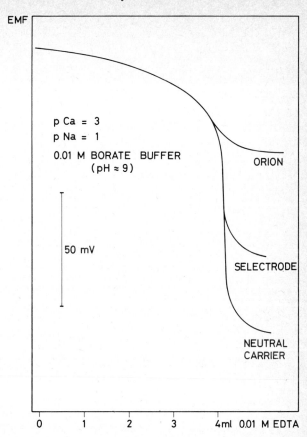

Fig. 10. Titrations of $1.0 \cdot 10^{-3}$ M $CaCl_2$ with EDTA at pH 9 using a neutral carrier electrode [45] (see Figs. 4 and 6), an Orion 92-20 electrode (see Fig. 6 in [16]) as well as a Selectrode (see Fig. 6 in [16]).

far as measurements in blood serum are concerned. In measurements of total calcium activities in blood serum, protein-bound Ca^{2+} can be replaced by Zn^{2+} ions and therefore high selectivities relative to Zn^{2+} are of interest [43].

The Ba^{2+}-selective liquid-membrane electrode [45] displays, except for magnesium, worse selectivies throughout than the Ba^{2+} sensor described by Levins [40]. However, since our electrodes show a much higher EMF stability with lifetimes longer than 11 months and the design of ligands selective for Ba^{2+} ions is still in progress, further improvement of this type of electrode is probable.

A Na^+-selective electrode based on a neutral carrier has been described recently [41] (see Fig. 5). The ligand presented in Figure 4 shows slightly different selectivities, increased solubility in the membrane phase and may be used for electrodes of considerably increased life-time [46]. Although the selectivities of sodium-responsive glass electrodes are usually superior [1, 31], except in respect to H^+, this and further improved sodium liquid-membrane electrodes will bring advantages, especially in blood serum measurements and in the preparation of microelectrodes (see below).

The first Li^+-selective electrode based on a neutral carrier is presented here [45,47]. Although its selectivity relative to Na^+ is not yet adequate for a direct monitoring of Li^+ in blood serum, e.g., in the therapy of maniacal depressive psychosis [48], it might be attractive as a reference electrode. In such applications a constant activity of Li$^+$ in the sample solution has to be generated.

Electrode Design

By dissolving the ion-selective ligands (Fig. 4) in an adequate solvent (see Figs, 2, 3 and 6 to 9) ion-selective liquids may be obtained for direct use as membrane components in microelectrodes (see Fig. 1). Multi-channel microelectrodes with tip diameters of about 1 µ and up to four channels have been prepared [49–51]. They are potentially useful for the measurement of activity ratios of different ions in extremely small sample volumes [50].

It has been pointed out repeatedly [52, 53] that the interference in the cation response by lipid-soluble sample anions is still a severe limitation of neutral-carrier liquid-membrane electrodes. A theoretical treatment [54, 55] shows, however, that there are means to eliminate or at least reduce such an anion interference by the permanent incorporation of lipophilic anions, e.g., tetraphenylborate, into the membrane phase. This tetraphenylborate may simultaneously be used to produce a thermodynamically reversible couple with silver, by covering silver with silver-tetraphenylborate. These surfaces may be coated with thin layers of ligand-impregnated PVC (containing tetraphenylborate) to obtain metal-contacted membrane electrodes [56] [57] of high EMF stability. They are suited for use as components in flow-through (see Fig. 11) and miniaturized electrode systems. For multi-ion monitoring several of the units shown in Figure 11 (approx. 10 µl dead volume per unit) may be stacked together.

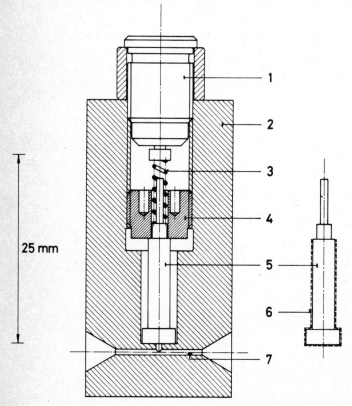

Fig. 11. Flow-through liquid membrane electrode [58]:
1: plug
2: plastic electrode body
3: metal spring
4: plastic screw to fix membrane carrier
5: membrane carrier (silver)
6: PVC membrane
7: sample channel

Future Prospects

There is no doubt that a more detailed study of the selectivity/structure relationship [25] will help in designing molecules with ion selectivities far superior to those obtained so far. In view of the analytical use of enzyme reactions [11–13], the measurement of NH_4^+ is of special interest and therefore the design of such electrodes is actively pursued. Since chiral ion-selective ligands can be prepared [59], we hope to be able to obtain enantiomer-selective liquid-membrane electrodes in due time.

Experimental

The preparation of the PVC membranes and the measuring technique have been described in detail elsewhere [41, 61] (see also [60]). Throughout, cells of the type

Hg; Hg_2Cl_2, KCl (satd)/electrolyte bridge/sample solution//
membrane//internal filling solution, AgCl; Ag

with double-junction reference electrodes (Philips R 44/2-SD/1) and electrode bodies (Philips IS 561) to mount the membranes (Philips liquid membrane electrodes IS 561-K^+, —Ca^{2+} and —NH_4^+ respectively) have been used at 25 °C. The internal filling solutions were aqueous 0.1 M KCl, 0.01 M NH_4Cl, 0.01 M $CaCl_2$, 0.01 M NaCl, 0.01 M $BaCl_2$ and 0.01 M LiCl to obtain the data presented in Figures 2, 3 and 6 to 9 respectively. The bridge electrolyte used was aqueous 1 M lithium acetate to obtain the data presented in Figures 2 and 3 and 0.1 M NH_4NO_3 for those given in Figures 6 to 9. Throughout, the separate-solution technique and the concentrations of sample solutions indicated in the figures were used to determine the selectivity factors.

The activity coefficients for NH_4^+, Li^+, Ca^{2+} and Ba^{2+} were estimated using *Debye-Hückel* equations [62–64] and for K^+ [65] and Na^+ [66] published values were used [65, 66].

Abbreviations used:

PVC: polyvinyl chloride
TPB: tetraphenylborate
DNP: dinonyl phthalate
TEHP: tris (2-ethylhexyl) phosphate
DBS: dibutyl sebacate
o-NPOE: o-nitro-phenyloctyl ether
DBE: dibenzyl ether

Acknowledgment:

This work was partly supported by the Schweizerischer Nationalfonds zur Förderung der wissenschaftlichen Forschung.

References

[1] *Eisenman, G.:* In: International Symposium on Modern Technology in Physiological Sciences, p. 245, ed. by *J. F. Gross, R. Kaufmann, E. Wetterer.* Academic Press, New York 1973
[2] *Rechnitz, G. A.:* Pure and applied Chem. 36 (1973) 457
[3] *Koryta, J.:* Anal. Chim. Acta 61 (1972) 329
[4] *Wuhrmann, H.-R., Morf, W. E., Simon, W.:* Helv. Chim. Acta 56 (1973) 1011
[5] *Buck, R. P.:* Anal. Chem. 46 (1974) 28R; 44 (1972) 270R
[6] *Covington, A. K.:* CRC Critical Reviews in Analytical Chemistry 3 (1974) 355
[7] *Pungor, E., Tóth, K.:* Pure and applied Chem. 36 (1973) 441
[8] *Pungor, E., Tóth, K.:* Vom Wasser, Vol. 42, p. 43. Verlag Chemie, Weinheim 1974
[9] *Ross, J. W., Riseman, J. H., Krueger, J. A.:* Pure and applied Chem. 36 (1973) 473
[10] *Ruzicka, J., Hansen, E. H.:* Anal. Chim. Acta 69 (1974) 129
[11] *Guilbault, G. G.:* In: Enzyme Engineering, p. 361, ed. by *L. B. Wingard,* Jr. Wiley, New York 1972
[12] *Weetall, H. H.:* Anal. Chem. 46 (1974) 603A
[13] *Moody, G. J., Thomas, J. D. R.:* Analyst 100 (1975) 609
[14] *Morf, W. E., Ammann, D., Pretsch, E., Simon, W.:* Pure and applied Chem. 36 (1973) 421
[15] *Pretsch, E., Ammann, D., Simon, W.:* Research/Development 25 (3) (1974) 20
[16] *Ruzicka, J., Hansen, E. H., Tjell, J. Chr.:* Anal. Chim. Acta 67 (1973) 155
[17] *Sharp, M.:* Diss., Univ. Umea, Umea 1973
[18] *Herman, H. B., Rechnitz, G. A.:* Science 184 (1974) 1074
[19] *Petránek, J., Ryba, O.:* Abstracts of IUPAC Meeting, Cardiff, Paper 13, April 1973
[20] *Ryba, O., Knižáková, E., Petránek, J.:* Coll. Czechoslov. Chem. Comm. 38 (1973) 497; *Ryba, O., Petránek, J.:* J. Electroanal. Chem. 44 (1973) 425
[21] *Moody, G. J., Thomas, J. D. R.:* Selective Ion-Sensitive Electrodes. Merrow, Watford, Herts/England 1971
[22] *Walker, J. L.* Jr.: Anal. Chem. 43 (1971) 89A
[23] *Rechnitz, G. A.:* Research/Development 24 (8) (1973) 18
[24] *Eisenman, G.:* In: Ion-Selective Electrodes, ed. by *R. A. Durst.* National Bureau of Standards, Spec. Publ. 314, Washington 1969
[25] *Simon, W., Morf, W. E., Meier, P. Ch.:* Structure and Bonding 16 (1973) 113
[26] *Morf, W. E., Kahr, G., Simon, W.:* unpublished results
[27] *Diebler, H., Eigen, M., Ilgenfritz, G., Maas, G., Winkler, R.:* II. I.C.C.C., Haifa, 1968; Pure and applied Chem. 20 (1969) 93
[28] *Dietrich, B., Lehn, J. M., Sauvage, J. P.:* Tetrahedron Letters 34 (1969) 2885, 2889
[29] *Schwarzenbach, G.:* Helv. chim. Acta 35 (1952) 2344
[30] *Morf, W. E., Simon, W.:* Helv. chim. Acta 54 (1971) 2683
[31] *Cammann, K.:* Das Arbeiten mit ionenselektiven Elektroden. Springer, Berlin 1973
[32] *Fiedler, U., Ruzicka, J.:* Anal. Chim. Acta 67 (1973) 179
[33] *Scholer, R.:* Diss. ETHZ Nr. 4940, Zürich 1972
[34] *Moore, W. E.:* In: Glass Electrodes for Hydrogen and Other Cations, ed. by *G. Eisenman.* Dekker, New York 1967
[35] *Davies, J. E. W., Moody, G. J., Price, W. M., Thomas, J. D. R.:* Laboratory Practice 22 (1973) 20
[36] *Guilbault, G. G., Smith, R. K., Montalvo, J. G.* Jr.: Anal. Chem. 41 (1969) 600
[37] *Pedersen, C. J.:* J. Amer. Chem. Soc. 89 (1967) 2495, 7017; 92 (1970) 386, 391
[38] *Rechnitz, G. A., Ehud Eyal:* Anal. Chem. 44 (1972) 370
[39] *Pioda, L. A. R., Stankova, V., Simon, W.:* Anal. Letters 2 (1969) 665
[40] *Levins, R. J.:* Anal Chem. 43 (1971) 1045; 44 (1972) 1544
[41] *Ammann, D., Pretsch, E., Simon, W.:* Anal. Letters 7 (1) (1974) 23
[42] *Orion:* Electrode Specification 1972
[43] *Christiansen, T. F.:* private communication
[44] *Eisenman, G.:* In: Advances in Analytical Chemistry and Instrumentation, Vol. 4, p. 213. Wiley-Interscience, New York 1965
[45] *Güggi, M., Ammann, D., Pretsch, E., Simon, W.:* in preparation
[46] *Ammann, D., Bissig, R., Güggi, M., Pretsch, E., Simon, W., Borowitz, I. J., Weiß, L.:* Helv. Chim. Acta 58 (1975) 1535
[47] *Boron-Rettore, P. M.:* Diplomarbeit ETH, Zürich 1974
[48] *Ceriotti, G.:* In: Clinical Biochemistry, p. 1566, ed. by *H. Ch. Curtis* and *M. Roth.* de Gruyter, Berlin 1974

[49] *Kessler, M., Hajek, K., Simon, W.:* International Workshop on Ion Selective Electrodes and on Enzyme Electrodes in Biology and in Medicine. Reisensburg near Ulm/Germany, Sept. 15–18, 1974
[50] *Simon, W., Pretsch, E., Morf, W. E., Güggi, M., Kahr, G., Kessler, M.:* The Society of Analytical Chemistry, Centenary Celebrations, Paper C 55, London, July 1974
[51] *Neher, E., Lux, H. D.:* Pflügers Arch. Suppl. 332 (1972) R 88
[52] *Lal, S., Christian, G. D.:* Anal. Letters 3 (1970) 11
[53] *Boles, J. H., Buck, R. P.:* Anal. Chem. 45 (1973) 2057
[54] *Morf, W. E., Kahr, G., Simon, W.:* Anal. Letters 7 (1) (1974) 9
[55] *Morf, W. E., Ammann, D., Simon, W.:* Chimia 28 (1974) 65
[56] *Cattral, R. W., Freiser, H.:* Anal. Chem. 43 (1971) 1905
[57] *Smith, M. D., Genshaw, M. A., Greyson, J.:* Anal. Chem. 45 (1973) 1782
[58] *Simon, W., Kahr, G., Oehme, M., Dohner, R. E.:* in preparation
[59] *Helgeson, R. C., Timko, J. M., Cram, D. J.:* J. Amer. Chem. Soc. 95 (1973) 3023
[60] *Craggs, A., Moody, G. J., Thomas, J. D. R.:* J. Chem. Education 51 (1974) 541
[61] *Pick, J., Tóth, K., Pungor, E., Vasák, M., Simon, W.:* Anal. Chim. Acta 64 (1973) 477
[62] *Robinson, R. A., Stokes, R. H.:* Electrolyte Solutions, p. 229. Butterworths, London 1968
[63] *Bates, R. G., Alfenaar, M.:* In: Ion-Selective Electrodes, p. 191, ed. by *R. A. Durst.* National Bureau of Standards. Spec. Publ. 314, Washington, 1969
[64] *Bates, R. G.:* In: International Symposium on Selective Ion-Sensitive Electrodes in Cardiff. p. 407, ed. by *G. J. Moody.* Butterworths, London, 1973
[65] *Staples, B. R.:* National Bureau of Standards, Certificate Nr. 2202 for Potassium, Washington, Febr. 1971
[66] *Staples, B. R.:* National Bureau of Standards Certificate Nr. 2201 for Sodium, Washington, Febr. 1971

Richard P. Buck

Application of Solid-State Principles to Response Interpretation and Design of Solid Membrane Electrodes

When silver wire is completely imbedded in a AgX or Ag_2S crystal, pressed pellet or cast membrane, a so-called "all-solid-state" selective membrane electrode is formed (*Farren* and *Staunton* 1970; *Staunton* 1970; *Vesely* and *Jindra* 1968) whose response is reversible to solution activities $a(Ag^+)$ and $a(X^-)$ or $a(S^=)$. Inert electronic conductors and metals whose salt formation free energies are positive relative to silver, e.g., C, Hg and Pt behave similarly (*Koebel* 1974; *Marton* and *Pungor* 1971; *Ruzicka* and *Lamm* 1971a; *Ruzicka* and *Lamm* 1971b; *Vesely* and *Nicolaisen* 1972). Electrodes of the second kind which consist of lightly-plated porous AgX film on the surface of silver wires also respond reversibly to ionic activities and serve as classical reference electrodes (*Ives* and *Janz* 1961; *Janz* and *Ives* 1968). The latter electrodes need not even be plated with salt since constant standard potentials are observed simply when silver wire is dipped into stirred, deaerated solutions saturated with AgX or Ag_2S powders. For reasons that are clear from the thermodynamic analysis, second-kind electrodes reach equilibrium most rapidly by procedures leading to saturation of the salts with silver and uniform solution activities within the electrode pores. These procedures include current reversal during anodization, thermal-induced-partial decomposition of the salt and spontaneous self electrolysis among shorted electrodes (*Ives* and *Janz* 1961; *Janz* and *Ives* 1968).

All-solid-state membrane electrodes, on the other hand, do not allow direct contact between bulk metal and solution, and observed standard potentials can differ significantly from electrodes of the second kind using apparently the same materials (*Trümpler* 1921). This unusual property is not restricted to metal-contacted, single salt membranes, but occurs with the Ag_2S – based electrodes such as commercially available PbS, CdS and $CuS-Ag_2S$ mixed sulfide, cation-selective compositions (*Hansen, Lamm* and *Ruzicka* 1972; *Koebel* 1974; *Ruzicka, Lamm* and *Tjell* 1972; *Ruzicka* and *Hansen* 1973; *Trümpler* 1921), the selenide and telluride-based electrodes (*Hansen, Lamm* and *Ruzicka* 1972; *Hirata* and *Higashiyama* 1971d) and complex mixtures of sulfides as electrodes (*Hirata, Higashiyama* and Date 1970; *Hirata* and *Higashiyama* 1971a; *Hirata* and *Higashiyama* 1971b; *Hirata* and *Higashiyama* 1971c). All-solid-state membrane electrode potentials for the halides can be irreversibly sensitive to photolysis to a degree not found with corresponding second-kind electrodes.

Some other reported electrodes may be either second kind or all-solid-state depending on the porosity and coating thickness (*Mesaric* and *Dahmen* 1973; *Williams, Piekarski* and *Manning* 1971). A number of electrodes involving metal/membrane contacts cannot be characterized at all without identification of the potential-determining processes. Recent examples were reviewed by *Buck* (*Buck* 1974), but many earlier examples involving glass/metal or glass/salt contacts are in the literature (*Beckman* Instruments, Inc. 1972; *Bender* and *Pye* 1938; *Brent* and *Georges* 1973; *Gebert* and *Friedman* 1973; *Guignard* and *Friedman* 1970; *Manfre* and *Vianello* 1973; *Niedrach* and *Stoddard* 1972; *Petersen* and *Matsuyama* 1971).

This paper has several purposes: to outline and to illustrate the thermodynamic reasons why silver-salt solid-state membrane electrodes differ from second-kind electrodes, to compare their responses and to suggest a new explanation of potential sensitivity to photolysis. The thermodynamic arguments use the electrochemical potential concept from which the interfacial potential differences, metal/salt and metal/solution may be described.

Relative interfacial potentials

The essential features that confer differences among the three electrode configurations: second kind (I), ionic contact membranes (II) and all-solid-state (III) used in electrochemical cells, are the different potential-determining processes at the interfaces.

Cu'| External Ref. Elect.| Test Sol'n.   Porous $Ag_{1\pm\delta}X$| Ag| Cu   (I)
$\quad\quad\quad\quad\quad\quad\quad\quad$ a(Ag$^+$), a(X$^-$)

Cu'| External Ref. Elect. | Test Sol'n. | $Ag_{1\pm\delta}X$| Inner Sol'n. | Ag| Cu   (II)
$\quad\quad\quad\quad\quad\quad\quad\quad$ a(Ag$^+$), a(X$^-$) $\quad\quad\quad\quad$ a'(Ag$^+$), a'(X$^-$)

Cu'| External Ref. Elect.| Test Sol'n. | $Ag_{1\pm\delta}X$| Ag, C, Pt, Hg | Cu   (III)
$\quad\quad\quad\quad\quad\quad\quad\quad$ a(Ag$^+$), a(X$^-$)

Rapidity of electron vs. ion exchange causes the membrane potentials to depend upon salt stoichiometry or extent of "complete" equilibrium of the salt and its elements (*Kröger* 1964). Salts which can support a thermodynamically significant quantity of defect elements in equilibrium with electrons and holes should and do show dependence of E^0 on stoichiometry when used in cells with electronic conducting contacts.

The symbol $Ag_{1\pm\delta}X$ implies that the solid phase contains distinguishable, equilibrated, uniform excess of Ag ($+\delta$) or X ($-\delta$). *Wagner* (1959) has deduced that only 5×10^{-13} mole fraction excess Ag in AgBr corresponds to unit activity (saturation), while 5×10^{-8} mole fraction excess Br is saturation with respect to bromine. *Valverde* (1970) has found the range $-3.5\times10^{-5}\leq\delta\leq1.8\times10^{-5}$ mole fraction defect (δ) in $Ag_{2\pm\delta}S$ at 150 °C. Extremes corresponding to S or Ag saturation, respectively, are wider at room temperature (*Yushina, Karpachev* and *Fedyaev* 1972). Other examples and methods for determination of defect compositions are reviewed by *Wagner* (1971).

Cell potentials I, II, and III are computed by summing the interfacial and diffusion components. Relative salt/solution interfacial potentials, established by ion exchange, are of the same form regardless of cell configuration provided the deviations from stoichiometry are small. Equilibrium is described by

$$[\bar{\mu}^0(Ag^+)+\bar{\mu}^0(X^-)]-[\mu^0(Ag^+)_{sol'n}+\mu^0(X^-)_{sol'n}]=RT\ln\frac{K_{sp}(AgX)}{\bar{a}(AgX)} \quad (1)$$

where super bars apply to salt and $\bar{a}^2(AgX)=\bar{a}(Ag^+)\bar{a}(X^-)$. Writing partial ion-exchange constants as

$$\ln k_i=[\mu^0_i-\bar{\mu}^0_i]/RT \quad (2)$$

the interfacial potential is given by

$$\bar{\Phi}-\Phi_{sol'n}=\frac{RT}{F}\left[\frac{\mu^0(Ag^+)_{sol'n}-\bar{\mu}^0(Ag^+)}{RT}+\ln\frac{a(Ag^+)_{sol'n}}{\bar{a}(Ag^+)}\right]=$$

$$=\frac{RT}{F}\ln\left[\frac{k_{Ag}+a(Ag^+)_{sol'n}}{\bar{a}(Ag^+)}\right] \quad (3)$$

Since the solid is presumed in equilibrium with its elements,

$$\bar{\mu}(Ag) = \bar{\mu}(Ag^+) + \bar{\mu}(e) \tag{4}$$

then,

$$\Phi - \Phi_{sol'n} = \frac{RT}{F}\left[\frac{\mu^0(Ag^+)_{sol'n} + \bar{\mu}(e) - \bar{\mu}^0(Ag)}{RT} + \ln\frac{a(Ag^+)_{sol'n}}{\bar{a}(Ag)}\right] \tag{5}$$

or for equivalent equilibrium with respect to anion,

$$\Phi - \Phi_{sol'n} = \frac{RT}{F}\left[\frac{\bar{\mu}^0(X_2)/2 + \bar{\mu}(e) - \mu^0(X^-)_{sol'n}}{RT} - \ln\frac{a(X^-)_{sol'n}}{\bar{a}(X)}\right] =$$

$$= \frac{RT}{F}\ln\left[\frac{k_{X^-}\, a(X^-)_{sol'n}}{\bar{a}(X)}\right] \tag{6}$$

Description of relative interfacial potential differences across metal/solution interfaces is based on two equalities using the electrochemical potential concept.

$$\tilde{\mu}(Ag^+)_{Ag} + \tilde{\mu}(e)_{Ag} = \tilde{\mu}^0(Ag) \text{ within silver metal} \tag{7}$$

$$\tilde{\mu}(Ag^+)_{Ag} = \tilde{\mu}(Ag^+) \text{ sol'n at the silver-solution interface} \tag{8}$$

For the simple second-kind cell (I), or a first-kind electrode where AgX is omitted,

$$E_{cell\,(I)} = \Phi(Ag) - \Phi(sol'n) - E_{Ext.\,Ref.} = \frac{RT}{F}\left[\frac{\mu^0(Ag^+)_{sol'n} - \mu^0(Ag) + \mu(e)_{Cu}}{RT} +\right.$$

$$\left. + \ln\frac{a(Ag^+)_{sol'n}}{a(Ag)}\right] - E_{Ext.\,Ref.} \tag{9a}$$

$$= E^0_{Ag^+/Ag} + \frac{RT}{F}\ln a(Ag^+)_{sol'n} - E_{Ext.\,Ref.} \tag{9b}$$

$$= E^0_{Ag^+/Ag} + \frac{RT}{F}\ln K_{sp}(AgX)$$

$$- \frac{RT}{F}\ln a(X^-)_{sol'n} - E_{Ext.\,Ref.} \tag{9c}$$

Equation 9b follows from the assumption of zero volts for the standard hydrogen electrode and cancellation of electron chemical potentials by use of common metals (Cu) at the extremes of complete cells. Thermodynamically, aged second-kind electrodes are the same as first-kind electrodes, and may function identically, e.g., ion exchange between Ag^+ (sol'n) and Ag^+ (metal) determines the interfacial potential. The porous AgX coating merely serves to establish heterogeneous equilibrium between solution species Ag^+ and X^-. Then defect structure and nonstoichiometry of AgX play no easily discernable role.

For Cell (II), the same overall result is found, although the potential contributions arise from different interfaces.

$$E_{cell\,(II)} = E_{Int.\,Ref.} + \frac{RT}{F}\ln\left[\frac{a(Ag^+)_{sol'n}}{a'(Ag^+)}\right] - E_{Ext.\,Ref.} \tag{10a}$$

$$= E^0_{Ag^+/Ag} + \frac{RT}{F}\ln a(Ag^+)_{sol'n} - E_{Ext.\,Ref.} \tag{10b}$$

This result has been verified both for solid crystal (*Durst* 1969) and heterogeneous supported-powder membranes (*Pungor* and *Toth* 1973). Citations of response data are in recent reviews (*Buck* 1972; *Buck* 1974; *Koryta* 1972; *Toren* and *Buck* 1970).

Potentials of all-solid-state electrode cells (III) depend critically on the metal/salt interfacial potential. If the process is exclusively electron exchange, the cell potential depends upon salt stoichiometry. If the process is ion exchange, the response is identical with Cells I and II. Intermediate cases are hypothetically possible.

Case A – Only Electron Exchange Between Salt/Metal is Reversible

Then

$$\Phi(Ag) - \overline{\Phi} = \frac{\mu(e)_{Ag} - \overline{\mu}(e)}{F} \tag{11}$$

and

$$E_{cell\,(III)} = \frac{RT}{F}\left[\frac{\mu^0(Ag^+)_{sol'n} - \mu^0(Ag) + \mu(e)_{Cu}}{RT} + \ln\frac{a(Ag^+)_{sol'n}}{\overline{a}(Ag)^*}\right] - E_{Ext.\,Ref.} \tag{12a}$$

$$= E^0{}_{Ag^+/Ag} + \frac{RT}{F}\ln\left[\frac{a(Ag^+)_{sol'n}}{\overline{a}(Ag)^*}\right] - E_{Ext.\,Ref.} \tag{12b}$$

The starred activity $\overline{a}(Ag)^*$ is the free metal value in the salt referred to pure Ag as unit activity. Regardless of the actual concentrations of Ag or X in AgX, their activities can be referred to the standard states of the elements. This means that AgX equilibrated with solid Ag or unit fugacity X_2 (or other standard state) has unit activity Ag or X. In the derivation of Equation 12a, b, the terms $-\overline{\mu}^0(Ag) RT \ln \overline{a}(Ag) + \overline{\mu}(e)$ occur. Referencing these to the pure metal standard state makes use of the equalities

$$\overline{\mu}^0(Ag) + RT \ln \overline{a}(Ag) = \overline{\mu}^{0*} + RT \ln \overline{a}(Ag)^* = \mu^0(Ag) + RT \ln a(Ag) \tag{13}$$

where normally

$$\overline{\mu}^0(Ag) = \mu^0(Ag) = RT \ln a(Ag) = 0 \tag{14}$$

The corresponding classical sequence for obtaining the potential formula is

Reference Reaction: $M' + Cl^- \rightarrow M'Cl + e^-$ (Ref. Elect. M'/M'Cl) (15a)

Salt Reaction: $e^- + AgX \rightarrow X^- + Ag^*$ (15b)

Case B – Only Silver Ion Exchange Is Reversible at Salt/Metal Interface

Then

$$E_{cell\,(III)} = E^0{}_{Ag^+/Ag} + \frac{RT}{F}\ln a(Ag^+)_{sol'n} - E_{Ext.\,Ref.} \tag{16}$$

and the overall reaction is

$$Cl^- + M' + AgX \rightarrow M'Cl + Ag\,(a=1) + X^- \tag{17}$$

An intermediate case involving formation of Ag at two activities seems hardly likely. Reactions at interfaces of pure form according to Cases A and B depend upon the contact metal or other conductor. Case A is *very hypothetical* for Ag contacts and could be observed in a transient measurement, if at all. The reason is that any AgX-containing Ag (a = Ag*) in contact with bulk Ag is not in equilibrium and eventually Ag diffuses from metal to salt until $\overline{a}(Ag)^* = 1$. Motion of Ag in AgX as neutral atoms or as $Ag^+{}_i$ and e^- is not known; although the diffusion coefficient of interstitial silver at 25 °C is approx. 10^{-5} cm^2/sec. (*Süptitz*

1969). An earlier theory of the membrane potentials for ionic contacts (Cell II) made use of the high mobility of interstitial silver ions to justify the assumption that the interior diffusion potential is zero (*Buck* 1968). It is clear that both Cases A and B give identical cell potentials when the salt phase is equilibrated unit activity silver.

When a salt crystal contacts a metal which is more noble than Ag, with respect to formation of its metal halide, Case A has special legitimacy. Electrodes using carbon contacts are advocated by *Ruzicka* (*Ruzicka, Lamm* and *Tjell* 1972), while mercury contacts have been used by *Marton* and *Pungor* (*Marton* and *Pungor* 1971). The importance of inert contacts has been stressed by *Koebel* (1974).

Case C AgX and Ag_2S Membranes Are Contacted by a Reactive Metal

This case is particularly important because it can describe practical situations such as the potential of AgX or Ag_2S contacted by Pb, Cd, Zn or solder. Both the initial state and the final, equilibrium state, when a finite amount of the contacting metal has been converted to salt with formation of Ag, can be treated. The potential of the initial transient problaby corresponds to Case A (only electron transfer occurs) and the potential depends on the initial Ag activity in the salt. However, at equilibrium, Ag activity will have increased to unity, while the contacting metal activity will have decreased to the value given by

$$(RT/2\,F)\ln a(M)^* = E^0_{M^{++}/M} - E^0_{Ag^+/Ag} - \frac{RT}{2\,F}\ln[K_{sp}(Ag_2S)/K_{sp}(MS)] \tag{18a}$$

$$= (RT/2\,F)(\Delta F^0\,cal./1364.3) \tag{18b}$$

for a Ag_2S membrane contacted by metal M. Similar equations using tabulated potentials, solubility products or free energy data can be found to describe the activities of Tl, Cu or Pb in contact with silver halides. Mercury does not attack the stoichiometric silver halides and so responds more nearly as a Case A contact. Mercury does react with halide electrodes which are presaturated with halogen.

Expected effects on E^0 of elemental defects in all solid state membrane electrodes

Equilibrium and metastable all-solid-state electrodes using metallic contacts belonging to Cases A, B and C should show E^0 values differing from electrodes of the first and second kinds, depending on the metal activity in the salt. The form of the response is

$$E^0\,(\text{for } a(Ag^+) = 1) = E^0_{Ag^+/Ag} - \frac{RT}{F}\ln a(Ag)^* \tag{19a}$$

$$E^0\,\text{for } a(M^{+n}) = 1) = E^0_{M^{+n}/M} - \frac{RT}{nF}\ln a(M)^* \tag{19b}$$

Since a (Ag)* and a (M)* vary between extremes set by external elemental activities, the range of E^0 variation follows from the formation constants

$$K_0\,(AgX) = \frac{\bar{a}\,(AgX)}{a(Ag)\,a(X_2)^{1/2}} \text{ or } K_0(MS) = \frac{\bar{a}\,(MS)}{a(M)\,a(S)} \tag{20}$$

by substitution into Equation 19a or b and the relation

$$E^0_{X_2/X^-} - E^0_{Ag^+/Ag} = \frac{RT}{F}\ln K_{sp}\,K_0(AgX) \tag{21}$$

or its equivalent for sulfide salts. The ranges of apparent E^0 values for unit activity silver ion or halide ion in solution have been deduced (*Marton* and *Pungor* 1971). For $a(Ag^+) = 1$, the E^0 range is:

$$E^0_{Ag^+/Ag} \text{ (for } \bar{a}(Ag)^* = 1) \text{ to } E^0_{Ag^+/Ag} + \frac{RT}{F} \ln K_0(AgX) \text{ (for } \bar{a}(X)^* = 1) \quad (22)$$

For $a(X^-) = 1$, the corresponding range is:

$$E^0_{Ag^+/Ag} + \frac{RT}{F} \ln K_{sp}(AgX) \text{ (for } \bar{a}(Ag)^* = 1) \text{ to } E^0_{X_2/X^-} \text{ (for } \bar{a}(X)^* = 1) \quad (23)$$

Intrinsic or stoichiometric all-solid-state AgX membrane electrodes will have values between these extremes. This value can be expressed for unit activity silver ion in solution as

$$E^0_{stoichiometric} = E^0_{Ag^+/Ag} + \frac{RT}{2F} \ln p^{eq.}_{X_2} + \frac{RT}{F} \ln K_0(AgX) \quad (24)$$

The dissociation pressure of X_2, $(2 K_0)^{-2/3}$, is not necessarily the equilibrium value for electronically intrinsic AgX, because the solubility equilibria with formation of holes is not included. *Matejec, Meissner* and *Moisar* (1973) suggest that at this calculated pressure, AgCl and AgBr already contain excess holes. Crystals grown form the melt, using ultra-pure starting materials to achieve freedom from divalent impurities, usually contain excess silver (*Slifkin* 1973). Consequently E^0 values near to or exactly equal to the standard Ag^+/Ag potential may be observed because halogen is lost by volatilization.

For a single metal sulfide electrode contacting its metal, results of form similar to Equations 22 to 24 are found. For example, when $a(M^{++}) = 1$, the range is $E^0_{M^{++}/M}$ for metal saturation, to $E^0_{M^{++}/M} + \frac{RT}{2F} \ln K_0(MS)$ for sulfur saturation. On the other hand, a mixed sulfide electrode, such as $CdS-Ag_2S/Cd$ cannot show as wide a range of E^0 variability, because the metal-saturated value is not thermodynamically stable. Cadmium metal activity in the mixed sulfide will be spontaneously (perhaps slowly) reduced to the value computed from Equation 18a, b. The range of E^0 values is

$$E^0_{Ag^+/Ag} + \frac{RT}{2F} \ln K_{sp}(Ag_2S)/K_{sp}(CdS) \text{ to } E^0_{Cd^{++}/Cd} + \frac{RT}{2F} \ln K_0(CdS) \quad (25)$$

for $a(Cd^{++}) = 1$ and $\bar{a}(Ag)^* = 1$ to $\bar{a}(S)^* = 1$.

There is both firm and indicative evidence for this thermodynamic analysis. *Wagner* (1959) showed conclusively by coulometric variation of Ag in Ag_2S that the cell potential varied as $\ln a(Ag)^*$ for the solid cell

$$Ag \mid AgI \mid Ag_2S \mid Ag \quad (IV)$$

The analysis presumed that the Ag/AgI interface transported Ag^+ only, or that AgI was equilibrated with Ag. Much earlier, *Trümpler* (1921) noted that nonstoichiometric sulfides, containing excess sulfur, deviated in E^0 from the 2nd second kind stoichiometric electrodes. *Koebel* (1974) has recently verified that the mixed sulfide responses for CuS, CdS and PbS mixed with Ag_2S agree with the theory described here. The copper case is complicated because of the many intermediate compounds in the phase diagram. Fortunately these have been identified and their formation free energies measured by *Matthieu* and *Rickert* (1972).

Calculated results

For compounds containing metals of a single valence, calculated all-solid-state E^0 values for unit activity of cations or anions follow directly from salt free energies, solubility products and standard potentials via Equations 19a, b and 25. Uncertainties arise because free energy data and solubility product data are not internally consistent. Metal activities for replacement reactions of mixed salt electrodes can be computed from either Equation 18a or 18b with differing results of as much as $\pm$ 50 mv. Use of Ringbom's solubility products (*Ringbom* 1953) leads to particularly discordant results. A summary of calculated results are given in Table I. The CuS-Cu_2S-Ag_2S case is complicated by the numerous compounds in the Cu-S system. *Koebel* (1974) has computed the unit activity Ag case using free energy data. Our computation is higher by $+$ 16 mv and is in better agreement with experimental results.

Experimental

As part of a larger program on the equilibrium electrochemistry and non-equilibrium transport properties of silver halide membranes at room temperature, responses of AgBr ionic contact (II) and all-solid-state (III) cells have been measured. The AgBr membranes were prepared or obtained according to:
a) Precipitation from $AgNO_3$, KBr solutions at stoichiometric, with washing. Pressed pellets are transparent and glassy.
b) Precipitation from $AgNO_3$, KBr solutions, 10% excess $AgNO_3$. Pressed pellets are cloudy.
c) Precipitation from $AgNO_3$, KBr solutions, 10% excess KBr. Pressed pellets are cloudy.
d) Precipitates dissolved in conc. HBr saturated with bromine. Reprecipitation on dilution with water
e) f) and g) Precipitates a), b) and c) stored in closed container with saturated bromine vapor

Electrodes were made from pressed disks (10 tons for 15 minutes) and sealed on to glass tubes. with epoxy resin. Inner reference probes were AgBr-coated silver wires dipping into 0.01 M KBr. Measurements were made at 25 °C in subdued light. Pellets were approx. 2 mm thick.

Results

Experimental results collected in Table II are a comprehensive survey of reported all-solid-state standard potentials and new values for AgBr obtained in this laboratory. Many of the data apply to ill-defined solid phases in the sense that no effort was made to saturate the solid salts with either metal or elemental anion. Numerous examples of agreement between responses of second-kind and solid state electrodes contacted with a parent metal for long periods of time (e.g., commercial all-solid-state responses) are not included. For the silver salts, there is no question of disagreement between E^0 values for Ag-saturated all-solid-state and second kind electrode responses. In most cases, the observed standard potentials in Table II lie between the extremes set thermodynamically for element-saturated membranes. Examples of controlled-composition electrode responses, using known contacting metals, show excellent agreement with theory.

Potentials of all-solid-state AgBr membranes deserve comment. Membranes pretreated with bromine using C and Pt contacts show E^0 values near to theoretical. Some loss of bromine occurs during the mounting and measuring step. Any effect of the ionic stoichiometry, by

Table 1. Estimated extreme values for standard potentials of single salt "all-solid-state" Electrodes at 25 °C (vs St'd Hydrogen Electrode)

Salt	pK_{sp} Ref. 1	pK_{sp} Ref. 2	$-pK_0$	$E^0\ \bar{a}\ (M)^* = 1$ $a(M^{+n}) = 1$	$E^0\ \bar{a}\ (X)^* = 1$ $a(M^{+n}) = 1$	$E^0\ \bar{a}\ (M)^* = 1$ $a(X^-)$ Ref. 1	$= 1$ Ref. 2	$E^0\ \bar{a}\ (X)^* = 1$ $a(X^{-n}) = 1$
AgCl	9.75	9.77	19.22	0.799 v.	1.935 v.	0.223 v.	0.222 v.	1.359 v.
AgBr	12.31	12.30	16.81	0.799	1.793	0.071	0.095	1.065
AgSCN	11.97	12.0	11.46	0.799	1.478	0.091	0.09	0.77
AgI	16.08	16.07	11.62	0.799	1.486	−0.152	−0.15	0.535
Ag_2CN_2	15.9	13.8	0 ± 3.	0.799	0.84	−0.141	−0.17	−0.1
Ag_2S	49.2	50.26	7.051	0.799	1.007	−0.656	−0.69	−0.48
			6.86		1.002			
PbS	26.6	28.15	16.24	−0.126	0.354	−0.90	−0.96	−0.48
ZnS	23.8 s	25.15	34.74	−0.762	0.265	−1.47	−1.51	−0.48
	22.8 w	22.8	32.39	−0.762	0.196		−1.44	
CdS	26.1	28.0	24.63	−0.402	0.326	−1.17	−1.23	−0.48
CuS	35.2	36.1	8.576	—	0.595	—	—	−0.48
Cu_2S	47.6	48.9	15.10	0.337 (Cu^{++})	—	−0.89	−0.93	—
HgS	52.4 r	—	8.55	0.854	1.106	−0.71	−0.93	−0.48
	51.8 b	53.8	8.10	0.854	1.09	−0.68	—	−0.48

Estimated extreme values for standard potentials of mixed salt "all-solid-state" Electrodes 25 °C (vs St'd Hydrogen Electrode)

salts	$E^0 \bar{a} (M)^* = 1$ $a(M^{+n}) = 1$	$E^0 \bar{a} (Ag)^* = 1$ $a(M^{+n}) = 1$	$E^0 \bar{a} (X)^* = 1$ $a(M^{+n}) = 1$
$PbS-Ag_2S$	—0.126 v.	0.151 v.	0.354 v.
$CdS-Ag_2S$	—0.402	0.123	0.326
$CuS-Ag_2S$	—	0.464	0.595
Cu_2S-Ag_2S	0.337	0.464	—

Ref. 1) *I. M. Kolthoff, E. B. Sandell, E. J. Meehan* and *S. Bruckenstein*, "Quantitative Chemical Analysis", 4th Ed., The Macmillan Co., 1969.

Ref. 2) *W. M. Latimer*, "The Oxidation States of the Elements", 2nd Ed., *Prentice-Hall*, Inc. N.J., 1952.

intercomparison of electrodes e), f) and g) is not significant. Loss of bromine is quite slow as indicated by the values after selected times. Use of silver contacts leads rapidly to the silver-saturated value, even when the pellet is presaturated with bromine. A time constant of 1 to 3 seconds is indicated for the potential change from carbon to silver contacts. Silver-contacted electrodes show variability in E^0 values according to preparation. Type a) electrodes show

Table 2. Experimental E^0 values compared with metall-saturated values, all-solid-state configuration

Salt	Theoretical Standard Potential Range (Metal to Anion Element Saturation) for $a(M^{+n}) = 1$ vs St'd. H_2 Electrode	Experimental Values recomputed to $a(M^{+n}) = 1$ (using Eq. 23 or equivalent)	Electrode Contact and Conditions
AgCl	0.799—1.935 v.	0.807 *Ruzicka* et. al. (1972)	$AgCl-Ag_2S/C$
		0.9 *Marton* and *Pungor* (1971)	Heterog./Hg
		0.841 *Marton* and *Pungor* (1971)	Heterog./Hg
AgBr	0.799—1.793	0.820 *Trümpler* (1021)	Ag/AgCl/AgBr/M KBr, Br_2-saturated
		0.805 *Ruzicka* at. al. (1972)	$AgBr-Ag_2S/C$
		0.9 *Marton* and *Pungor* (1971)	Heterog./Hg
		0.856 *Marton* and *Pungor* (1971)	Heterog./Hg
		1.770	AgBr/C Prec. e) M KBr
		1.776	AgBr/C Prec. f) M KBr
		1.767	AgBr/C Prec. g) M KBr
		1.767 to 1.711 (12 Hrs.)	AgBr/C Prec. d) M KBr
		1.777 to 1.656 (1 Hr.)	AgBr/Pt Prec. d) M KBr
		0.856	AgBr/Hg Prec. d) M KBr
		0.797 $\pm$ 0.003	AgBr/Ag Prec. a) M KBr
		0.902 to 0.798 (1 Hr.)	AgBr/Ag Prec. b) M KBr
		0.784	AgBr/Ag Prec. c) M KBr
AgI	0.799—1.486	0.825 *Trümpler* (1921)	AgI/Ag/M KI, I_2sat'd.
		0.805 *Ruzicka* et. al. (1972)	$AgI-Ag_2S/C$
		0.9 *Marton* and *Pungor* (1971)	Heterog./Hg
		0.876 *Marton* and *Pungor* (1971)	Heterog./Hg

Table 2. (Continued)

Salt	Theoretical Standard Potential Range (Metal to Anion Element Saturation) for a $(M^{+n}) = 1$ vs St'd. H_2 Electrode	Experiment Values Recomputed to a $(M^{+n}) = 1$	Electrode Contact and Conditions
Ag_2S	0.799—1.006 ± 0.004	0.800 *Koebel* (1974)	Ag_2S/Ag
		0.80—0.83 *Vesely* et al. (1972)	Ag_2S/Ag
		0.80—0.865 *Trümpler* (1921)	Ag_2/Ag
		0.994 *Koebel* (1974)	Ag_2S/C (S-saturated)
		1.010 *Vesely* et al. (1972)	Ag_2S/C (S-saturated)
		0.82 to 0.86 *Noddack* (1955)	Unspecified
		0.82 *Sato* (1960, 1966)	Unspecified
		0.84 *Berner* (1963), *Sato* (1966)	Unspecified
Cu_2S $Cu_{1.8}S$	0.337—0.504 v.	0.337—0.535 (unstable) *Trümpler* (1921)	Cu
		0.49 (Stable) *Trümpler* (1921)	Cu
		0.436 *Hirata* et al. (1970)	$Cu_{1.8}S$/Inert
		0.51 *Sato* (1960, 1966)	Unspecified
		0.47 *Noddack* (1955)	Unspecified
CuS $Cu_{1.8}S$	0.504—0.595	0.49—0.61 *Trümpler* (1921)	Cu or Pt
		0.58 *Noddack* (1955)	Unspecified
		0.57 *Sato* (1960, 1966)	Unspecified
PbS	—0.126—0.354	0.274—0.354 *Tammann* (1920)	Cu
		—0.126 to 0.20 *Trümpler* (1921)	Pb
		0.19—0.366 *Piontelli* (1949)	Pb
		0.274—0.454 *Noddack* (1955)	Unspecified
		0.364 *Sato* (1960)	Unspezified
PbS-Ag_2S	0.151 (Ag-contact) to 0.354 (S-saturated)	0.346 *Ruzicka* et al. (1972)	Carbon
		0.175 ± .005 *Koebel* (1974)	Ag
		0.375 ± .005 *Koebel* (1974)	Carbon (S-saturated)
		0.236 *Hirata* et al. (1971c)	Inert
CdS-Ag_2S	0.123 (Ag-Contact) to 0.326 (S-saturated)	0.326 *Ruzicka* et al. (1972)	Carbon
		0.168 ± .005 *Koebel* (1974)	Ag
		0.369 ± .010 *Koebel* (1974)	Carbon (S-saturated)
		0.342 to 0.356 *Ruzicka* and *Hansen* (1973)	Carbon S-saturated)
Cu_XS-Ag_2S	0.464 (Ag-contact) to 0.595 (S-saturated) or 0.337 (Cu-saturated)	0.464 *Matthieu* and *Rickert* (1972)	$Cu_{1.95}S$
		0.478—0.545 *Koebel* (1974)	Ag
		0.569—0.606 *Koebel* (1974)	Carbon (S-saturated)
		0.614 *Hansen* et al. (1972) *Ruzicka* et al. (1972)	Carbon (S-saturated)
		0.515 *Hansen* et al. (1972)	*Beckman*
		0.541 *Hansen* et al. (1972)	Orion

almost exactly tabulated values: 797 $\pm$ 2 mv, Type b) average 798 mv after 1 hour, while type c) electrodes are low.

Silver bromide membranes contacted with mercury show an effect of presaturation with bromine. Application of phase rule argument suggests that mercury acts as an inert contact over a range of silver activities in AgBr from unity to the value set by the overall equilibrium among the two elements and their bromide salts. For Br_2-saturated AgBr, mercury will react to decrease bromine activity and increase silver activity to its equilibrium value. The expected E^0 values for the half-cells Ag^+ / AgX / Hg_2X_2 / Hg are 0.844, 0.867 and 0.910 v for AgCl, AgBr, and AgI from free energy values. The *Marton* and *Pungor* results (*Marton* and *Pungor* 1971), 0.841, 0.856 and 0.878 v are strikingly similar although their method of preparation was not disclosed. Our result for Hg-contacted, bromine-presaturated electrodes is also 0.856 v.

Discussion

The experimental evidence summarized here confirms the thermodynamic analysis of all-solid-state electrode potentials, based on metal/salt electron exchange reversibility. Standard potentials depend upon elemental defect activities in the solids. Second kind electrodes give the same E^0 values as the metal-saturated all-solid-state electrodes. This result does not imply that second-kind electrodes must be metal-saturated, because their ion exchange equilibria determine the responses. However, where the salt touches the metal, saturation is a spontaneous process and instability might occur by local cell action involving electron transfer until the salt composition is uniform. As listed above, empirical procedures leading to metal-saturation do stabilize second-kind electrode potentials.

Two further topics are relevant to the relation of second-kind and all-solid-state electrodes. These are effects of light and solution-soluble oxidants and reductants. Both are beyond the main thrust of this paper and will not be discussed in detail. In the most recent review by *Janz* and *Ives* (1968) effects of light on second-kind electrodes were still unsettled. Behavior of ionic contact membrane cells (II) and all-solid-state cells (III) under irradiation are less well known or understood. There are potential shifts which follow an on-off light cycle (*Van Peski, De Vooys* and *Engel* 1969), and other more slowly varying shifts (*Marton* and *Pungor* 1971; *Sanders* and *Kolthoff* 1940). It is possible that the solid-state defect equilibria, so important for the interpretation of E^0 values, will prove to be a means for understanding these effects.

References

Beckman Instruments, Inc.: Solid-State Ion-Sensitive Glass Electrode. Brit. Patent 1,260,065 (Jan. 12, 1972)

Bender, H., Pye, D. J.: Glass Electrodes for Use in Hydrogen Ion Concentration Determination. U.S. Patent 2,177,596 (May 17, 1938)

Berner, R. A.: Electrode Studies of Hydrogen Sulfide in Marine Sediments. Geochim. cosmochim. Acta 27 (1963) 563

Breant, M., Georges, J.: Twinned Mercury-filled Glass Electrode System for Acid-base Titrations in a Number of Nonaqueous Solvents. Talanta 20 (1973) 914

Buck, R. P.: Theory of Potential Distribution and Response of Solid-State Membrane Electrodes. Anal. Chem. 40 (1968) 1432

Buck, R. P.: Ion Selective Electrodes, Potentiometry and Potentiometric Titrations. Anal. Chem. 44 (1972) 270R

Buck, R. P.: Ion Selective Electrodes, Potentiometry and Potentiometric Titrations. Anal. Chem. 46 (1974) 28R

Durst R. A. (Ed.): Ion Selective Electrodes. National Bureau of Studies Spec. Pub. 314, U.S. Government Printing Office 1969

Farren, G. M., Staunton, J. J.: Electrode to

Determine Ion Activity in Solution. Ger. Patent 1,940,353 (Feb. 19, 1970)

Gebert, G., *Friedman, S. M.:* Implantable Glass Electrode Used for pH Measurement in Working Skeletal Muscle. J. appl. Physiol. 34 (1973) 122

Guignard, J. P., *Friedman, S. M.:* Construction of Ion-selective Glass Electrodes by Vacuum Deposition of Metals. J. appl. Physiol. 29 (1970) 254

Hansen, E. W., *Lamm, C. G., Ruzicka, J.:* Selectrode-universal Ion-selective Solid-state Electrode II. Anal. Chim. Acta 59 (1972) 403

Hirata, H., *Higashima, K.:* Analytical Study of a Cadmium Ion-selective Ceramic Membrane Electrode. Z. anal. Chem. 257 (1971a) 104

Hirata, H., *Higashima, K.:* New Type of Lead (II) Ion-selective Ceramic Membrane Electrode. Anal. Chim. Acta 54 (1971b) 415

Hirata, H., *Higashima, K.:* Analytical Study of the Lead Ion-selective Ceramic Membrane Electrode. Bull. Chem. Soc. Japan 44 (1971c) 2420

Hirata, H., *Higashima, K.:* Ion-selective Lead Selenide and Lead Telluride Membrane Electrodes. Anal. Chim. Acta 57 (1971d) 476

Hirata, H., *Higashima, K., Date, K.:* Copper (I) Sulfide Ceramic Membranes as Selective Electrodes for Copper (II). Anal. Chim. Acta 51 (1970) 209

Ives, D. J. G., *Janz, G. J.:* Reference Electrodes. Academic Press, New York 1961

Janz, G. J., *Ives, D. J. G.:* Silver, Silver Chloride Electrodes. Ann. N.Y. Acad. Sci. 148 (1968) 210

Koebel, M.: Standard Potentials of Solid-state Metal Ion-selective Electrodes. Anal. Chem. 46 (1974) 1559

Koryta, J.: Theory and Applications of Ion-selective Electrodes. Anal. Chim. Acta 61 (1972) 329

Kröger, F. A.: The Chemistry of Imperfect Crystals. North Holland Publ., Amsterdam (Holland) 1964

Manfre, G., *Vianello, D.:* Metals Wires Coated with Glass or Glass Ceramics. Ger. Patent 2,235,171 (Feb. 1, 1973)

Marton, A., *Pungor, E.:* Standard Potential of Heterogeneous Precipital-based Membrane Electrodes. Anal. Chim. Acta. 54 (1971) 209

Matejec, R., *Meissner, H. D., Moisar, E.:* Solid State Chemistry of the Silver Halide Surface. In: Progress in Surface and Membrane Science, Vol. 6, p. 1, ed. by *J. F. Danielli, M. D. Rosenberg* and *D. A. Cadenhead.* Academic Press, New York 1973

Matthieu, H. J., *Rickert, H.:* Elektrochemisch-thermodynamische Untersuchungen am System Kupfer-Schwefel bei Temperaturen T = 15–90 °C. Z. Phys. Chem. N.F. 79 (1972) 315

Mesaric, S., *Dahmen, E. M. F.:* Ion-selective Carbon-paste Electrodes for Halides and Silver (I) Ions. Anal. Chim. Acta 64 (1973) 431

Niedrach, L. W., *Stoddard, W. H.* Jr.: Ion-selective Electrodes. Ger. Patent 2,220,841 (Nov. 9, 1972)

Noddack, W., *Wrabetz, K.:* Über das elektrochemische Verhalten einiger Schwermetallsulfide. Z. Elektrochem. 59 (1955) 96

Petersen, A., *Matsuyama, G.:* Glass Electrode. Ger. Patent 2,050,050 (Nov. 25, 1971)

Piontelli, R.: Studies on the Reactions of Sulfides and Electrolyte Solutions. I: Galena. Bull. Soc. chim., Fr. 16 (5) (1949) D 197

Pungor, E., *Toth, K.:* Precipitate-based Ion-selective Electrodes. Pure appl. Chem. 34 (1973) 105

Ringbom, A.: Solubilities of Sulfides. Report to the Analytical Section. IUPAC 1953

Ruzicka, J., *Hansen, E. H.:* Selectrode. Universal Ion-selective Electrode IV. Solid-state Cadmium (II) Selectrode. Anal. Chim. Acta 63 (1973) 115

Ruzicka, J., *Lamm. C. G.:* New Type of Solid-state Ion-selective Electrodes with Insoluble Sulfides or Halides. Anal. Chim. Acta 53 (1971) 206

Ruzicka, J., *Lamm, C. G.:* Selectrode – The Universal Ion-selective Solid-state Electrode – I: Halides. Anal. Chim. Acta 54 (1971) 1

Ruzicka, J., *Lamm, C. G., Tjell, J.:* Selectrode, Universal Ion-selective Electrode III: Concept, Constructions, and Materials. Anal. Chim. Acta 62 (1972) 15

Sanders, H. L., *Kolthoff, J. M.:* Photovoltaic Behavior of Pure Silver Bromide. J. Phys. Chem. 44 (1940) 936

Sato, M.: Oxidation of Sulfide Ore Bodies II. Oxidation Mechanism of Sulfide Minerals at 25 °C. Econ. Geol. 55 (1960) 1202

Sato, M.: Half-cell Potentials of Semiconductive Simple Binary Sulfides in Aqueous Solution. Electrochim. Acta 11 (1966) 361

Slifkin, L.: Crystal Growth Facility. Materials Research Center, Univ. of North Carolina, Chapel Hill/N.C., Private Comm. 1973

Staunton, J. J.: Nonpolarized Ion Selective Electrodes. Ger. Patent 2,002,676 (Nov. 5, 1970)

Süptitz, P.: Die Diffusion von Fremdionen in AgCl- und AgBi-Kristallen. In: Reactivity of Solids, p. 29, ed. by *J. W. Mitchell.* Wiley, New York 1969

Tammann, G.: Über den Ionenaustausch an der Oberfläche von Mineralien. Z. anorg. allg. Chem. 113 (1920) 149

Toren, E. C., Buck, R. P.: Potentiometric Titrations. Anal. Chem. 42 (1970) 284R

Trümpler, G.: Zur Kenntnis des elektromotorischen Verhaltens von Metallverbindungen mit Elektronenleitung. Z. Phys. Chem. 99 (1921) 9

Valverde, N.: Coulometrische Titrationen zur Bestimmung des Homogenitätsbereiches von festem Silbersulfid, Silberselenid und Silbertellurid. Z. Phys. Chem., N.F. 70 (1970) 113

Van Peski, A. C. H., DeVooys, D. A., Engel, D. J. C.: Transient Changes in the Potential of Ag/AgCl-Electrode Induced by a Light Pulse. Nature 223 (1969) 177

Vesely, J., Jensen, O. J., Nicolaisen, B.: Ionselectrodes Based on Silver Sulfide. Anal. Chim. Acta 62 (1972) 1

Vesely, J., Jindra, J.: Ionic Electrodes on Single Crystal Basis and with Solid Internal Contacts. Proc. IMEKO Conf., Akad. Kaido, Budapest/Hung. 1968, 66–67

Wagner, C.: Ionen- und Elektronenleitung in Silberbromid und Abweichungen von der idealen stoichiometrischen Zusammensetzung. Z. Elektrochem. 63 (1959) 1027

Wagner, C.: The Determination of Small Deviations from the Ideal Stoichiometric Composition of Ionic Crystals and Other Binary Compounds. In: Progress in Solid State Chemistry, Vol. 6, p. 1, ed. by *H. Reiss* and *J. O. McCaldin.* Pergamon Press, New York 1971

Williams, T. R., Piekarski, S., Manning, C.: Potentiometric Titrations with a Mercury-Mercury Sulfide Electrode. Talanta 18 (1971) 951

Yushina, L. D., Karpachev, S. V., Fedyaev, Y. S.: Solutions of Silver in Ag_2S. Elektrokhimiya 8 (1972) 1513

W. E. Morf and W. Simon

Theoretical Treatment of the Dynamic Response of Ion-Sensitive Membrane Electrodes

(2 Figures)

One of the critical limiting factors in the use of ion-sensitive membrane electrodes, especially in routine analysis, is their so-called speed of response. In order to investigate the parameters affecting the dynamic characteristics of a sensor, a theoretical model was evolved (*Morf* and *Simon* 1974). The aim of the present contribution is to summarize the framework of theoretical results in view of their practical importance.

The EMF-response of a membrane electrode assembly to primary ions I^{z_i}, in the absence of any interfering species, may be described by the Nernst equation:

$$E(t) = E_i^0 + s \log a_i' \qquad (1)$$

$E(t)$ is the cell potential at the time t, E_i^0 signifies the reference potential and s is the slope of the response function (59.2 mV/z_i at 25 °C). The ion activity a_i' clearly refers to the sample boundary which is in equilibrium with the membrane surface. The usual form of the Nernst equation:

$$E(\infty) = E_i^0 + s \log a_i \qquad (2)$$

is valid only after long equilibration periods when the equilibrium state holds throughout the bulk of sample solution (activity a_i). The existence of time-dependent deviations between measured boundary activity a_i' and bulk activity a_i, which are related to diffusion processes, is the main reason for a finite speed of response.

A rather simple diffusion model (*Markovic* and *Osburn* 1973; *Morf* and *Simon* 1974) may be applied to membranes that incorporate charged ion-exchange sites (*solid-state-, glass-, or liquid ion-exchange membranes*). Since the composition of these membranes remains approximately constant in the absence of interfering ions, it is the ionic diffusion within the aqueous boundary layer (thickness δ; diffusion coefficient D') that is directly rate-controlling. A mathematical analysis of the diffusion problem leads to the following approximate description of the dynamic response:

$$E(t) - E(\infty) = s \log \left[1 - \left(1 - \frac{a_i^0}{a_i} \right) e^{-t/\tau'} \right] \qquad (3)$$

with the time constant:

$$\tau' = \frac{4 \delta^2}{\Pi^2 D'}$$

The symbol a_i^0 signifies the activity of the conditioning solution which is replaced at $t = 0$ by the sample solution of activity a_i. According to (3), the major factors affecting the response time are the boundary-layer thickness (strongly reduced by stirring) and the direction of sample-activity change. The calculated response vs. time profiles in Figure 1, which are in

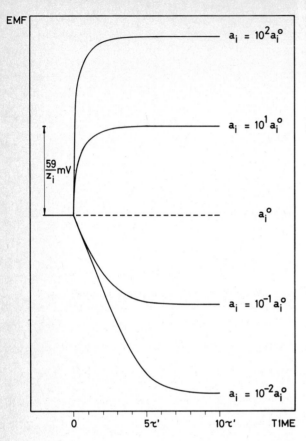

Fig. 1. Theoretical EMF-response vs. time profiles for *ion-exchange* membrane electrodes, calculated according to Equation 3.

qualitative agreement with experimental curves (*Toth* and *Pungor* 1973), demonstrate clearly that the speed of response is considerably faster when changing from a low activity a_i^0 to a high activity a_i, as compared to a change in the opposite direction.

A more complicated theoretical procedure (*Morf* and *Simon* 1974) is necessary in the case of membranes based on electrically neutral cation-selective ligands (*neutral carrier membranes*). For this membrane type the concentration of dissolved electrolyte (cationic complexes and anions) is variable with space and time and, therefore, diffusion processes reaching into the membrane interior have to be considered. The following relationship was derived for the dynamic response of carrier membrane electrodes:

$$E(t) - E(\infty) = s \log \left[1 - \left(1 - \frac{a_i^0}{a_i} \right) \frac{1}{\sqrt{t/\tau} + 1} \right] \tag{4}$$

with the time constant:

$$\tau = \tau' \frac{\Pi D}{4 D'} K^2$$

D is the diffusion coefficient of the electrolyte within the membrane and K describes the partition of ionic forms between sample solution and membrane (for practical purposes:

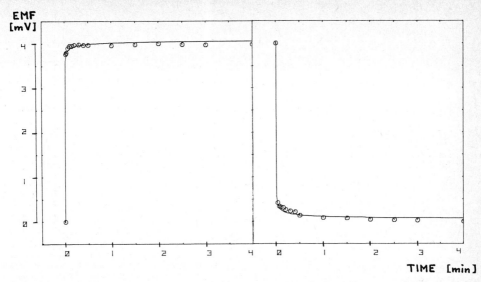

Fig. 2. EMF-response vs. time profiles for a K$^+$-sensor based on the *neutral carrier* valinomycin. Solid lines: theoretical curves obtained from Equation 4 with $E(\infty) = E(0) \pm 4$ mV and $\tau = 20$ ms; circles: experimental points (*Lindner* 1974).

$K \ll 1$ (*Boles* and *Buck* 1973). Accordingly, in addition to the parameters of the measurement mentioned above, it is the extraction capacity of a carrier membrane that is mainly decisive for its speed of response. The use of a highly nonpolar membrane (low K-values) in a flow-through cell (high stirring rates) is preferable in the context of most analytical applications. As indicated in Figure 2, response times in the order of a few seconds may be obtained thereby.

Acknowledgement

This work was partly supported by the Schweizerische Nationalfonds zur Förderung der wissenschaftlichen Forschung.

References

Boles, J. H., Buck, R. P.: Anion Responses and Potential Functions for Neutral Carrier Membrane Electrodes. Anal. Chem. 45 (1973) 2057

Lindner, E.: Private Communication of Experimental Results. Eidg. Techn. Hochschule, Zürich (Switzerland) 1974

Markovic, P. L., Osburn, J. O.: Dynamic Response of Some Ion-Selective Electrodes. Amer. Inst. Chem. Eng. J. 19 (1973) 504

Morf, W. E., Simon, W.: Ion-Selective Electrodes. In: Electrodes Based on Antibiotics and Other Ion-Selective Organic Ligands. ed. by *M. S. Frant* and *J. W. Ross*. 1974

Morf, W. E., Lindner, E., Simon, W.: Theoretical Treatment of the Dynamic Response of Ion-Selective Membrane Electrodes. Anal. Chem. (1974)

Toth, K., Pungor, E.: Recent Results on the Dynamic Response of Precipitate-Based Ion-Selective Electrodes. Anal. Chim. Acta 64 (1973) 417

R. Wodick

Compensation of Measuring Errors Produced by Finite Response Time with Ion-selective Electrodes

The inertia of ion-selective electrodes causes measuring errors, especially when the response of the electrode does not take place considerably faster than the processes to be investigated. We developed a method which allows for the elimination of the influence of inertia of ion-selective electrodes.

We proceed on the assumption that the response processes are based on process law which is homogeneous and linear, such is the case as with the equation for diffusion

$$\frac{\partial z}{\partial t} = D \text{ div grad } z. \tag{1}$$

We assume that the values $z(t)$ are a one-to-one relation

$$z(t) = F(a(t)) \tag{2}$$

of the ion activity $a(t)$ to be found. With these general assumptions we can show that the values, $y(t)$, can be described as a convolution integral

$$y(t) = \int_{-\infty}^{t} z'(\tau) s(t-\tau) d\tau, \tag{3}$$

where $y(t)$ is the measuring value dependent on time, $s(t)$ is the response function of the electrode for a sudden change in activity. It is possible that we have to consider two different functions, $s^+(t)$ and $s^-(t)$, according to an increase ($s^+(t)$) or decrease ($s^-(t)$) in ion activity. Eq. (3) – an inhomogeneous Volterra integral equation of first order – can be generally solved only with a relatively high numerical expenditure. In all cases in which it is possible to describe $s(t)$ approximately by an exponential function

$$s(t) = 1 - e^{-At}, \tag{4}$$

eq. (3) can be solved in closed form

$$z(t) = y(t) + \frac{1}{A} y'(t). \tag{5}$$

If $s(t)$ can be described only by a series of exponential functions having further summands different from zero, then we obtain a differential equation of higher order. By using the response time, T_{90}, (T_{90} is the time which elapses until 90% of the final response is measured in the case of sudden change in activity), we obtain

$$z(t) = y(t) + 0.43 \, T_{90} \, y'(t). \tag{6}$$

Equations (5) and (6) proved useful in practice to compensate the errors caused by the inertia of the electrodes (*Wodick* 1972, 1973a, 1973b).

References

Wodick, R.: Compensation of Measuring Errors Produced by Finite Response Time in Polarographic Measurements with Electrodes Sensitive to Oxygen and Hydrogen. Pflügers Arch. 336 (1972) 327–344

Wodick, R.: Bestimmung des Einstellverhaltens von sauerstoffempfindlichen Platinelektroden durch Messungen im lebenden Gewebe. Pflügers Arch. 339 (1973a) 49–58

Wodick, R.: Eine Methode zur Bestimmung des Einstellverhaltens bei wenig trägen wasserstoff- bzw. sauerstoffempfindlichen Platinelektroden. Basic Res. Cardiol. 68 (1973b) 595–613

E. Pungor, K. Tóth and G. Nagy

Some Aspects of the Application of Ion-selective Electrodes in Flowing Systems

(16 Figures)

Introduction

More and more reliable and relatively fast analytical methods are required for both laboratory and industrial purposes. To satisfy these requirements analytical techniques and apparatus have recently been developed for carrying out analysis in streaming solutions. This trend can be attributed to two factors: on the one hand for controlling continuous technological processes a monitor can be constructed in the most simply by building the sensor into the appropriate part of the flow-through reaction chamber; on the other hand, in an automatic analyzing unit, sampling and washing are more simple if the analysis is carried out in streaming solutions, e. g., in Technicon analyzers.

Every new method of concentration signal transformation offers a potential possibility for the development of devices for continuous monitoring and automatic analysis. Since much work has been carried out in developing apparatus of this kind, e. g., in the field of optical analysis, it is not surprising that even in the early stages of research on ion-selective electrodes, efforts have been made to develop detectors based on ion-selective electrodes for use in streaming solutions.

Aside from the reviews covering only the large amount of work done in the field of continuous pH measurements, the first review of the industrial application of ion-selective electrodes with continuous techniques is given by *Light* (1969a, 1969b).

For the determination of the fluoride content of natural waters a continuous technique using a fluoride-selective electrode was described by *Babcock* and *Johnson* (1968). For the continuous monitoring of various ionic species with ion-selective electrodes several techniques have been reported in the literature (*Cornish* and *Simpson* 1971; *Di Martini* 1970; *Liberti* and *Mascini* 1970).

Ion-selective electrodes have been used to advantage for measuring enzyme activity in devices containing flow-through cells. *Llenado* and *Rechnitz* (1973) give a comprehensive report on the Technicon Auto-Analyzer system in enzyme analysis. Their report deals with the application of iodide- and cyanide-selective electrodes in determining enzyme activity.

Flow-through analytical reactors with ion-selective electrodes were constructed by Guilbault and coworkers for the determination of activities of the enzymes rhodenase and cholinsterase. As has been said earlier, there is now a trend to use the Technicon Autoanalyzer or similar apparatus in connection with ion-selective electrodes (*Llenado* and *Rechnitz* 1973; *Oliver* et al. 1970; *Ruzicka* and *Tjell*, 1969). Another trend in continuous analysis using potentiometric sensors is the construction of multiple-ion monitors with flow-through cells especially in the field of water – (*Pungor*, unpubl. results, *Slevogt* et al. 1974) and clinical analysis (*Havas* 1973; *Ruzicka* and *Tjell* 1969).

Aside from practical applications, some aspects of the use of ion-selective electrodes in streaming solutions have also been investigated (*Fleet* and *Rechnitz* 1970; *Fleet* and von *Storp* 1971 a, b; *Pungor* and *Toth* 1974; *Thomson* and *Rechnitz* 1972).

Applying the monitors discussed earlier, the basis for the evaluation of the results is the electrode potential signal. Ion activities and/or concentrations are determined according to the usual method of direct potentiometry with calibration or standard addition techniques. However, direct potentiometry has a serious drawback in comparison with potentiometric titration techniques; this is the lesser accuracy.

To overcome this, several attempts have been made to develop apparatus suitable for potentiometric titration in streaming solutions. The devices developed so far are based on a linearly increasing reagent administration rate achieved with a specially designed method using the gradient principle (*Eichler* 1970; *Fleet* and *Ho* 1974).

In our laboratory the research aimed toward continuous analysis goes back to the early fifties, when the elution processes in chromatographic columns filled with ion exchangers were monitored by an oscillometric detector cell (*Pungor* 1965). Later on, flow-through potentiometric detector cells were developed and applied to continuous analysis (*Pungor* et al. 1970; *Feher* et al. 1973; *Váradi* et al. 1974). On one hand, they have been used in continuous monitoring of the concentration of an electrically active species, and on the other hand the so-called injection technique was developed on a potentiometric basis.

Also, the behavior of two kinds of detector cells, i.e., detector cells working in laminar (*Pungor* et al. 1970; *Feher* et al. 1973) and turbulent (*Váradi* et al. 1974) conditions, respectively, were studied thoroughly.

The injection technique was found to be very favorable among other reasons because of its easy calibration and its characteristic fast results in convenient form.

Recently, we have been studying some of the theoretical aspects of the application of ion-selective electrodes in streaming solutions. the most important results are briefly summarized in this paper.

Materials and Methods
Instruments

The electrode potential was measured in all cases with an expanded scale mV meter Type OP-205, Radelkis, Budapest; Type 604 Differential Electrometer, Keithley; Digital PH meter, Radiometer Type PHM 64. For recording the electrode potential a compensograph Type OH-814/1, Radelkis and an XY plotter (Type 26000 A3, Bryans) were used. For the programed coulometric titration technique a potentiostat (Tacussel, Type ASA 100-1) was used as a current generator, while the appropriate programing was done with a signal generator (Philips, Type PM-5168). The streaming of the solution was assured by a peristaltic pump (LKB VARIOPERPEX Type 12000, and Unipan, Type 304).

Reagents

All reagents used were of analytical grade.

Electrodes

Silicone rubber iodide-, cyanide- and chloride electrodes of small surface area made in our laboratory were used in this study. As fluoride electrodes, Radelkis and ORION (Type 96-09) electrodes were applied.

As reference electrodes S.C.E.-s were used. To avoid deterioration of the sample by leakage of ions through the junction, the reference electrode was always placed downstream from the indicator electrode.

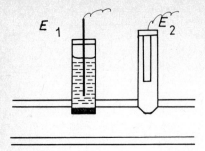

Fig. 1. Flow-through cell for continuous monitoring. E_1: ion-selective electrode; E_2: reference electrode.

Cells

The same types of potentiometric flow-through cells with different ion-selective electrodes were used Fig. 1. However, the reference electrode – in some cases – was placed in a vessel holding solution flown off from the main stream.

Measuring techniques. Determination of fluoride in air

The flow rate of gas was measured by a rotameter. The TISAB solution used contained NaCl, Na-citrate and acetic acid-sodium acetate buffer to adjust the pH. The absorbing solution was a 1:1 mixture of TISAB and distilled water. Standard NaF solutions were prepared by mixing pure NaF solutions with equal volumes of TISAB.

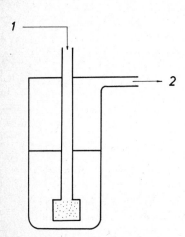

Fig. 2. Gas-absorbing vessel.
1: Gas inlet
2: Gas outlet

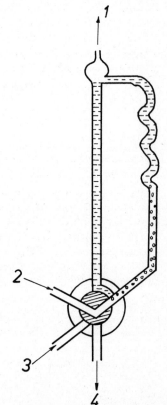

Fig. 3. Recirculation – type gas absorber.
1: Gas outlet
2: Gas inlet
3: reagent inlet
4: outlet to the measuring cell

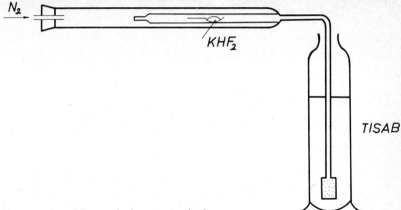

Fig. 4. Measuring set-up used for producing gas standards.

Figures 2 and 3 show diagrams of the two different absorbers used. The first one was essentially a gas washing bottle with a specially prepared teflon bubbler in place of the sintered glass bubbler, and the second one a recirculation-type absorber. The operation of the latter is based on the principle of the mammoth pump: the gas introduced into one of the two communicating vessels reduces the density of the liquid on one side and lifts the liquid on the basis of the difference in density. Thus the absorbing liquid is continuously circulated in the system whereas the gas passes through only once. The first type of absorber was found to be sufficiently good at flow rates of 3 to 40 l/hour, and the recirculation – type at 3 to 20 l/hour. The system may be considered as quasi-continuous. The absorption efficiency was studied by means of home-made gas standards. Gaseous HF was prepared by slowly heating KHF_2 in a special device (Fig. 4) and the HF evolved was carried with a stream of nitrogen into the absorber.

Injection technique

A diagram of the apparatus used is shown in Figure 5.
A constant-flow rate was provided by a peristaltic pump (S). The flow rate of the solution could be varied within a range of 0,3–40 ml/min by adjusting the pump. The individual samples of small volume were injected with a Hamilton syringe through the elastic wall of the tube carrying the solution (marked J) A magnetic stirrer (R) was installed between the place of the injection and the detector cell to provide a homogeneous distribution of the sample injected perpendicularly to the flow. The volume of solution between the points of injection and detection was kept as small as possible.
The potential difference between the ion-selective electrode (E_1) and the reference (E_2) was measured with a mV-meter (M) and registered by a recorder (IP).
The ionic strength of the continuously streaming solution was adjusted to 0,1 M by adding a suitable amount of potassium nitrate.

Programed coulometric titration technique

The so-called programed coulometric titration technique was used for the determination of the chloride content of streaming water samples. As titrant silver ions coulometrically

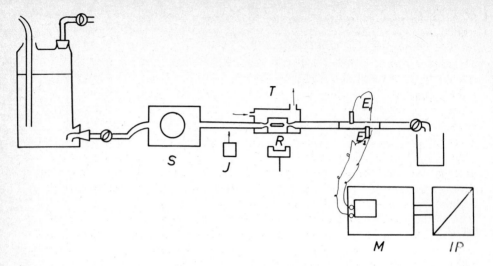

Fig. 5. Block diagram of the experimental set-up constructed for the injection study.
S: peristaltic pump; J: site of injection; T: water jacket; E_1: ion-selective electrode; E_2: reference electrode; M: mV-meter; IP: recorder

generated in situ were added to the sample solution. The change in chloride or silver ion activities – as a result of the fast precipitation reaction and programed silver ion generation – was monitored – with a chloride-selective membrane electrode built into the tube carrying the sample solution. Figure 6 shows the apparatus used in the measurements.

The pump (P) (LKB, VARIOPERPEX type 12000) provided a constant flow rate of the sample solution.

The flow through the generator cell (C_1) contained the silver anode (G_1) and the platinum cathode (G_2). The sphere of the cathode was separated from the flow-through anodic section

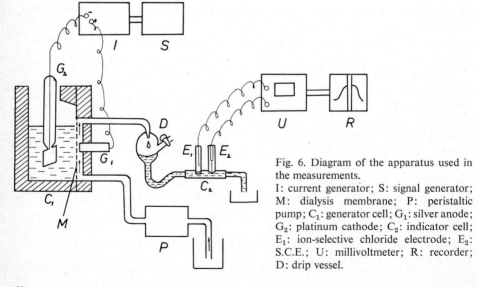

Fig. 6. Diagram of the apparatus used in the measurements.
I: current generator; S: signal generator; M: dialysis membrane; P: peristaltic pump; C_1: generator cell; G_1: silver anode; G_2: platinum cathode; C_2: indicator cell; E_1: ion-selective chloride electrode; E_2: S.C.E.; U: millivoltmeter; R: recorder; D: drip vessel.

of the cell by a dialysis membrane (M) in order to avoid gas bubbles. Potassium nitrate was added to the sample and to the standard solutions before they entered the generator cell to increase the conductivity. The concentration of potassium nitrate was 0.5 M.

The current generator (Tacussel; type ASA 100-1) (I) controlled by a signal generator (S) (Philips; type PM 5168) was used to provide the electric current for the silver ion generation. For every measurement an appropriate single, pulse triangular signal was given according to which the current intensity in the generator circuit changed from 0.0 to 0.1 A and symmetrically back to 0.0. In this way the silver ion generation and thus its addition to the streaming sample solution were controlled by an appropriate program.

The ion-selective chloride electrode (E_1) was placed in the flow-through detector cell (C_2). The potential difference between the indicator electrode (E_1) and a saturated calomel electrode (E_2) was measured by a millivoltmeter (U) (Keithley Differential Electrometer, type 604) and registered by a recorder (R) (Bryans, Type 26000 A 3).

Between the generator and the detector cell the stream was interrupted by a special drip-vessel (D) to electrically insulate the two cells and to achieve a homogeneous distribution of the generated silver ions throughout the cross-section of the tube. The measurements were evaluated on the basis of the time interval between the equivalent points of the two symmetrical titration curves.

Results and discussion

Monitoring of fluoride in air

The determination of fluoride in polluted air is a very serious problem, particularly in areas with aluminum industry. Fluoride in air must be absorbed and concentrated in a suitable aqueous solution prior to measurement with fluoride ion-selective electrodes. In the first type of absorber (Fig. 2) a liquid layer of about 20 cm heigth was sufficient to achieve practically complete absorption of gaseous HF, since the gas was distributed in small bubbles with a diameter of 0.1–0.5 mm. In the second type (Fig. 3) the absorbing vessel was a plastic tube of 2 m length wound into a spiral. Here, bubbles of about 4 mm diameter were formed which needed a long path for the HF to be completely absorbed. The advantage of the first type was its simple construction; a disadvantage was that the solution retained in the pores of the bubbler had a higher fluoride concentration than the bulk of the solution. This difficulty could be overcome by washing out the bubbler and mixing the solution obtained, but this made the operation lengthy. This problem did not arise with the second recirculation-type absorber, with which no subsequent washing and homogenization were necessary to obtain about 100% of the HF introduced into the absorbing solution. The liquid was efficiently homogenized by the gas stream itself. These advantages made the latter type of absorber more favorable than the former. The absorption efficiency was studied with both types as function of the flow rate and HF concentration of the gas. The results of some of these experiments are collected in Tables I and II.

The standard deviations of the measurements are 4.4% and 5.5% respectively.

The electrodes were calibrated with standard solutions of known concentration and the calculations were made on the basis of a calibration graph taking the gas flow rate and volume of absorbing solution into consideration.

With this system only average HF levels could be determined over a certain period of time. During this time a volume of gas containing enough HF to make fluoride in the absorbing

Table I. Absorbtion efficiency measurements on the gas washing bottle absorber.

HF concentration in gas mg/l	Gas flow rate l/hour	Fluoride concentration in absorbing solution moles/l		Error %
		Measured	Calculated	
0,19	20	$3,63 \cdot 10^{-4}$	$3,58 \cdot 10^{-4}$	+ 1,3
0,30	10	$3,17 \cdot 10^{-4}$	$3,30 \cdot 10^{-4}$	- 3,9
0,50	6	$4,37 \cdot 10^{-4}$	$4,34 \cdot 10^{-4}$	+ 0,8
0,58	40	$2,00 \cdot 10^{-3}$	$2,00 \cdot 10^{-3}$	0
0,81	10	$8,70 \cdot 10^{-4}$	$9,02 \cdot 10^{-4}$	- 3,5
0,84	3	$2,76 \cdot 10^{-4}$	$2,78 \cdot 10^{-4}$	- 0,7
0,43	10	$2,51 \cdot 10^{-3}$	$2,70 \cdot 10^{-3}$	- 7,0
0,59	6	$1,21 \cdot 10^{-3}$	$1,19 \cdot 10^{-3}$	+ 1,8
1,67	6	$1,09 \cdot 10^{-3}$	$1,18 \cdot 10^{-3}$	- 7,5
2,40	10	$2,35 \cdot 10^{-3}$	$2,35 \cdot 10^{-3}$	0
4,15	3	$1,26 \cdot 10^{-3}$	$1,26 \cdot 10^{-3}$	- 7,3

liquid at least 10^{-5} M had to pass through the system. At low levels of fluoride in air only discontinuous measurements could be accomplished with our device; continuous measurements can only be carried out a fairly high fluoride concentrations, although *Liberti* and *Mascini* (1970) have reported on successful continuous measurement of low HF concentrations in air.

On the basis of our experience we rather suggest a continuous fluoride analysis carried out after an appropriate accumulation process.

On the basius of the experiments reported on here, an automatic apparatus for the continuous monitoring of air pollutant fluoride is under construction in our laboratories. The design of the apparatus is shown in Figure 7.

Table II. Absorbtion efficienhy measurements on the recirculation absorber.

HF concentration in gas mg/l	Gas flow rate l/hour	Fluoride concentration in absorbing solution moles/l		Error %
		Measured	Calculated	
0,15	20	$1,69 \cdot 10^{-3}$	$1,83 \cdot 10^{-3}$	- 7,7
0,19	12	$1,48 \cdot 10^{-3}$	$1,48 \cdot 10^{-3}$	0
0,22	18	$3,24 \cdot 10^{-3}$	$3,26 \cdot 10^{-3}$	- 0,6
0,23	14	$2,05 \cdot 10^{-3}$	$2,00 \cdot 10^{-3}$	+2,5
0,29	10	$1,20 \cdot 10^{-3}$	$1,16 \cdot 10^{-3}$	+3,4
0,64	3,5	$1,60 \cdot 10^{-3}$	$1,72 \cdot 10^{-3}$	- 7,0
0,93	2,5	$5,50 \cdot 10^{-4}$	$5,83 \cdot 10^{-4}$	- 5,6
0,96	5,5	$1,42 \cdot 10^{-3}$	$1,32 \cdot 10^{-3}$	+7,5

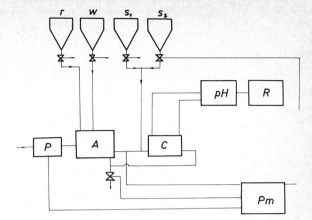

Fig. 7. Monitor for air pollutant fluoride (block diagram).
R: reagent; W: destilled water; S_1 and S_2: standards; P: pump; A: absorber, C: measuring cell; P_m: programer.

Injection technique

General aspects

As has been pointed out in the introduction, the injection technique worked out in our laboratory using potentiometric detectors was found very suitable for the analysis of electrically active components (*Feher* et al. 1973; *Nagy* et al. 1970). The technique and the apparatus required for this method are simple to use and perform the analysis very rapidly. Furthermore, it seems to be quite likely that monitors working on the same principle can be constructed.

Our intention to use the injection technique for potentiometric purposes was influenced by two factors. Firstly, we wanted to see whether the advantages of the injection technique could also be used favorably in potentiometry. Secondly, the logarithmic signal transformation of a potentiometric sensor may offer some additional advantages.

Generally speaking, with the injection technique a small volume of sample is injected into a solution streaming with a constant flow rate. After leaving the mixing chamber built into the tube carrying the solution, the solution enters the flow-through detector cell. As an effect of a single injected dose, alterations appear in the concentration of the solution flowing through the cell which can either be a result of a simple dilution or the result of a chemical reaction between the injected sample and an ingredient of the streaming solution. If this process is monitored by an appropriate sensor, the signal recorded after an injection changes its value with respect only to the streaming solution (base-line) and as the dose leaves the cell the signal gradually falls back to the value of the base-line. Thus, the signal recorded in time after a single shot injection has a peak-type shape with a well-defined area.

In potentiometry the area under the peak has a well defined physical meaning, namely, the electric charge flown through the cell as voltammetric current after a dose of injected solution (of course its sign can be positive or negative). The charge can be expressed in coulombs.

However, if potentiometric sensors are used with the injection technique, the recorded peak area (this is the integral of the electrode potential in time) can not be given in common terms. It may be considered as action since its dimension is erg. sec. In spite of this, the peak areas can be taken as readings and used for analytical purposes.

Another property of the registered potential signal which could also be used as a basis for signal transformation is the peak height.

E. Pungor, K Tóth, G. Nagy

When the alteration of concentration after an injection is monitored with an ion-selective electrode, the potential change can of course be positive or negative depending on whether the ion activity increases or decreases. The potential increases or decreases in comparison with the electrode potential corresponding to the continuously streaming solution ("base-line"). To obtain a constant "base-line" potential with an ion-selective electrode even in streaming solution, one should of course have a constant level of the ion, to which the electrode reacts reversibly.

A given injected dose of an appropriate solution gives rise to a concentration change (ΔC_t). The correlation between the concentration change and the amount injected (M) is given by the following general equation:

$$\int_0^\infty K \Delta C_t V \, dt = M \tag{1}$$

where
  K is stoichiometric constant
  ΔC_t is equal to ($C_t - C_0$)
  C_t is the actual concentration
  C_0 is the concentration in the continuously streaming solution
  V is flow rate (ml/min)

When potentiometric sensors are used, the transformation equation is

$$\Delta E_t = S \ln \left[1 + \frac{\Delta C_t}{C_0 + \sqrt{L} + \sum_{i=0}^{n} k_i a_i} \right] \tag{2}$$

considering that only ions of valency 1 take part in the electrode reaction

where
  S is the slope of the electrode
  ΔE_t is the potential change es an effect of injection
  a_i is the activity of interfering ion i
  k_i is the selectivity constant
  L is the solubility product of the precipitate incorporated into the electrode

The exact relations between $E_{t\,max}$ (peak height) and A (peak area) and the parameters of the experiment can only be deduced if the concentration profile in time can be expressed with a suitable model. The simplest approximation can be given by considering ideal conditions, i.e.:

1. The rate of injection is constant, i.e., no concentration gradient is formed in the direction of flow inside the "slug" of liquid containing the material injected
2. The mixing of the solution in the stirrer is instantaneous.

Under these conditions the concentration distribution in time is given by the following equations (*Nagy* et al. 1970) if $0 \leq t < \tau$

then

$$\Delta C_t = \frac{M}{V \tau} \left[1 - e^{-\frac{V}{W} t} \right] \tag{3}$$

while
if $t \geq \tau$

then

$$\Delta C_t = \frac{M}{V\tau}\left[1-e^{-\frac{V}{W}\tau}\right]\cdot e^{-\frac{V}{W}(t-\tau)} \tag{4}$$

where   W is the volume of the mixing chamber
   τ is the time required by the "slug" of solution containing the injected material to pass the entrance section of the mixing chamber. If "tailing" is ignored (assumption 1) then τ is equal to the time of injection.

By substituting Equations 3 and 4 into Equation 2 one obtains the value E_t expressed by the parameters. The integral of the equations expressing E_t gives the peak area. However, we could solve this equation so far only by approximation with a computer.

From Equation 3 one can obtain the concentration value (C_τ) at the height of the peak when $t = \tau$ is assumed. On the basis of Equation 2 the relationship between peak height (E_τ) and the parameters of the experiments could be obtained:

$$C_\tau = \frac{M}{V\tau}\left(1-e^{-\frac{V}{W}\tau}\right) \tag{5}$$

$$E_\tau = E_0 + S \ln\left[C_0 + \sqrt{L} + \sum_{i=0}^{k} k_i\, a_i + \frac{M}{V\tau}\left(1-e^{-\frac{V}{W}\tau}\right)\right] \tag{6}$$

Experimental Findings

The injection technique was tested with a cyanide- and an iodide-selective electrode. The signals obtained were of the peak type as expected (Fig. 8).
The effects of the following parameters on the peak area and the peak height were studied:
   – flow rate
   – amount of ion injected

Effect of flow-rate

It was found that in the normal activity range of the electrode function the flow rate of a streaming solution with constant activity and constant and relatively high ionic strength had no influence on the signal (i.e., electrode potential).

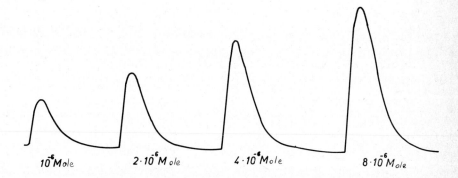

Fig. 8. The signals of an ion-selective electrode recorded after injection
Solution injected: $1\cdot 10^{-6} - 8\cdot 10^{-6}$ mole KI
Streaming solution: 10^{-4} m KI

With the injection technique the flow rate had a marked effect both on peak area and peak height. The experimental results are shown in Figures 9 and 10. From Figure 9 it can be seen that an inverse proportionality exists between the flow rate and the peak area. The standard deviation of the individual readings was found to be within ± 1,9 %.

The peak area can be expressed as a function of the contact time (the time during which the electrode is in contact with the sample solution differing in activit from the continuously streaming solution) in the following way:

$$A = \text{constant} \cdot t_c = \text{const.} \frac{1}{V}$$

where t is the contact time.

In Figure 10 it can be seen that the relation between peak height and flow rate is not as simple as that between peak area and flow rate.

Effect of concentration

In the cases of cyanide and iodide injection it was found that well-defined relationships exist between the peak area and the amount injected as well as between the peak height and the

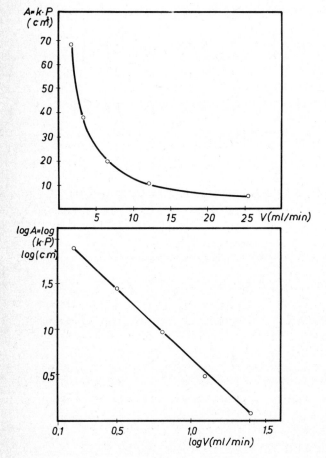

Fig. 9. Dependence of the peak area on the flow-rate.
Injected solution: $2 \cdot 10^{-5}$ mole KI
streaming solution: 10^{-4} M KI

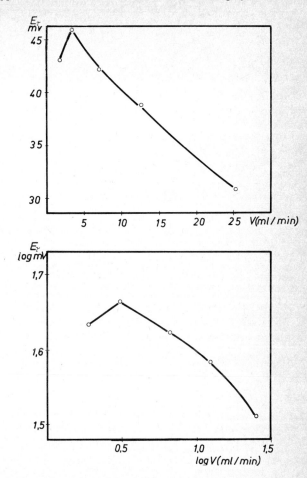

Fig. 10. Dependence of the peak potential on the flow rate. Injected solution: $2 \cdot 10^{-5}$ mole KI
Streaming solution: 10^{-4} M KI

amount injected (Figs. 11 and 12). The relationships are linear in a wide range of the amount injected; however, the curves deviate from linearity when very small amounts are injected. This can be explained by a decrease of the slope of the electrode in the concentration range near the detection limit.

The advantage of the injection technique is that the detection limit approaches the theoretical value when a suitable flow-rate is applied.

We have tried to compare our experimental findings with the results of the theoretical considerations discussed earlier, and it was found that they show complete agreement only under certain conditions. However, the trends of the findings were in agreement with the theoretical expectations. To make the model more complete efforts are being made to consider the deviations from the ideal conditions.

For the practical description of the curves, curve fitting methods can also be applied.

The application field of the injection technique can be expanded for solving various problems. It can be used for monitoring a species injected which undergoes a fast chemical reaction with the streaming reagent.

As an example, the determination of relatively concentrated cyanide sample solutions is mentioned here. In this case the reagent, i.e., silver nitrate, flows at a constant rate and the

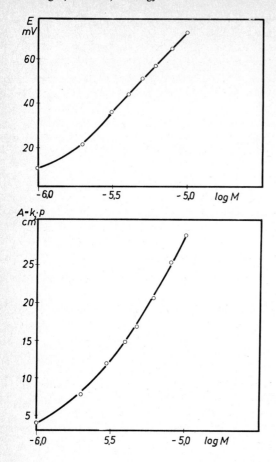

Fig. 11. Calibration graph for iodide.
Flow rate: 2,8 ml/min
Streaming solution: 10^{-4} M KI

cyanide is injected in a given interval into the stream. The most important advantage of this technique of cyanide monitoring is that the life time of the electrode is greatly lengthened since the electrode is not poisoned by the streaming solution.

Another advantage of the injection technique is that it can be used for the determination of the concentration of the streaming solution, since the signal transformation is logarithmic. As mentioned earlier this possibility does not exist in the case of linear signal transformation, e. g., potentiometry. The exploitation of this possibility offered by the non-linear signal transformation was one of the aims of this study.

It is obvious that if a certain amount of an ion is injected into a streaming solution containing the same ion in a given concentration, the peak area and the peak height depend to a great extent on the concentration in the streaming solution. This can provide a basis for using the injection technique to determine the concentration in the streaming solution with calibration. The calibration graph is made by injecting the same amount of the primary ion into streaming standard solutions.

Furthermore, ion activity can be determined in a streaming sample solution by injecting standards. The concentration of the streaming solution can be determined from the peak area or the peak height in the following way.

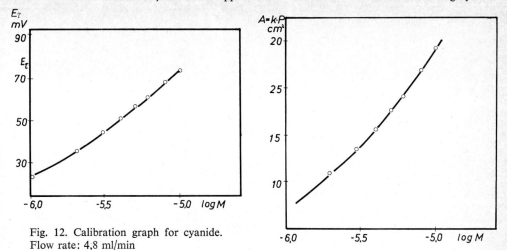

Fig. 12. Calibration graph for cyanide.
Flow rate: 4,8 ml/min
Streaming solution: 10^{-4} M KI + 10^{-3} M NaOH + 10^{-1} M KNO_3

From Equation (1) and (2) one can derive

$$M = C_0 \int_0^\infty \left[e^{\frac{\Delta E_t}{S}} - 1 \right] V \, dt \qquad (7)$$

If a peak-type recording is available, the amount of M_1 can be approximated by an expression derived on the basis of Equation (7), namely,

$$M_1 \cong V \cdot C_0 \, \Delta t \sum_i^n \left[e^{\frac{E_1(t_i)}{S}} - 1 \right] \qquad (8)$$

where C_0 and S are unknown values
Δt is the time interval, on the basis of which the approximation of the peakarea is made.

If another amount of sample (M_2) is injected, a similar equation can be written for M_2:

$$M_2 \cong V \cdot C_0 \, \Delta t \sum_i^n \left(e^{\frac{E_2(t_i)}{S}} - 1 \right) \qquad (9)$$

On the basis of these two equations S and C_0 can be determined. To obtain more accurate results it is advisable to inject more than two standards. Of course, M_1, M_2 and M_3 should not be equal to each other.
The calculation of C_0 and S values from the Equations (8) and (9) can be done with a computer program the block diagram of which is shown in Figure 13. The concrete program used in our work is given as follows:

```
'BEGIN
    'COMMENT
    ALGOL 1204 PROGRAM FOR COMPUTING THE ELECTRODE SLOPE AND THE
    CONCENTRATION OF FLOWING SOLUTIONS ( TWICE STANDARD ADDITION
    BY INJECTING — POTENTIOMETRICAL DETECTING );
```

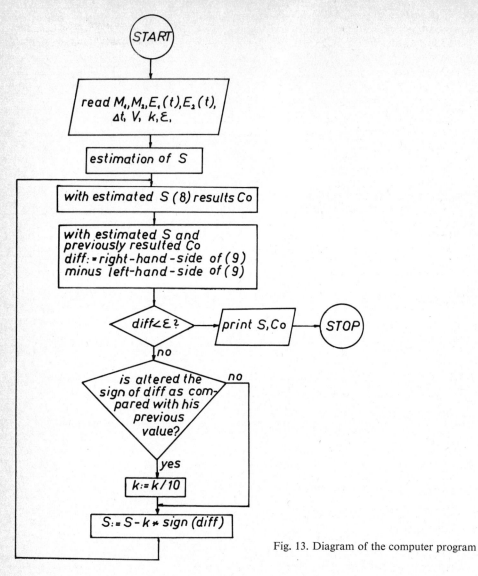

Fig. 13. Diagram of the computer program

```
 INTEGER N;
READ(N);
'BEGIN
  'INTEGER I,IT,P;
  'REAL F1,F2,DT,M1,M2,V,S,S1,K,EPS,DIFF,DIFF1,SUM,R1,R2,C,C1;
  'ARRAY DE1[1:N],DE2[1:N];
  READ(F1,F2,DT,DE1,DE2,M1,M2,V,S,K,EPS);
  'FOR I:=1 'STEP 1 'UNTIL N 'DO
    'BEGIN
    DE1[I]:=DE1[I]*F1;
    DE2[I]:=DE2[I]*F2;
    'END;
```

```
    K:=ABS(K);
    P:=SIGN(S);
    S:=ABS(S)/2.303;
    R1:=M1/V/DT;
    R2:=M2/V/DT;
    IT:=0;
A1: IT:=IT+1;
    SUM:=0.0;
    'FOR I:=1 'STEP 1 'UNTIL N 'DO
      SUM:=SUM+EXP(DE1[I]/S);
    C:=R1/(SUM-N);
    SUM:=0.0;
    'FOR I:=1 'STEP 1 'UNTIL N 'DO
      SUM:=SUM+EXP(DE2[I]/S);
    DIFF:=R2—C*(SUM—N);
    'IF IT=1 'THEN
      'BEGIN
      DIFF1:=DIFF;
      S1:=S;
      C1:=C;
      'GOTO A1;
      'END;
    'If ABS(DIFF/R2)<EPS 'AND ABS((S—S1)/S)<EPS 'AND ABS((C—C1)/C)
      <EPS 'THEN 'GOTO A2;
    'IF DIFF1*DIFF<0 'THEN K:=K/10;
    S:=S—K*SIGN(DIFF);
    DIFF1:=DIFF;
    S1:=S;
    C1:=C;
    'GOTO A1;
A2: LINE(6);
    S:=P*S*2.303;
    PRINT("TWO STANDARD ADDITION RESULTS: $);
    LINE(2);
    PRINT("INJECTED STANDARD QUANTITIES: $);
    FORMAT("!!!!!—1.2ⓐ—12!!AND!!—1.2ⓐ—12!!MOLES??$);
    PRINT(M1, M2);
    FORMAT("NUMBER!OF!ITERATIONS!=!—123456?$);
    PRINT(IT);
    FORMAT("FEASIBLE!RELATIVE!ERROR!OF!THE!EQUATIONS!=!—0.12345?$);
    PRINT(EPS);
    FORMAT("THE!CALCULATED!ELECRTODE!SLOPE!=!—12.12!MV?$);
    PRINT(S);
    FORMAT("THE!CALCULATED!CONCENTRATION!=!—1.123ⓐ—123!MOLE/LIT?$);
    PRINT(C);
    'END;
'END;
?
?
```

Some values for C_0 and S were calculated on the basis of data pairs ($E(t_i)$; t_i) taken from the potentiometric recordings, which corresponded to the injection of two standards (M_1 and M_2). These results are listed below:

TWO STANDARD ADDITION RESULTS WITH DIFFERENT FEASIBLE RELATIVE ERROR VALUES (INJECTION OF KI SOLUTION IN FLOWING KI SOLUTION;
the concentration of the latter is unknown)

INJECTED STANDARD QUANTITIES: M1=4.0ⓐ—6 moles AND M2=8.0ⓐ—6 moles;

1.
number of iteration steps = 13
feasible relative error of equations = 0.001
the calculated electrode slope S = 59.76 mV
THE CALCULATED CONCENTRATION Co = 1.2448ⓐ—4 mole/lit
2.
number of iteration steps = 19
feasible relative error of equations = 0.0005
the calculated electrode slope S = — 59.87 mV
THE CALCULATED CONCENTRATION Co = 1.253ⓐ—4 mole/lit
3.
number of iteration steps = 19
feasible relative error of equations = 0.0001
the calculated electrode slope S = — 59.87 mV
THE CALCULATED CONCENTRATION Co = 1.253ⓐ—4 mole/lit

The rapid convergence proves the efficiency of the method of calculation used.

Programed coulometric titration technique

In our work an argentometric titration with programed coulometric reagent administration and potentiometric detection using ion-selective electrodes has been worked out and has been applied for the determination of chloride ion concentration in flowing natural water samples.

General considerations

A complete titration can be carried out in a solution streaming laminarly in a tube if the titrant is added continuously with a linearly increasing rate in a certain period of time. This is true if it is supposed that the mixing of the titrant is homogeneous in the direction perpendicular to the flow and when the reaction is quantitative and relatively fast and no "tailing" takes place in the tube.

The titration curve can be recorded by an appropriate detector, eg., an ion-selective electrode placed in the tube downstream from the site of reagent administration.

After achieving an adequate degree of overtitration the reagent administration can be decreased at the same rate at which it was increased. In this case two symmetrical titration curves connected to each other are obtained.

The time interval between the equivalent points of the two titration curves is well defined and depends only on the concentration of the streaming sample solution if the other conditions are kept constant.

For the time interval (Q) existing between the two equivalent points the following can be derived:

$$Q = 2\tau - 2\frac{C V_M}{a n} \tag{10}$$

where a stands for the stoichiometric constant of the reaction
C represents the concentration, while
V_M is the flow rate of the sample
n is constant,
τ is the time of reagent administration rate increasing and decreasing.

Experimental Findings

The addition of the silver ions for the argentometric titration was done coulometrically as was mentioned in the experimental part.

Two curves recorded after single-shot reagent generation are shown in Figures 14 and 15. As expected, each recording can be divided into two well defined parts namely, two potentiometric titration curves. The first one is a result of the chloride titration, while the second is that of the linearly decreasing silver ion generation rate. According to the theoretical expectations the two parts should be the mirror pictures of each other. The distortion is due to the tailing effect taking place in the tube.

The titration curves were found to be almost symmetrical to each other in the case of high chloride concentrations ($C \geq 10^{-2}$ M) and small "n" values.

However, the tailing seemed to be more pronounced at low chloride concentrations and at higher "n" values. Furthermore, the distortion of symmetry appeared to be more significant in the lower range of sample flow rate.

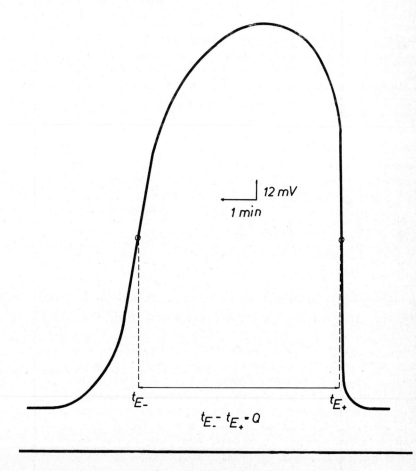

Fig. 14. Electrode potential vs. time curve after single-shot, triangular reagent generation. Pulse frequency: $5 \cdot 10^{-3}$ cps; max. current 0,1 A; chloride concentration: $2 \cdot 10^{-3}$ M

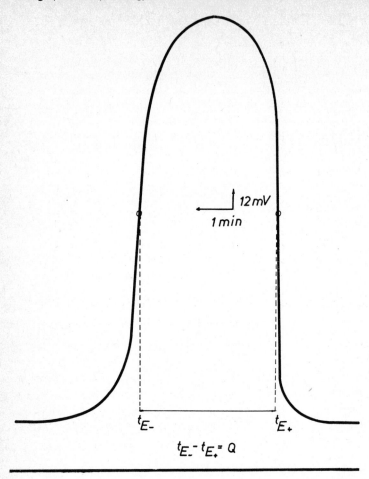

Fig. 15. Electrode potential vs. time curve after single shot triangular reagent generation. Pulse frequency: $5 \cdot 10^{-3}$ cps; max. current 0,1 A; chloride concentration: $5 \cdot 10^{-3}$ M

Owing to the huge change in the electrode potential registered at the inflection points the time interval "Q" existing between them can easily be determined. Accordingly calibration curves can be made by plotting the "Q" values vs. the actual chloride ion concentration of standards. As an example, Figure 16 shows a calibration curve which is almost linear and provides an excellent way of accurately determining chloride in streaming sample solutions. By the variation of parameters τ, n, V_m the sensitivity and the accuracy of the measuring method based on the determination of Q values can be adjusted to match the sample concentration and the requirements.

In addition, it is worth mentioning that in our laboratory promising experiments are in progress to investigate further the properties and the possible applications of this programed coulometric titration technique in streaming solutions.

Amperometric, biamperometric and different potentiometric detector cells have already been constructed and applied in this work.

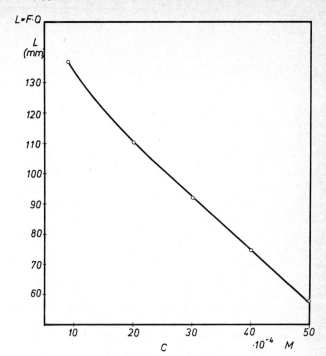

Fig. 16. A calibration curve for the determination of chloride. Pulse frequency: $5 \cdot 10^{-3}$ cps; max. current 0,1 A; $L = F \cdot Q$ where F = 19 mm/min

Acknowledgement

The authors are indebted to Dr. *A. Pall* for her contribution in the fluoride determination, and to Mr *M. Gratzl* for his computer work and Dr. *Zs. Fehér*, who has contributed so much to this work.

References

Babcock, R. H., *Johnson*, K. A.: Selective Ion Electrode System for Fluoride Analysis. J. Amer. Water Works Ass. 60 (1968) 953–61

Cornish, D. C., *Simpson*, R. J.: Considerations in the Use of Ion Selective Electrodes for Monitoring and Control. Meas. Contr. 4 (1971) 308–311

Di Martini, R.: Determination of Nitrogen Dioxide and Nitric Oxide in the ppm Range in Flowing Gaseous Mixtures by Means of Nitrate Specific Ion Electrode. Anal. Chem. 42 (1970) 1102–1105

Eichler, D. L.: The Gradient Titration. Advances in Automated Analysis. Technicon International Congress 1969. Vol. II. p. 51–59. Mediac Inc., New York 1970

Fehér, Zs., *Nagy*, G., *Pungor*, E.: Studies on the Druglevel and Flow Rate of Blood Streaming in Living Organisms by Means of a Silicone-rubber Based Graphite Electrode. Hung. Sci. Instr. 26 (1973) 15–20

Fehér, Zs., *Pungor* E.: The Application of Hydrodynamic Voltammetry in Chemical Analysis. Anal. Chim. Acta 71 (1974) 425–432

Flett, B., *Ho*, A. Y. W.: Gradient Titration. A Novel Approach to Continuous Monitoring Using Ion-selective Electrodes. Anal. Chem. 46 (1974) 9–11

Fleet, B., *Rechnitz*, G. A.: Fast Flow Studies of Biological Reactions with Ion-selective Membrane Electrodes. Anal. Chem. 42 (1970) 690–694

Fleet, B., *von Storp*, H. L.: Analytical Evaluation of a Cyanide-ion-selective Membrane Electrode under Flow-Stream Conditions. Anal. Chem. 43 (1971a) 1575–1581

Fleet, B., *Von Storp*, H. L.: The Determination of Low Levels of Cyanide Ion Using a Silver

Responsive Ion Selective Electrode. Anal. Lett. 4 (1971 b) 425–435

Havas, J.: The Use of Radelkis Biological Microanalyzer for the Direct Potentiometric Determination of the Chloride-ion Concentration and pCl Value of Human whole Blood. Hung. Sci. Instr. 27 (1973) 47–50

Hussein, R. W., von Storp, L. H., Guilbault, G. G.: Determination of Rhodenase Activity with Cyanide-ion selective Electrodes under Flowstream Conditions. Anal. Chim. Acta 61 (1972) 89–97

Liberti, A., Mascini, M.: Specific Ion Membrane Electrodes for the Continuous Measurement and Monitoring of Atmospheric Pollutants. Lecture presented at the 2nd Int. Air Pollution-Prevention Associations, Washington/D. C. 1970

Light, T. S.: Industrial Analysis and Control with Ion- selective Electrodes. In: Ion-selective Elctrodes, pp. 349–74, ed. by R. A. Durst. N.B.S. spec. publ. (1969 a)

Light, T. S.: Ion Selecting Electrodes. Industrial Application. Ind. Water Eng. 6 (1969 b) 33–37.

Llenado, R. A., Rechnitz, G. A.: Ion-Electrode Based Autoanalysis System for Enzymes Anal. Chem. 45 (1973) 826–33

Nagy, G., Fehér, Zs., Pungor, E.: Application of Silicon Rubber-based Graphite Electrodes for Continuous Flow Measurement Part II. Anal. Chim. Acta 52 (1970) 47–54

Oliver, R. T., Lenz, G. F., Frederick, W. P.: The Determination of Fuoride in Vegetation and Gases by Automatic Direct Potentiometry. Advances in Automated Analysis Vol. II. Medial Inc. (1970) 3091–314

Pungor, E.: unpublished results

Pungor, E.: Oscillometry and Conductometry. Int. Series of Monographs in Anal. Chem. Vol. 21. pp. 206–207. Pergamon Press, New York 1965

Pungor, E., Fehér, Zs., Nagy, G.: Application of Silicone Rubber-based Graphite Electrodes for Continuous Flow Measurements Part I. Anal. Chim. Acta 51 (1970) 417–424

Pungor, E., Tóth, K.: Die Anwendung von ionenselektiven Elektroden in der Wasseranalyse. In: Vom Wasser, ed. by W. Husmann. Verlag Chemie, Weinheim 42 (1974) 43–71

Ruzicka, J., Tjell, J.C.: Ion-selective Electrodes in Continuous-flow Analysis, Determination of Calcium in Serum. Anal. Chim. Acta 47 (1969) 475–482

von Storp, L. H., Guilbault, G. G.: New Assay for Cholisterase, Potentiometric Determinations in Flow-streams. Anal. Chim. Acta 62 (1972) 425–430

Slevogt, K., Seelos, E., Schraidt, W.: Ionselektive Elektroden und deren Erprobung zur kontinuierlichen Konzentrationsbestimmung in Trink- und Abwasser. In: Vom Wasser, ed. by W. Husmann, Verlag Chemie, Weinheim 42 (1974) 1–42

Thompson, H. I., Rechnitz, G. A.: Fast Reaction Flow System Using Crystal Membrane Ionselective Electrodes Anal. Chem. 44 (1972) 300–305

Váradi, M., Fehér, Zs., Pungor, E.: Determination of Purin Bases by Means of Chromatovoltammetry. J. Chromatography 90 (1974) 259–265

F. G. K. Baucke

Investigation of pH Electrode Membrane Glasses by Means of an Ion Sputtering Method. — Discussion of Microelectrode Membranes

(7 Figures)

1. Introduction

The development of micropotentiometry in biological and medical work demands glass electrodes of progressively reduced size. Sealed pH-glass microelectrodes with tip diameters below 1 μm have been applied (*Khuri* 1969). With decreasing dimensions, however, the thickness of the glass membrane becomes increasingly comparable to the extension of the leached glass surface layer, and the membrane properties are controlled by this surface region to a growing extent. It is thus of high interest, if not fundamental, for the fabrication of smaller electrodes and development of new glasses to obtain information on the structure and properties of leached glass surface layers.

We have obtained such information by applying a new method of measuring concentration profiles of certain cationic components within glass surface regions to several lithium-containing pH-electrode glasses. The technique is based on the continuous measurement of the photon emission induced during ion-sputtering the specimens and is distinguished by a depth resolution of 30 to 50 Å. For a detailed description the reader is referred to the literature (*Bach* 1971; *Bach* and *Baucke*, 1974). Leached glass surface layers (*Baucke* 1974), cation migration (*Baucke* 1971; *Baucke*, 1975), and the reaction between the glasses and humid atmospheres (*Bach* and *Baucke*, 1974) have been investigated by means of the sputtering method. In the following, the results of this work will by summarized, as far as necessary, and will serve as the basis for the discussion of the compositional and electrical structure, the useful life-time, and the lower limit of thickness of glass microelectrode membranes. Besides, some recommendations for microelectrode storage and pretreatment, which result from this work, will be given.

Summary of Results

Leached glass surface layers formed in aqueous solutions

During contact of the glasses with aqueous solutions the lithium concentration profile at the glass surface changes significantly (Fig. 1A). The change rate decreases with increasing loss of lithium and becomes zero within approximately one day (Fig. 1B) which compares well with the "conditioning time" usually recommended for glass electrodes. The subsequently constant profile demonstrates a steady state given by equal rates of formation and dissolution of the leached surface layer*). Steady state concentration profiles are characterized by an

---

*) The dissolution rate of the particular glass referred to was measured to be 0.1 Å h^{-1} (25 °C) and 1.0 Å h^{-1} (50 °C) in 0.1 n H_2SO_4 and six times as high in 0.1 n KOH.

outer "leached layer" with low lithium concentrations and small and constant concentration gradient and a "transition layer" with higher lithium concentrations and non-uniform concentration gradients. They are dependent on solution pH and temperature and change reversibly when the glass is submitted to new conditions for sufficiently long time (Fig. 2).

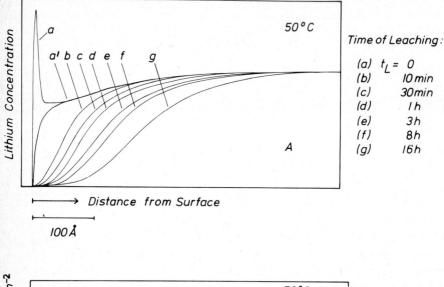

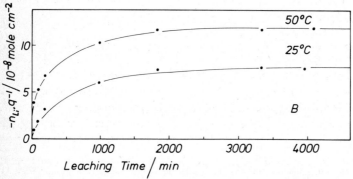

Fig. 1. Lithium concentration profiles in electrode membrane glass surface after different periods of leaching in 0.1 n H_2SO_4 at 50 °C (A) and the corresponding lithium loss as a function of time (25, 50 °C) (B). Profile (a): untreated specimen after storing at humid atmosphere, profile (a′): after wiping with moist tissue.

Lithium ions leaving the transition layer are replaced by an equal number of hydrogen ions. The interdiffusion coefficient has a minimum within the range of maximum lithium concentration gradient. The corresponding maximum local resistivity exceeds the bulk glass resistivity by a factor of 340 at 25 °C. A range with high resistivity within the leached surface layer has been reported (*Wikby* 1972; *Wikby* 1973) and had been proposed earlier (*Brand* and *Rechnitz* 1969).

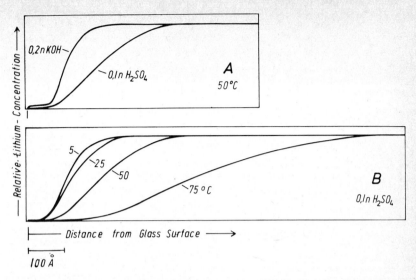

Fig. 2. Reversible steady state Li-concentration profiles developed at different pH (A) and temperature (B), respectively.

Cation migration in membrane glasses

Cation migration was studied under nonblocking and nonpolarizing conditions by subjecting membranes in contact with electrolyte solutions to electric fields. The moving boundary between lithium ions drifting towards the cathodic surface and guest ions entering the glass anodically was subsequently detected by the ion sputtering technique. As shown by infrared spectra, protons, and not hydronium ions, are transferred from aqueous acid solutions into the glass and migrate in the host network. The drifting boundary between lithium ions and following guest protons is fairly sharp (Fig. 3) and demonstrates the migration of distinct layers of different ions, as assumed by *Haugaard* (*Haugaard* 1938; *Haugaard* 1941). The transport number of protons in 100% protonated glass and that of lithium ions in the original glass are unity. The corresponding mobilities of both ions as determined from the distance dependence of the boundary velocity and the activation energies of the transport processes are given in Table I.

By applying appropriate anodic solutions any concentration ratio of lithium ions and protons can be transferred into the glass phase simultaneously, and the anodic glass can be made purely lithium ion-, proton- or mixed ion-conducting. The mobilities of the ions are concentration-dependent. The transfer conditions suggest the discussion of a new pH-electrode mechanism.

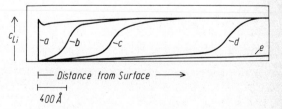

Fig. 3. Li-concentration profiles at anodic membrane surfaces before (a) and after 30 min. (b), 90 min, (c), 240 min (d), and 17 h (e) of applying an electric field of 70 kV cm^{-1} at 50 °C.

Table 1. Li$^+$ and H$^+$ mobilities (Å s^{-1}/Vcm^{-1}) and activation energies (Kcal mol^{-1}) in a rate-cooled pH-electrode glass

T/°C	u_{Li}	u_H	u_{Li}/u_H
25	6.3×10^{-7}	1.8×10^{-10}	3500
50	6.0×10^{-6}	2.4×10^{-9}	2600
75	4.4×10^{-5}	2.7×10^{-8}	1600
E_A	17.4	20.7	

Conclusions for the structure of leached surface layers

While the outer leached layer has constant or fairly uniformly changing properties with respect to interdiffusion (Fig. 2), the transition layer has a resistivity maximum where the lithium concentration is 30 to 50% of the site concentration. It is probably caused by the low concentration-dependent mobilities of both ions, particularly protons, within this range of the transition layer, which seems to maintain a highly glass-like network structure. Adjacent to this there is probably a gradual transition to the structure characterizing the outer leached layer. The transition layer seems to be the seat of the diffusion potential proposed earlier (*Karreman* and *Eisenman* 1962).

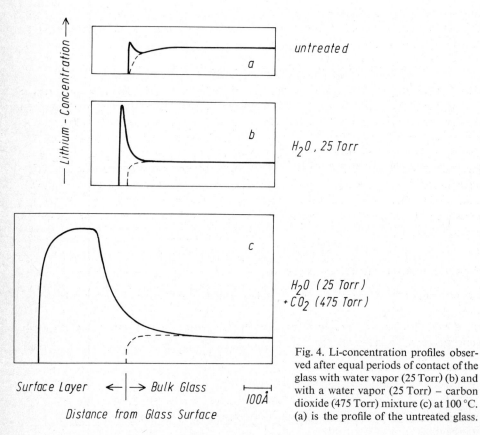

Fig. 4. Li-concentration profiles observed after equal periods of contact of the glass with water vapor (25 Torr) (b) and with a water vapor (25 Torr) – carbon dioxide (475 Torr) mixture (c) at 100 °C. (a) is the profile of the untreated glass.

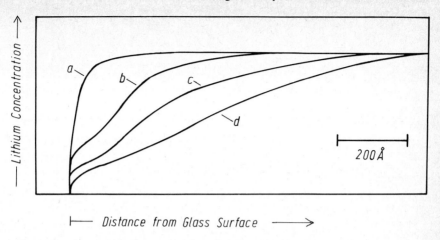

Fig. 5. Li-concentration profiles within glass after removing the surface layers developed in a CO_2-H_2O vapor mixture during 120 h (b), 144 h (c), and 216 h (d) at 35 °C. (a) untreated specimen.

Reaction of glasses with humid atmospheres

Contact of the glasses with water vapor results in interdiffusion of lithium ions and protons supplied by adsorbed water molecules. The resulting reaction products are deposited at the glass surface, for instance:

$$\equiv \text{SiOLi(glass)} + H_2O \text{ (surface)} \rightarrow \equiv \text{SiOH (glass)} + \text{LiOH (surface)}. \tag{1}$$

Such deposits of lithium hydroxide, oxide or carbonate are usually found after storing the glass in the atmosphere (Fig. 1A). If the reaction product forms a uniform layer, the mobilities of the interdiffusing cations in both phases determine whether diffusion within the glass or within the layer controls the reaction. Formation of a uniform lithium carbonate layer at a high rate is thus observed when carbon dioxide and water vapor are present simultaneously at elevated temperature (Fig. 4). Since a steady state in the surface layer, as observed in aqueous solutions, does not develop, the protons exchanged for more mobile lithium ions diffuse into progressively deeper glass regions (Fig. 5). The ion exchange and corresponding change of glass composition continues as long as water vapor is supplied to the surface.

Discussion of pH-microelectrode membranes

The information on pH-electrode glasses summarized above has some consequences for the development of new and the choice of available glasses to be used for making microelectrodes with further reduced dimensions. They are also of some interest for electrode fabrication and treatment.

Limitations of membrane thickness

The membrane resistance, whose maximum tolerable value is dictated by the measuring equipment, is given by the glass resistivity at the measuring temperature and the cell constant of the membrane. For a certain glass, consequently, the lower limit of reducing the dimensions of a microelectrode is imposed by a minimum membrane thickness given by the necessary mechanical stability. The high local resistivity within the leached surface layers observed adds a further limitation since the total membrane resistance, which is the sum of bulk glass and leached layer resistances, is determined by the bulk glass resistance, and thus a function of the membrane thickness, only as long as the condition.

$$\rho_{gl} [d - (d_{s1} + d_{s2})] > 2 \rho_{max} \bar{d}_{max} \qquad (2)$$

is observed, where ρ_{gl} and ρ_{max} are the bulk glass and maximum layer resistivity, respectively, d is the membrane thickness, $(d_{s1} + d_{s2})$ the sum of the two thicknesses of the entire leached layers, and $\bar{d}_{max}$ the average thickness of the high-resistivity range. There is thus a certain minimum membrane thickness for each electrode glass below which the membrane resistance is determined only by the transition layer resistance within the leached surface layers. This minimum may be much larger than that demanded for the mechanical membrane stability. For the glass of Table I it is in the order of 0.5 μm if the high-resistivity range is assumed to be $\bar{d}_{max} = 10$ Å thick. The corresponding resistance of a membrane with 100 μm² surface area is in the order of 16^6 MΩ at 25 °C ($\rho_{gl} = 7 \times 10^{10}$ Ω cm, $\rho_{max} = 10^{13}$ Ω cm (*Baucke* 1974)). The high durability of this glass can thus not be utilized for making membranes thinner than the critical value 1 μm although it would promise the fabrication of extremely thin membranes needed for ultramicroelectrodes of sufficiently long lifetime*).

The application of glasses with lower bulk and simultaneously lower transition layer resistivities, on the other hand, is limited by the high dissolution rates and the formation of thick leached layers of such glasses. On soda-calcia-silica glasses we have observed leached layers as thick as several μm; even several tens μm have been reported (*Bocksay, Bouquet* and *Dobos* 1967). These high glass dissolution and leaching rates result not only in short lifetimes of membranes but may also affect the internal buffer solution, whose volume is reduced much faster with decreasing electrode dimensions than the inner membrane surface area, and may cause a drift of the glass electrode standard potential.

Fabrication, conditioning, and storage of microelectrode membranes

Care must be taken not to alter the composition of the electrode glass during the fabrication of membranes and electrodes. Microelectrodes, particularly sensitive because of their size, should thus not be pulled in a flame, as has been reported (*Khuri* 1969), since, in addition to unavoidable losses of alkali oxide at high temperatures, the high water and carbon dioxide contents of the hot flame atmosphere may cause a serious ion exchange and corresponding changes of the glass composition (Figs. 4 and 5).

Microelectrodes should be conditioned in solutions with low pH. Dilute (10^{-2} or 10^{-1} n) sulfuric acid is recommended since it generates well-defined leached layers (Figs. 1 and 2), causes a minimum glass dissolution by keeping the pH in the membrane vicinity low, and contains no chloride ions, which may be incorporated into surface layers of some glasses at low pH (*Schwabe, Dahms, Nguyen* and *Hoffmann* 1962).

---

*) The glass dissolution rate yields the membrane lifetime 200 d/1000 Å at 25 °C corresponding to 20 d/1000 Å at 50 °C.

Investigation of pH Electrode Membrane Glasses

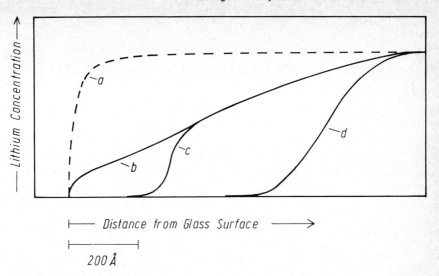

Fig. 6. Li-concentration profiles before (a) and after subjecting the glass to H_2O vapor $-CO_2$ atmosphere and removing the surface deposit formed (b); see fig. 5d. Subsequent leaching (c, d) can cause a steady state only within a range with uniform Li-content (d).

Storage of glass microelectrodes in a humid atmosphere has been recommended in order to protect the surface from "dehydration" (*Khuri* 1969). During extended contact of the membrane with water vapor, however, protons penetrate into progressively deeper glass layers (Fig. 5), so that subsequent conditioning of a membrane thus affected does not generate a steady-state leached surface layer until the unchanged bulk glass with uniform composition is reached (Fig. 6). During this period, an electrode potential drift may be caused by the continuous concentration changes within the transition layer range, probably the seat of the diffusion potential of the membrane. Fresh membranes, consequently, should be stored over an extremely efficient desiccant. Rigorously dried magnesium perchlorate is a most effective material (*Trusell* and *Diehl* 1963) and is being used in our laboratory with good results.

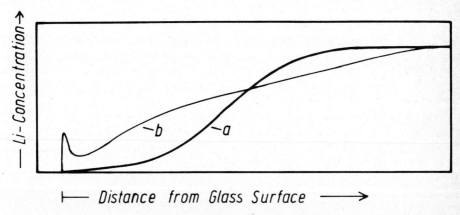

Fig. 7. Li-concentration profile within a membrane surface after leaching in 0.1 n H_2SO_4 (a) and subsequent storing in the atmosphere for two weeks (b).

Protons of a leached layer interdiffuse with lithium ions of the bulk glass and penetrate into deeper glass layers if a steady state of the layer is not maintained by storing the membrane in aqueous solution (Fig. 7). Used (leached) membranes, consequently, should only be stored in solution, if possible at low pH. The lifetime of the membranes can be increased considerably by storing them at low temperatures, which decreases glass dissolution rates as well as interdiffusion coefficients within the glass (Fig. 2).

Acknowledgement

The author is indebted to his co-workers Ing. (grad.) *G. Röth* and Mrs. *B. Mohr* for their careful experimental contributions.

References

Bach, H.: Eine Methode zur Bestimmung von Mikrokonzentrationsgradienten an Glasoberflächen mit Ionenätzung. Intern. Congr. on Glass, Versailles, Sept. 1970, Vol. I, pp. 155–170

Bach, H., Baucke, F. G. K.: Investigation of Reactions between Glasses and Gaseous Phases by Means of Photon Emssion Induced During Ion Sputtering. Physics and Chemistry of Glasses, 15 (1974) 123

Baucke, F. G. K.: Determination of Cation Mobilities in Glasses by Direct Measurement of Drift Velocities (Moving Boundary). Z. Naturforsch. 26a (1971) 1778

Baucke, F. G. K.: Investigation of Surface Layers, Formed on Glass Electrode Membranes in Aqueous Solutions, by Means of an Ion Sputtering Method. J. Non-Cryst. Sol. 14 (1974) 13–31

Baucke, F. G. K.: Cation Migration in Electrode Glasses. In: Materials Sci. Rev. 9: Mass Transport Phenomena in Ceramics, pp. 337–354, ed. by A. R. Cooper and A. H. Heuer, Plenum New York 1975

Boksay, Z., Bouquet, G., Dobos, S.: Diffusion Processes in the Surface Layer of Glass. Phys. and Chem. of Glasses 8 (1967) 140–144

Brand, M. J. D., Rechnitz, G. A.: Mechanistic Studies of Ion-selective Glass Membrane Electrodes. Anal. Chem. 41 (1969) 1788–1793

Haugaard, G.: Studies on the Glass Electrode. Compt. rend. Lab. Carlsberg, Ser. Chim. 22 (1938) 199–204

Haugaard, G.: The Mechanism of the Glass Electrode. J. Phys. Chem. 45 (1941) 148–157

Karreman, G., Eisenman, G.: Electrical Potentials and Ionic Fluxes in Ion Exchangers: I: "n-Type" Non-ideal Systems with Zero Current. Bull. Math. Biophys. 24 (1962) 413–427

Khuri, R. N.: Cation and Hydrogen Microelectrodes in Single Nephrons. In: Glass Microelectrodes, pp. 272–298, ed. by *M. Lavallee, O. F. Schanne*, and *N. C. Herbert*, Wiley, New York 1969

Schwabe, K., Dahms, H., Nguyen, Q., Hoffmann G.: Vergleichende Untersuchungen der elektromotorischen Eigenschaften und des chemischen Verhaltens von Glaselektroden, II: Untersuchungen im Gebiet des Säurefehlers. Z. Elektrochem. 66 (1962) 304–309

Trusell, F., Diehl, H.: Efficiency of Chemical Desiccants. Anal. Chem. 35 (1963) 674–677

Wikby, A.: The Surface Resistance of Glass Electrodes in Neutral Solutions. J. Electroanal. Chem. 38 (1972) 429–440

Wikby, A.: The Resistance of the Surface Layers of Glass Electrodes. Extended Abstracts: Symposium on Mass Transport in Amorphous Solids, Electrochem. Soc. (London) and Soc. Glass Technol., Sheffield/England, January 1973, pp. 1–5

Discussion of Session I

Chairmen: *W. Simon*

(2 Figures)

Discussion of Paper by Eisenman, Krasne, Ciani: Further Studies on Ion Selectivity

Baum: Does it make any difference whether the matrix for the carrier is a phospholipid or a glycolipid?

Eisenman: I have to answer you in two ways. As far as equilibrium selectivity is concerned, the lipid should make no difference at all; and, as far as we can see, this is so. But if you are asking what happens to the "apparent" permeability ratios, and how important the kinetic effects are, then the lipid makes a big difference. In fact this was the sole variable when we compared trinactin in asolectin vs. glyceryl dioleate bilayers.
In this particular case we think we know what is going on and we believe that what happens has nothing to do with the phospholipid head group directly, but merely that a glyceryl dioleate bilayer has about 150 milivolts higher internal negativity compared to the aqueous phase, than does asolectin due to a dipole effect.

Brown: Firstly, does the kinetic expression include a term for unstirred layers? Secondly, what is the explanation for the position of thallium in the selectivity sequences?

Eisenman: As to question 1, our expression did not include such a term. However, in our experimental range there is no significant polarization in the aqueous phase. As to your second question concerning the thallium (I) selectivity:
We have been trying to use thallium as a kind of a probe for ligand type, because it is a more "polarizable" (and "polarizing") ion than the like-sized group Ia cations. Empirically, if you examine selectivities in molecules having ether ligands (or carboxylic groups) one finds for Tl^+ a selectivity in excess of that for like-sized group 1A cations (Rh^+ and K^+). *Krasne* and I have called this "supra 1A" selectivity; and we interpret it as reflecting non-coulomb interactions.

Kessler: How constant or how variable is the deviation from the Nernst behavior caused by kinetic effects in biological membranes?

Eisenman: For most biological membranes of bioelectric interest we think the permeation uses channels and not carriers. Since the model discussed is one specific to a carrier system, the results cannot be used directly, although there certainly are analogies between the rate of loading and unloading of a carrier and of the jumping of anions from one site to another in a channel.

Buck: Your successive levels of analysis and refinement of the permeability ratio theory parallels the equivalent theory for potential-dependent rates of electron transfers at metallic electrodes. The first model gives a gross potential dependence as the Tafel equation relates flux to overall potential difference. When potential barriers at the surfaces are added into the model you have an analog of the *Butler-Volmer-Erdey-Gruz* rate equation. The flux is proportional to a local potential-dependent rate constant and surface concentration (or bulk concentration by removing surface concentration via the *Frumkin* correction). I wonder, if it is legitimate to compare electron transfer rate models and their concentration dependences, with the ion exchange rates? For instance, can you find that the permeability rate constant is independent of concentration?

Eisenman: Not in the "kinetic domain". Remember that although the permeability ratio is expressed as a function of voltage, what was really varied experimentally was the concentration in the mixture, which causes the voltage to vary. The expression contains an implicit dependence on the concentration.

Buck: You cannot separate a potential-dependent rate constant from the effect of potential on the distribution of ions at the surface.

Eisenman: Absolutely not. There are a number of equivalent models.

Pressman: Those of us who are biologically orientated would certainly regard the number of membranes with operative carriers as being greater than the number of membranes with operative channels. Behind each membrane with an operative channel there must have been a carrier system loading the gradient across the membrane prior to channel function.

Eisenman: As an electro-physiologist, what I really meant was the number of electrically "interesting" membranes! To an electro-physiologist, these are ones in which the ionic movements are not electrically silent. I would guess that most of the carrier systems are electrically silent.

Pungor: How did you measure the selectivity ratios? Was it with single ions in the solutions, or have you used the appropriate ion to which the electrode responds and the ion for which the selectivity constant is to be determinded in the same solution?

Eisenman: Selectivity was measured by membrane potential in two-cation mixtures. for example, we start with sodium chloride on both sides of a bilayer membrane and then add potassium nitrate, for example, to one solution so as to measure the potential in mixtures of varying potassium nitrate and sodium chloride concentration. By the way, this is an important point because when people measure solely "bi-ionic" potentials, they throw away all possibility of detecting that they have a kinetic effect.

Christian: What effect or information do you feel you might obtain in a dehydrating medium, for example, in highly concentrated salts like aluminium chloride?

Eisenman: We have never seen any effect for ionic strengths varying between 10^{-3} and 3 molar.

Discussion of Paper by Simon, Morf, Pretsch: Synthetic Neutral Carriers for Cations

Herbert: What is the exact composition of the carrier in the PVC membrane? How critical is this when you make up these membranes?

Simon: It is not really critical. A typical-composition of a PVC membrane would be about 65% of a plasticizer such as dipentyl-phthalate or o-nitro-phenyl-octyl-ether, 30% of PVC and 5% of carrier.

Discussion of Session I

Khuri: How much interference would you expect by anionic phosphates, both organic and inorganic?

Simon: We checked the possible interference by inorganic phosphate and did not observe any, even without adding tetraphenylborate to the membrane phase. We have no experience with organic phosphates. Increasing interference will have to be expected with increasing lipophilicity of the anion.

Eisenman: What sort of resistivities did you get in your neutral carrier systems?

Simon: A typical PVC membrane electrode as described having an exposed circular membrane with a diameter of about 2 mm will have a restistance of 1 MΩ or less at a membrane thickness of 0.1 mm.

Eisenman: Have you used porous polyethylene?

Simon: No, we have not.

Eisenman: Have you an idea as to the applicability of this material to microelectrodes?

Simon: Since adequate solutions of the membrane components mentioned are of rather low viscosity, one should simply try to introduce them into the glass tips of microelectrodes.

Lehmenkühler: What is the respone time of these electrodes?

Simon: It depends on different parameters. You will see details in respect to the response time of these electrodes in the paper by Dr. *Morf*.

Scarpa: Can you give us an idea of the complex formation constants?

Simon: Because of technical problems in the measurement of formation constants only few numbers are available. For a Ba^{2+} carrier we got values of about 3 mol^{-1} lit. in water at 25 °C. We have to assume that values of approximately 10^5 mol^{-1} lit. in methanol at 25 °C are about ideal in view of the use of neutral carriers in liquid membrane electrodes.

Christian: Have you considered the use of polarography to measure these formation constants?

Simon: Yes, we considered it, but we have not yet done any measurements. There is a group in Czechoslovakia (*J. Koryta* et al.) which applies this technique to other carriers such as the crown compounds.

Pungor: May I add something to this question. I think the polarographie examination of the stability constants has many limitations. One must be very critical in accepting evidence from such a technique.

Discussion of Paper by Buck: Application of Solid-State Principles to Response Interpretation and Design of Solid Membrane Electrodes

Simon: You mentioned a paper published in 1921 describing solid saltmetal contacted systems which usually has not been cited by other workers in the field of ion-selective electrodes.

Buck: This paper is by *Trümpler*. His paper was an important early contribution in the field; although there is another pioneering publication by *Tammann* in 1920. Their results and other published nearly 30 years later are cited in table II.

Herbert: With respect to selectivity of these electrodes, apparently there is electron transfer in all of these electrodes, which you described. Would it be safe to say that any oxidation-reduction couple in solution would interfere with these electrodes?

Buck: I can give a qualified "Yes", depending on the redox potential and reaction kinetics. I didn't mention this topic because we are still measuring effects of redox reagents on AgBr membranes in both membrane and "all-solid-state" configurations. One of the predictions you can make from the theory is that (aa) elemental anion, for instance, dissolved bromine in bromide against a silver bromide electrode should perturb the membrane potential. One does not know, in advance, at what concentration of bromine a measurable pertubation will occur. *Jaenicke* has observed the effect and concludes that bromine generates holes. The holes diffuse with the silver interstitial ions to modify the diffusion potential and affect the overall membrane potential. The effect is observed also if permanganate in a acidified dilute silver solution is used. It is not known whether this response involves hole injection or direct oxidation of AgBr. An upper limit on the positive potential shift can be estimated from a flux balance of oxidant to the surface with Ag^+ diffusing away. The surface reaction may involve a rate constant (*Buck*, 1968) or reach equilibirum and therefore obey a mass transport law analogous to that describing cyanide attack. Iron (III) at high concentration also disturbs the potentials of SCN and I^- electrodes. Strong reductants, such as those used in photographic development, render the silverbased electrodes useless.

Ref.: W. Jaenecke, Phys. Stat. Sal. *3*, 31 (1963)
R. P. Buck, Analyt. Chem. *40*, 1437 (1968)

Lübbers: I have a question which I think is related to a certain aspect of your work. In biology we sometimes use reference electrodes, e. g., silver or platinum fused in glass. These electrodes operate well for several days und give very stable potentials. They however suddenly change their characteristics. Could you comment on this?

Buck: Yes, I can indicate a possible explanation by giving examples of other electrodes with a common problem, viz., lack of a stable or reversible potential-determining process at the inner surface. There are many reported electrodes including *Freiser*'s plastic-coated wires and those with glass to metal contacts which have undefined potentials (see figure 1.). I don't mean to say that the inner interfacial potential is not defined by nature, but we don't know what process nature has chosen to determine the potential or its stability and reversibility. There are quite a few of these unusual electrodes in addition to these in the figure which includes only those reported in the past two years. Very little is specifically known about the potential-determining processes except in a general way: At an interface there must be an electron or ion exchange and a reservoir of involved species. We have studied the mercury/pH glass interface by the AC impedance methods and found that the mercury/glass contact involves a reversible process. Electrons stay in the mercury and mecury ions seem to move the glass. There are evidently enough mercury ions in the glass surface "reservoir" to permit reversibility. I would not have expected that result. I cannot explain the reversibility or stability of the platinum/glass interface. Perhaps the Pt/PtOx couple defines the potential. It is an important point and should be straightened out experimentally.

Conductor	Coated support	Ingredient or second layer	Inside interface response Outide	interface response	Ref.
Pt	Stearic acid	Methyltri-n-octylammonium stearate	Proton exchange with PtOH?	Proton exch.	(1221)
Pt	Polyvinyl-chloride	Quaternary ammonium salt	?	Anion exch.	(508)
Pt	Polyvinyl-chloride	Ca didecylphosphate in dioctylphosphonate	?	Ca^{2+} exch.	(192)
Determination of nitrogen oxides using coated-wire nitrate electrode					(566)
Patent	Sulfonated polystyrene	Ion exchange material	?	Ion exchange	(509)
Pt	Perchlorate salts of radical cations		Electron exch.	ClO_4^- exch.	(1054–1056)
Pt	Pb, Cu, Ph$_4$As salts of radical anions		Electron exch.	Ion exch.	(1053)
Patents on radical ion-based electrodes					(1052)
Pt or Ag	Polyvinyl-chloride	KBPh$_4$	?	Ion exch.	(251)
Wire	Octadecyldimethylbenzyl ammonium chloride or thiocyanate				(1110)
Oxidized metal	Block copolymer siloxane, carbonate	p-Dodecyldinitro phenol, etc	Proton exch.	Proton exch.	(848)
Oxidized metal	Block copolymer siloxane, carbonate	Gramacidin, valinomycin, etc.	Proton exch.	Ca^{2+} or K^+ exch.	(848)
Graphite	Polyvinyl-chloride	Ca^{2+} ion exch.	Proton exch.?	Ca^{2+} exch.	(38)
Graphite	Xylene	Cu(II) dithizonate	Proton exch.?	Cu^{2+} exch.	(176)
Graphite	Silver salts, etc.		Electron exch.	Ag^+ exch.	(1003)
Graphite	Liquid ion exchangers or neutral carriers in solvents		Proton exch.?	Ion exch.	(1003)
C-paste	AgCl–Ag$_2$S mix		Electron exch	Ag^+ exch.	(774)
Pd, Pt–Rh, Cr, Fe	Glaze of ion-selective glass		Electron exch.?	H^+, K^+ exch.	(844)
Cu, Ag, Co, Cd	Glass + metal halide	Ion selective glass	Electron exch.	H^+ exch.	(102, 903)
Hg		Glass membrane	Hg^+ ion exch.?	H^+ exch.	(161)
Cu, Hg, Fe Al (3 μ diam.)	Oxide?	Borosilicate glass	Electron exch.?	H^+ exch.	(729)
Ag	Ag-film	Corning 015 glass	Electron exch.?	H^+ exch.	(345)
Pt	Hg	HgS	Electron exch.	Hg^{2+} or S^{2-} exch.	(1243)
Pd	PdO			$H \rightleftharpoons H^+ + e^-$	(386)

Table 1. Electrode Configurations with Salt/Metal Interfaces

Lübbers: Also the diffusion of the metal ions in the glass would be an important thing. It is well known that silver atoms can rather easily diffuse into the glass surface.

Buck: Yes, that factor may be important.

Hebert: The silver-silver chloride potential variability might be related to the fact that when you use this electrode you should have a saturated silver chloride solution.

Buck: I have always assumed that the worker saturated his solutions first with the electrode material. Otherwise the system is out of equilibrium to begin with!

Herbert: These metal salt responses are very similar to the antimony oxide electrode which is used for pH measurement. Have you had any experience with this electrode? There are many papers describing the use of this electrode for pH measurement and apparently quite a bit of this work is in error because the electrode will respond to oxygen and a variety of buffers such as phosphate and sulphate. I wonder if you have any experience with this?

Discussion of Session I

Buck: No, I have only read about it as one of several alternatives to the glass membrane electrode. I can recommend John Stock's review of a few years ago. Since this electrode is of the second kind, and two oxidation states of antimony are possible, it is susceptible to soluble redox interferences, especially air oxidation. I would be very leary of that system just for the reason that there a mixed mechanism is possible. In addition anions forming less soluble antimonyl salts than the oxides would distort the potential.

Herbert: I wonder if the distortions of potential may not be related to the solubility product of antimony oxide versus antimonyl sulphate and antimonyl phosphate. One could probably formulate selectivity constants for this electrode with respect to these various solutions.

Buck: Yes, that assumption is the place you start. The theory of interferences by metathetical reagents and experimental verification for silver salt electrodes show that the potentiometric selectivity is (to a good approximation) the ratio of Ksp values. However the functional dependence may be a good deal more complicated in this case.

Baucke: With respect to the standard potentials of silver-silver chloride electrodes of the second kind and membrane electrodes responding to the same ion may I add the following? We have recently determined standard potentials of electrodes of the second kind by means of membrane electrodes. In order to avoid contact of the identical solution with the electrode of the second kind and membrane electrodes, we have determined $E°$ values which came out to be identical within one millivolt to the values determined by Bates and Bower.

Buck: I think that the silver halide salts become saturated upon contact with silver metal very rapidly. Alternatively, pure silver halide salts undergo photolytic generation of silver by loss of the halogen to the atmosphere. In both cases you have essentially a silver halide-silver surface presented to the solution. The two types of electrodes should give identical $E°$ values except for possible small strains or surface energy differences.

Discussion of Paper by Morf and Simon: Theoretical Treatment of the Dynamic Response of Ion-sensitive Membrane Electrodes*)

Kessler: What is the shortest respone time which can be achieved with a membrane electrode using, for instance, a PVC matrix and the ligand valinomycin assuming that the electrode membrane is made as thin as possible?

Morf: In the case of valinomycin-based bulk membranes the time to come to within 0.2 mV of the final state is about 10 seconds or less for an EMP-increase of 59 mV. This corresponds to a 95%-response time of about 50 milliseconds. Such a high speed of response is obtained, for instance, by using silicone rubber membranes, which have been pioneered by Prof. *Pungor*, or by using very non-polar membrane solvents. For thinner membranes, the dynamic characteristics are nearly the same except that the final steady-state is established more rapidly.

Kessler: Is this about the limit that can be achieved?

---

*) and: *Wodick:* Compensation of Measuring Errors produced by Finite Response Time with Ion-selective Electrodes.

Morf: It depends strongly on the stirring of the sample solution. We have only stirred at rates of about 15 Hz. Of course, we can apply higher stirring- or flow-rates but there is evidence for the existence of a lower limit for the boundary layer thickness. Prof. *Pungor* has shown in his experiments with flow-through cells that the EMF vs. time profile may become rather independent of the flow-rate.

Pungor: The so-called flow-through cell meant here an arangement similar to the immersion technique.
We can find big variations in the response time depending on how the surface of the electrode is prepared. We have found response times in the range from 10 msecs to 600 msecs. I was convinced previously that the response time will lead to some understanding of the processes involved in the response of the electrodes, but now I am convinced that this is not the case except for the carrier membranes presented here by Prof. *Simon* and Dr. *Morf*. We are now preparing a paper about these findings. The measurement of the response time is anyway useful from a practical point of view.

Simon: I would like to make a general statement here as to the response time. Different people mean different things when they speak about the response time. Some indicate the time needed for 95%, 66% or 50% of the response of a sensor and some specify the time needed to get to, e.g., 0.1 mV to the final value. This can give differences in the response time by serveral orders of magnitude. We really have to specify clearly what we mean.

Eisenman: I was under the impression that you can reduce the unstirred layer almost to zero thickness under turbulent flow condition. The response time would then be essentially that of mechanical equilibrium at the interface.

Morf: If you use very high stirring rates, then you must consider some other processes which may become rate-limiting. These are, for instance, the equilibration processes at the phase boundary I have not considered in my simple diffusion model, because they are very complicated to treat. In this case, we cannot use the Nernst equation or similar equations, which are based on the assumption of a phase equilibrium, to describe the cell potential. Another process which may become limiting is due to the fact that any ion-containing membrane constitutes a combination of resistance and capacity; such a system has of course a certain response time. This also holds for the measuring equipment.

Eisenman: It is well to recall that when one is dealing with a change in concentration of an ion to which one's electrode is responding in Nernstean manner, one has a situation where only a chemically undetectable flow of charge occurs. Therefore, it seems to me you should get a response time of the order of 1 msec in a flowing situation as was seen by *Friedman* in Na^+ and K^+ selective glass electrodes.

Buck: I would like to comment that if you imagine as an ideal interface one which has an atomically smooth surface with infinitely fast ion exchange kinetics, then the ultimate short response time constant for a step concentration charge is set by the product of the ionic resistance of the membrane times the geometrical capacitance. The geometrical capacitance comes into the time constant, because in order to establish a potential you have to build up the ionic space charge on the solution sides of the interface. The space charge determines the curvature in the local potential. We measured high frequency time constants for glasses and

Discussion of Session I

AgCl membranes. They agree very well with the product of geometrical capacitances calculated simply from the dielectric constant, thickness and area and the measured resistance. The time constant values are in the low millisecond range. The ultimate time constant for real systems will be longer when the surfaces are rough or pores are present. In addition, slow surface kinetics may introduce an intermediate time constant less than the slowest diffusion-controlled time constant described by *Morf*. *J. Ross MacDonald*, in the Journal of Chemical Physics (1971–1974) has explored this topic extensively theoretically.

Lübbers: I would like to ask Dr. *Wodick* a question. For your solution you have used an exponential response curve. Could you also find an easy solution for the squarlaroot law? If I understood this presentation correctly, you gave us a formula with which you can overcome the difficulties which arise from the finite response time.

Wodick: Clearly, if you have a square-root formula for this function, you can approximate it with a series of exponential functions.

Morf: This is right from the theoretical standpoint. But why should we approximate a simple square-root expression by a series of exponential terms for practical purposes? We can apply the given potential-time relationship directly for an easy extrapolation of the final value. In addition, *Wodick's* method using the time-derivative of the potential may lead to a high variance of the extrapolated value due to noise.

Discussion of Paper by Pungor, Tóth, Nagy: Some Aspects of the Application of Ion-selective Electrodes in Flowing Systems

Fuchs: What is the advantage of using measurements by peak heights as compared to normal potential measurements in the steady state?

Pungor: I mentioned that in my lecture, but it was perhaps not clear enough. If you use a measuring system for many hours without calibrating, the varying factors of the systems, i.e., E_0, slope etc. result in poor analytical results. With the method I mentioned you can calibrate the electrode assembly without disconnecting it from the system to be investigated.

Betz: How did you place the measuring electrode and the reference electrode? Did both electrodes see the same concentration of the injected sample at the same time or did they see different concentrations?

Pungor: The two electrodes are placed in the flow system so that the measuring electrode is the first in the streaming solution, and the reference electrode is "downstream". The reference electrode solution flows very slowly. Generally we use the saturated calomel electrode as the reference.

Betz: Are you sure of a uniform distribution through this small pipe?

Pungor: Yes, we have used a mixing chamber before the cell. Perhaps you saw in the diagram that this chamber was placed after the injection point. If we do not have it, it is impossible to get satisfactory results.

Betz: We very often see that the solution is somewhat mixed but that the mixing is not uniform.

Pungor: The experimental finding is that mixing is complete. This is assured by proper arrangement of the mixing chamber. The solutions are mixed with a magnetic stirrer with an air bubble in the upper part of the chamber above the inlet and outlet points.

Kessler: Did you observe streaming potentials in your system or did you avoid them by stirring?

Pungor: In the calibration system which we have used here these do not have any influence.

Buck: Does your method of calculation make use of the *Gran's* plot?

Pungor: We can use it but did not, because we have written a computer program which does not need it.

Discussion of Paper by Baucke: Investigation of pH Electrode Membrane Glasses by Means of an Ion Sputtering Method. Discussion of Microelectrode Membranes

(2 Figures)

Eisenman: Have you got anything for aluminium silicates?

Baucke: No I have not, we have concentrated on pH glasses. We think we must know more about them.

Simon: I could not read the scale on one of your slides. For a typical lithium pH glass, what is the approximate thickness of the swelling layer?

Baucke: It depends very much on the temperature. At 50 °C we have about 300 Å which is 3×10^{-6} cm.

Hebert: Can you elaborate on resistance versus minimum thickness?

Baucke: This is repeating parts of my paper! The resistance of the membrane is the sum of bulk glass and two layer resistances. When you decrease the thickness of the membrane you must get into a thickness below which the resistance of the membrane is no longer a function of this thickness because the layer resistances are so high.

Hebert: You are speaking about two hydrated layers sandwiching a dry layer.

Baucke: I call it the bulk glass.

Jensen: Representing one of your competitors, I was of course touched by your last remark on how you used electrodes having them kept dry as you recommend. Inside the electrode you have a liquid. Wouldn't it create an asymmetric potential if you keep the outside dry and still have this filling?

Baucke: I was talking about membranes only, without any filling. Keep both sides dry, and you can store them for years. After use keep them in an aqueous solution of moderately low pH. That is what I meant.

Eisenman: Is yours a lithium barium silicate?

Baucke: Yes, with more ingredients.

Discussion of Session I

Eisenman: There are several phenomena which *Walker*, *Sandblom* and I described in 1967 in NAS_{27-4} which I mention because I would like to encourage you to look at the sodium aluminum silicates and also because they may bear on some phenomena seen by users of potassium-selective glass microelectrodes made from Corning K^+ selective tubing, which is made of my NAS_{27-4} composition. Several things occur in hydration. The first is that the resistance decreases by about 4 orders of magnitude in wet glass compared to dry glass (this is opposite to what you reported in your different composition). Secondly, we found an analogous maximum in the resistance in mixtures of sodium and potassium. Thirdly, one property of the completely hydrated film unsupported on dry glass was different from the property of a hydrated film on the surface of the dry glass of the usual glass-electrode. In the latter a step change from pure Na solution, Na^+ to K^+ leads to a pure step response of the electrode. In contrast, in the former we found the response is a step plus a subsequent creep having the time course of diffusion. I have always wondered whether or not the mol fraction dependence would be as severe in the hydrated film by dry glass, as we found in the completely hydrated membrane which was free to swell and shrink in the plane of the membrane, which of course, the supported film cannot. Our experiments could resolve this.

Baucke: One thing has never been clear to me. In the modified Nernst equation do you mean the mobility within each layer or with-in the bulk glass?

Eisenman: In this particular glass the mobility is throughout the hydrated layers, which are uniform to surprisingly great depths as judged by the linearity of radio active tracer uptakes with the square root from 15 seconds to 3 days.

General Discussion Monday a.m.

Eisenman: In principle, I don't see why you couldn't overcome the unstirred layer effect by superposing an AC field (e.g., 1 v/cm at 10^6 Hertz).

Buck: There is an exact time-dependent solution for the transient potential due to diffusion of two ions through a gel-glass layer. Also the time-dependent concentration at all points in the layer after a step concentration change on one side is given (J. Electroanalyt. Chem. 1968). The form and time constant are nearly the same as *Morf* gives. Only the position at the surface giving the plane response differs. We used the model to give a basis for the paper by *Rechnitz* and *Hameka* and our own results on the drift of Na^+-selective glass electrodes. Removal of the potential transient is theoretically possible by data treatment. In our laboratory Prof. *Reilley* does this with a fast Fourier transform of the transient potential data dividing out the system response function and reinverting to the time domain. You may expect a flat response or some other function that is more easily interpreted. One must know the form of the system response function which is the Fourier transform of the response to a delta function. It may also be easily computed from *Morf's* equation.

Lux: Many electro-physiologists are concerned with fast-reacting systems such as nerve cells. Determining transient changes of potassium activity outside such cells depends on a setup with a fast response time. We therefore measured response times of potassium microelectrodes in the vicinity of a point source from which potassium was ejected by a electrophoretic current step, and compared the halftimes of the response with the time predicted by diffusion laws. We found that down to 2 msecs the response time could be very well predicted from

Discussion of Session I

diffusion equations. Thus, it appeared to us that attaining a high response time is an electric problem, since the chemical response appears to be very fast. This applies to liquid ion-exchanger electrodes.

Baum: There are two factors in response time that I have found experimentally for liquid ion-exchange membranes: one is the concentration of charge carriers in the membrane, and second the ionic strength of the solution also has some rôle. At low ionic strength it just takes a long time to get a stable reading. I wondered whether any of your treatments included some of these factors.

Morf: In a more detailed treatment, we also considered an additional diffusion potential within the aqueous boundary layer. This potential term arises from the diffusion of all the ions within this layer and, therefore, depends on non-interfering ions, too. For instance, *Pungor* has shown experimentally that ionic-strength buffers may thus have an important effect on the response time. Another point is that "low ionic strength" often means "going down to a low activity"; in this case, an increased response time is clearly predicted by the equations presented here.

Pungor: I would like to add something. I think an important question is the turbulent flow which was mentioned by Prof. *Eisenman*. If you use a laminar flow system for the measurements where the solutions of various concentration are injected, the concentration profile is not always very well defined. We have used a system where the solution was flowing against the surface of the electrode, and we have investigated the condition under which we always get the same response time with the same solution varying the flow rate. With the application of turbulent flow we can reach a flow rate above which the response time does not vary. I think this method can be applied generally for response time measurement. Returning to the lecture of Dr. *Morf*, I would like to point out that it was really a very clear presentation of the problems connected with response time.

Lübbers: Is there an effect of the membrane matrix on the ion selectivity of the system?

Simon: Yes, there is a considerable contribution from such supports. We had to discard most of the earlier measurements of selectivities obtained with filter paper supports, because different batches led to different selectivities. No such problems are observed when using a PVC matrix.

Lübbers: Because there is a very big difference in the response time of the setup Dr. *Lux* is using and the response time of the carrier membrane you are applying, could you predict if the carrier without a membrane matrix would have the same response time?

Morf: Since the percentage of solvent retained on the matrix is rather high (60–70%), there are only slight differences in diffusion rates.

Kessler: I think that in all systems where diffusion is involved, the diamter of the electrode plays an important rôle. From work with platinum electrodes we know that the response time can be decreased by the use of very small electrodes.

Coleman: Dr. *Simon*, in your comparison of the selectivities of the various calcium exchangers was there a common exchanger support?

Simon: In all the measurements of the selectivities I have discussed we have been using a PVC matrix.

Discussion of Session I

Reneau: I would like to question Dr. *Morf* and Dr. *Wodick*. The equations indicate an assumption of homogeneous diffusion conditions with respect to distance in both the boundary layer and the membrane. A voltage boundary layer, a concentration boundary layer and, under stirring conditions, a momentum boundary layer are present. For an exact description one would need to include all three of these boundary layers and the thicknesses of each. Quitting the momentum boundary layer, do you assume an identical thickness for the electrical boundary layer and the concentration boundary layer?
Would this have an effect on the results?

Morf: For simplicity's sake, I have only considered the unstirred boundary layer introduced by Nernst and a homogeneous membrane type. For the first period of equilibration in carrier membranes it is the boundary layer of the membrane that is responsible for the response behavior, whereas after very long times the interior of the membrane becomes important. Inhomogeneous properties of the membrane may therefore lead to transient phenomena.

Reneau: I agree with you that solution difficulties are formidable. Inside the membrane you are using a constant diffusion coefficient which changes with a charged or noncharged membrane. Recent work has shown that significant diffusional interactions can occur between diffusing ions and the membrane or membrane charges. It is suggested that this can be accounted for with interactional diffusion coefficients. Have you seen any evidence of this phenomenon?

Buck: There are many sets of impedance-frequency data on diverse electrochemical systems, which cannot be completely interpreted. When these data are plotted by the Cole & Cole method (the imaginary impedance against the real impedance) the semicircles are depressed. If the system contains only RC elements, one or more semicircles with centers on the real axis are possible depending on the time constants of the system.
The data can be fitted by an empirical parameter called α. This effect has not been explained for impedance data on a molecular level, but the present feeling is that the depression is due to a distribution of diffusion coefficients within the material. Furthermore, *Macdonald* has succeeded in showing that α is not connected with the height of the barriers, but is connected with possible jump distances. It is an entropic effect which seems to be very common. The difficulty is simply that one cannot solve boundary value problems with variable diffusion coefficients to compare data with assumed concentration or space-dependences. It is also very difficult to make the measurements of the diffusion coefficients themselves.

Reneau: Yes, but with numerical analysis techniques sometimes it is possible to solve this condition, but it is quite involved. May I continue with one last question which is very short, on this same subject. With respect to K, the partition coefficient, is it assumed to be equal on both sides of the membrane?

Morf: For identical solutions: yes.

Eisenman: To the extent that it even becomes important to reduce the time constant of diffusion in the bulk of your membrane. I would remind you about the ultrathin bilayer systems, where the diffusional time constants are of the order 10^{-5} to 10^{-4} sec.

Discussion of Session I

General discussion – Monday afternoon

Buck: I would like Dr. *Baucke*'s opinion on whether or not, on the basis of his work, there is a need for reconsideration of the theory of glass membrane potential taking into account the surface layers which have not been taken into account in great detail before.

Baucke: We have to think about what is the real mechanism of the potential formation; that is, what happens at the immediate interface between leached layer and solution. There is evidence that the dissociation of the SiOH group, which, as I said, controls the entrance of lithium ions and protons into the glass phase, is the response mechanism of the pure pH electrode. I cannot say anything about the response mechanism of lithium or of other cations at present. We have done one experiment which should be mentioned. We have taken pH glass electrodes and have measured their response. Then we have completely replaced the lithium ions by protons at the surface within a layer thickness up to 4 or 5000 Å. Subsequently, the electrodes had exactly the same response and approximately the same potentials, also in alkaline-free buffer solutions – as before their protonation.

Eisenman: Certainly your conclusion is consistent with the ion-exchange interpretation which has generally been accepted since 1962, when I suggested that it was silanol sites that have the marked preference for protons compared to cations like lithium. You also have sufficient data to calculate the potential, but you will have to generalize the model to include mol fraction-dependent mobilities (*Eisenman, Sandblom,* and *Walker,* Science *155,* 965 (1967)) which is going to be a bit nasty. You can measure the ion exchange equilibrium directly because you can measure the proton/lithium displacement through the profiles, and it is inconceivable to me that, when you analyse this, you will not come out with the classical ion-exchange interpretation extended to include mol fraction dependence of the conductances.

Baucke: I would like to show two slides regarding this point. Firstly, lithium concentration profiles which were measured after applying an electric field to the membrane (the anodic side is on the left of the figure). Different ratios of lithium (below the curve) and protons (above) were supplied to the anodic surface. The lithium/proton ratio transferred into the glass

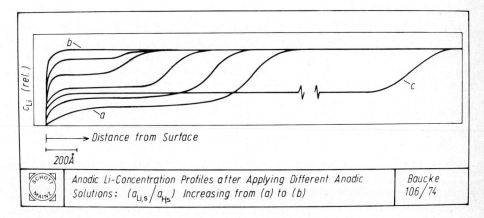

Fig. 1.

Discussion of Session I

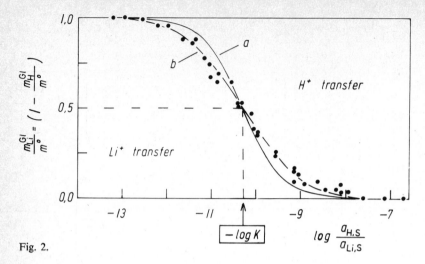

Fig. 2.

depends on the pH controlling the dissociation of ≡SiOH groups at the glass surface and, again, by the proton activity and the lithium ion activity of the solution. Referring now to the second slide, we plotted the mol fraction of lithium admitted to the glass as a function of the activity ratio in solution. We have derived an equation which I wanted to show but, because of the limited time, I was not able to. This equation describes the experiments fairly well and gives, at the point of 0.5 ion ratio, the dissociation constant of silicon groups. In addition, the ideal curve is known so that the derivation from the ideal curve tells you something about the distribution of dissociation constants of SiOH groups at the surface.

Eisenman: The hydrogen to lithium selectivity, from the ion exchange theory should be of the order of 10^9 od 10^{10} or something like that. Do you remember the values you estimated?

Baucke: Yes and more – 10^{11}.

Thomas: This is a preview of something I am going to be talking about in the very last talk this evening, but I think its relevant now. I have found with pH-sensitive microelectrodes and sodium microelectrodes that the pH electrode seems to work only after about 3 weeks soaking at room temperature, or 2 hours boiling, and it suggests to me that the glass needs to be completely hydrated. Whereas the sodium-sentitive electrodes are best as soon as they are filled. After use for a day or two they get slower and their resistance increases as if the hydrated layer has a high resistance. If you then etch the glass with alkaline EDTA you can restore them almost to their original low resistance. So the two glasses NAS 11–18 and Corning 0150 behave completely oppositely. Hydration is good for pH, but bad for NAS 11–18.

Eisenman: I would just like to add one thing to complete *Roger Thomas'* statement. For WAS 27–4, the potassium electrode, again hydration is desirable, because for this composition the resistance drops by four orders of magnitude upon hydration and in a few days you have hydrated all the way through more than a micron of thickness.

Lübbers: I would like to ask Professor *Pungor* a question about the calibration process he mentioned. If I have understood this correctly, his injection method is used to calibrate continuously functioning electrodes. Does this imply that the calibration curve which is not linear must be rather stable?

Pungor: I mentioned during my lecture that we inject equal amounts of calibration solution two or three times. We then can calculate the slope and also the E_0.

Session II

Microelectrodes

Chairmen: *I. A. Silver*

Y. Saito, H. Baumgärtl, D. W. Lübbers

The RF Sputtering Technique as a Method for Manufacturing Needle-shaped pH Microelectrodes

(4 Figures)

Microelectrodes of different ion-sensitive glasses can be used to measure the H^+, Na^+, and K^+ activity continuously in small amounts of fluid or in tissue. The technical construction and the performance of microcapillary electrodes (*Khuri, Agualian, Harik* 1968; *Uhlich, Baldamus, Ullrich* 1968; *von Stackelberg*, 1970) and recessed electrodes (*Thomas* 1970) seem to be satisfactory. The manufacturing of needle-type microelectrodes still gives rise to considerable technical difficulties (*Carter, Rector, Campion, Seldin* 1967; *Gebert* 1971; *Hinke* 1967) if a size of only a few microns is necessary. It is possible to produce pH-sensitive conical glass tips of 1 μm diameter or less, but the insulation of the upper parts of the glass tip to prevent interference with the measurement at the tip remains an unsolved problem. Since 1969 we studied methods of depositing insulating layers of slight thickness on ion-sensitive glasses by evaporating or sputtering different materials in vacuum. Using SiO_2 or insulating glasses evaporation with the electronic beam and a DC-sputtering device gave layers of insufficient resistance. Hence, the RF-sputtering technique remained as a last possibility. Since the equipment necessary for this technique is infortunately rather expensive, we made our first experiments in the laboratory of Kontron, München*), and of Balzers, Liechtenstein*).

To find suitable materials for the deposition we discussed our problems in 1972 with Schott & Gen., Mainz*, who provided targets of pH glass and lead glass.

We approached our problem from three different angles.

1. We used the sputtering technique to insulate the shaft of a needle-shaped pH electrode filled with buffer solution.
2. The insulation was deposited on a metal-filled pH electrode.
3. pH glass was deposited on a metal surface which was fused into an insulating glass.

All three methods proved successful.

In the following we will describe mainly the manufacturing and testing of the buffer-filled pH electrode, since we have most experience with this electrode.

Fig. 1 shows the different steps in manufacturing micro pH electrodes of the needle-shaped type. In the symposium the different steps were demonstrated by a film.

A. *Production of the lead glass cone (1–5).*

Lead glass tubes (type 124, OSRAM, München; outer diameter 6.2–6.5 mm; wall thickness 0.55–0.7 mm) are cut to pieces of about 160 mm length and then cleaned by ultrasonics

---

*) We would like to thank the companies for their help. We are also most grateful to Dr. *H. Frischat*, Lehrstuhl für Glas und Keramik, Institut für Steine und Erden, Technische Universität Clausthal, who advised and helped us with the more general problems of the formation and behavior of the glass layers.

Y. Saito, H. Baumgärtl, D. W. Lübbers

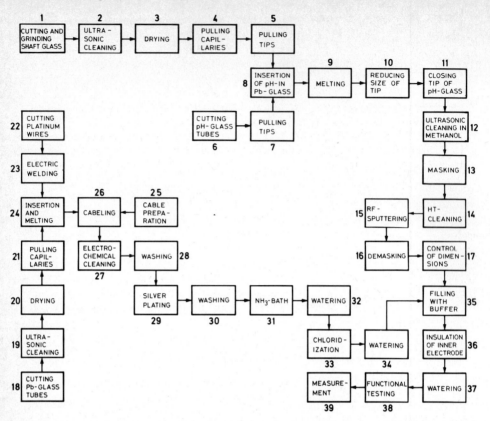

Fig. 1. Diagram of the different steps in manufacturing a needle-shaped micro pH electrode.

(40 KHz) in a 1:1 mixture of methanol and 2% RBC solution (*C. Roth*, Karlsruhe). Thereafter the tubes are carefully rinsed and dried at 50 °C. The tubes are drawn to capillaries by placing their middle part in a heating coil (ca. 650 °C) then they are cut in the middle. The capillary with the drawn tip is again put in a heating coil and drawn by gravity to a small tip of about 1 μm.

B. *Manufacturing of the pH glass cone (6–7).*

As pH glass we use capillaries of 0150 glass (Corning Glass; outer diameter 1 mm, wall thickness 0.25 mm). The glass is stored in methanol. Open conical tips of ca. 7 mm length and less than 1 μm diameter are drawn from the capillaries by gravity using a heating coil.

C. *Fusing of lead glass and pH glass (8–9).*

The tip of the lead glass tube is broken off under microscope control so that an opening with a diameter of 20–25 μm is produced. Then the pH glass cone is inserted into the lead glass tube so that it protrudes by about 0.5 mm. The contact zone of the pH glass with the lead glass is fused together over a length of about 100 μm. During this procedure the tip of the pH glass is inserted into a micropipette in order to avoid mechanical movements of the tip. The glass parts are brought in contact with a light pipe to achieve good illumination by conducting the light in the glass wall of the lead and pH glass. The fusing must be carried out on a table which is as free of vibrations as possible (Gestag, Fellbach).

D. *Manufacturing of the micro tip (10–11).*

The tip of the pH cone is fused to a thin platinum wire so that the electrode is hanging freely. The tip is heated by a micro coil and drawn by gravity to an open tip of 0.5–2 μm diameter. The tip is closed under microscope control (magnification 450 ×) by approaching the tip to a heated platinum wire with a micromanipulator.

E. *Masking of the electrode tip (12, 13).*

The electrode is cleaned in a methanol bath by ultrasonics. The masking is done under microscope control (magnification 450 ×). The tip of the electrode is inserted into a small droplet of Fotoresist (AZ 111, Shipley) and withdrawn. The electrode is moved and turned around so that the droplet of Fotoresist which remains on the electrode becomes a oblong protruding over the electrode tip (Fig. 2a). All these manipulations are carried out with micromanipulators.

F. *RF-Sputtering (14–15).*

RF-sputtering is carried out in a SPURTON II (Balzers, Liechtenstein) by varying the power between 0.5 and 1.5 Kw, and the time of sputtering between 0.5 and 1.5 hours for a single layer. We use the normal technique (*Gaydou* 1967; *Sager* 1971). The electrode is about 20 cm distant from the target. We used different targets (see *Lübbers, Baumgärtl, Saito* 1974). The sputtering is carried out with oxidic materials in a mixture of Argon and Oxygen at pressures of about $5 \cdot 10^{-4}$ and $3 \cdot 10^{-4}$ Torr. Sputtering of Si_3N_4 is carried out at an N_2

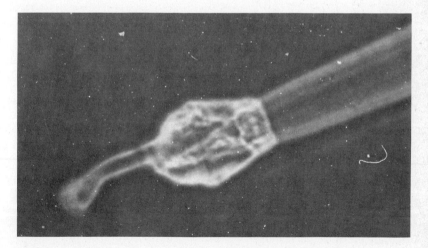

Fig. 2a. Light micrograph of the tip of a pH needle micro electrode masked with Fotoresist.

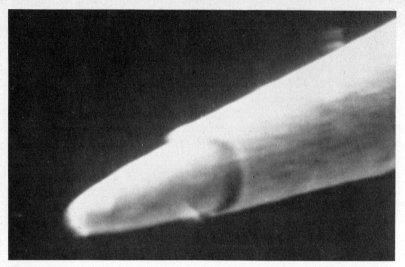

Fig. 2b. Scanning electron micrograph of the pH micro needle electrode with an insulation layer of Al_2O_3.
(Length of the tip 10 μm; diameter of the tip 2.0 μm; diameter at the beginning of the insulation ayer: 6.5 μm).

partial pressure of $4 \cdot 10^{-4}$ Torr plus Ar of $4 \cdot 10^{-4}$ Torr. The maximal thickness of the layers is about 1 μm. The temperature of the electrode rises to a maximum of about 80 °C. Higher temperatures are avoided.

G. *Demasking and control (16–17).*

The tip of the Fotoresist is slightly damaged by touching under microscope control. The droplet is solved and destroyed in an aceton bath by 1 minute of ultrasonics (about 40 kHz) and half an hour standing. Then the electrode is ready to be used and the tip is controlled by microscope (magnification 450 ×). Fig. 2b shows an electron scanning micrograph of an electrode manufactured as described above. The length of the tip is about 10 μm, the diameter 6.5 μm at the beginning of the insulation layer. The diamter of the tip itself is about 2 μm.

H. *Manufacturing of the internal reference electrode (18–34).*

A tube of lead glass (type 8095, Schott & Gen. Mainz; outer diameter about 4.0 mm) is carefully cleaned by ultrasonics, dried and drawn to a capillary (18–21). Platinum wires of 0.5 mm and 0.1 mm diameter are electrically welded together. The 0.5 mm platinum wire is inserted into the capillary tip so that it protrudes by about 10 mm. The platinum wire is fused in the capillary (22–24). The 0.1 mm platinum wire is soldered to a coaxial cable the forepart of which is freed of insulation, pressed into the upper end of the lead glass tube and glued with UHU plus (H. & M. Fischer, Bühl, Baden) (25–26). The protruding part of the platinum wire is electrochemically cleaned in concentrated HNO_3 against a platinum sheet as cathode (current density 1.2 mA/cm²; 30 min) and rinsed (27–28). A silver layer is deposited electrolytically on the platinum wire (solution: 10 g K Ag (CN)₂ in 1 Liter H_2O for 6 hours at a current density of 0.4 mA/cm²). The wires are carefully rinsed in bidestilled water,

The RF Sputtering Technique as a Method for Manufacturing Needle-shaped pH Microelectrodes

washed with NH_3 and left in NH_3 for six hours. Then the electrodes are kept in bidestilled water for three days. They are andodically chloridized on 0.1 m HCl (current density 0.4 mA/cm^2; 30 min) in the dark and rinsed again (29–34).

I. *Assembly of the electrode (35–37)*.

The glass part of the electrode is filled with 1 ml methanol and carefully evacuated with an aspirator vacuum pump until the methanol fills the tip. The methanol is replaced by buffer solution (1 part standard acetate of pH 4.62 and 2 parts saturated KCl solution). A tightly fitting Teflon tube of about 10 mm length is pressed over the upper end of the Pb-glass tube. The reference electrode is inserted into this Teflon tube. The space between the Teflon wall and the wall of the reference electrode is filled with pitch. The pitch is heated from outside by high-frequency induction so that it fills the space completely and without air bubbles. The extremely high resistance required is achieved by this procedure. Then the finished electrode is watered. Fig. 3 shows a diagram of the electrode described.

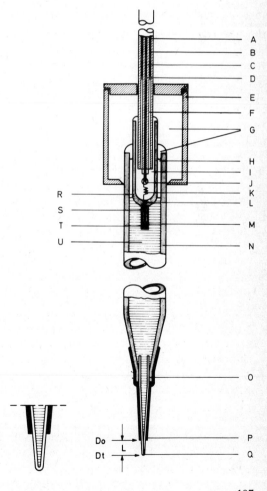

Fig. 3. Glass needle electrode with internal buffer solution to measure H$^+$ activity.
A = Teflon insulation; B = screen; C = polyethylene; D = copper wire; E = Teflon tube; F = air; G = pitch; H = air; I = UHU plus; J = soldered contact; K = platinum wire diameter 0.1 mm; L = Platinum wire of the internal reference electrode fused into lead glass, R.; M = platinum wire diameter 0.5 mm; N = lead glass capillary, O = place of melting; P = insulating layer produced by RF sputtering; Q = pH-sentitive glass tip; R = lead glass of the internal reference electrode; S, T = Ag/Ag/AgCl layer; U = internal buffer solution

J. *Testing of the electrode (38–39).*

The electrode is tested by measuring the resistance, pH sensitivity and sensitivity to K^+ and Na^+ ions. The reaction of the electrode is observed during several months. Fig. 4 shows the behavior of the electrode during calibration with different buffer solutions. We obtained a pH sensitivity of 58.1 mV/pH (25 °C). The pH sensitivity is restricted to the tip of the electrode. This is proven by inserting the tip into a layer of agar agar with a pH of 7.0 and then bringing the other parts of the electrode in contact with a solution of pH 4.62. This procedure does not change the signal of the electrode. The asymmetry potential of our electrodes may vary between + 42 mV and 2.3 mV. The influence of K^+ and Na^+ ions is small. By changing the concentration of K^+ or Na^+ between 0.001 and 1 molar the steepness of the pH calibration curve can vary by about 2 mV/pH. The absolute deviations are in the range of 2 mV. With Na^+ they are higher at alkaline pH (at pH 9, e.g., 8 mV). This shows that the glass is satisfactorily preserved during the manufacturing process. More details on the behavior of the resistance are reported elsewhere (*Lübbers, Baumgärtl, Saito* 1974).

In a second approach we investigated the behavior of pH micro needle electrodes having an internal platinum electrode as reference. Platinum wires of 200 μm diameter were etched with AC current so that a conical tip was formed. Tip diameters between 0.5 and 1 μm (*Baumgärtl* and *Lübbers* 1973) were obtained. These platinum wires were fused in capillaries of pH glass 0150 of Corning Glass over a length of 1 mm. The electrodes showed a sensitivity of 52.9–54.4 mV/pH (25 °C), a 90% response time of 0.2–3 s and a linear calibration curve in the range investigated of 4.0 to 9.0 pH. The drift amounted to 3–30 mV/h, the electrical resistance was $0.25–1.1 \cdot 10^{-9}$ Ohm. At pH 7, the voltage of this pH glass platinum wire electrode was 270 to 340 mV more positive than the potential of an Ag/AgCl reference electrode in a buffer of pH 4.62. Small pH electrodes with rapid response could be produced with this technique. In further experiments we will try to reduce the length of the pH-sensitive tip by sputtering layers of high insulation resistance on this type of electrode. Other metals such as silver or gold should also be tested (*Friedmann* et al. 1958, *Gebert* and *Friedmann* 1973).

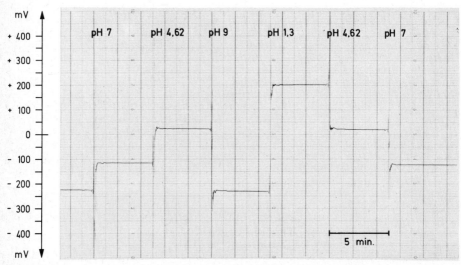

Fig. 4. Calibration of a needle-shaped micro pH electrode the shaft of which has been insulated by RF sputtering. (Original record). pH sensitvity: 58.1 mV/pH (25 °C), 90% response time: 25 s.

Instead of sputtering insulating materials we also tried to sputter ion-sensitive glasses directly. Material from a pH glass target was sputtered on the polished surface of a 200 μm platinum wire fused in a tube of lead glass (type 8095, Schott & Gen., Mainz). The pH glass film had a thickness of about 0.7 μm. In the range investigated between 4.0 and 9.0 pH the calibration curve of this electrode was linear. After 20 days' watering the pH sensitivity was 54.5 mV/pH (25 °C), the electrical resistance amounted to $8.8 \cdot 10^9$ Ohm and the 90% response time was 45 s.

Summary

Our experiments have shown that by RF-Sputtering it is possible to deposit thin layers of insulating materials on ion-sensitive glasses and to insulate the shaft of the electrode adequately. Very satisfactory pH microelectrodes of conical shape can be obtained with this technique. Most of the experiments were done with buffer-filled pH micro needle electrodes, but we could show that the technique can also be applied to metal-filled pH electrodes. In a similar way ion-sensitive glasses can be deposited on metal or other surfaces to produce ion-sensitive microelectrodes. In our experience RF sputtering has proven to be a very useful technique for manufacturing ion-sensitive microelectrodes.

References

Baumgärtl, H., Lübbers, D. W.: Platinum Needle Electrode for Polarographic Measurement of Oxygen and Hydrogen. In: Oxygen Supply, Theoretical and Practical Aspects of Oxygen Supply and Microcirculation of Tissue, pp. 130–136, ed. by *M. Kessler, D. F. Bruley, L. C. Clark jr., D. W. Lübbers, I. A. Silver, J. Strauss,* Urban & Schwarzenberg, München 1973

Carter, N. W. Rector, F. C., Campion, D. S., Seldin, D. W.: Measurement of Intracellular pH with Glass Microelectrodes. Fed. Proc. 26 (1967) 1322–1326

Friedman, S. M., Jamieson, J. D., Hinke, J. A. M., Friedmann, C. L.: Use of Glass Electrode for Measuring Sodium in Biological Systems. Proc. Soc. Exp. Biol. Med. 99 (1958) 727–730

Gaydou, F. P.: Ein Verfahren zur Herstellung dünner Schichten durch Ionenzerstäubung im 10^{-4}-Torr-Bereich. Mikroelektronik 2 (1967) 183–192

Gebert, G.: Die Messung von Na^+-, K^+- und H^+-Aktivitäten im Gewebe mit Glasmikroelektroden. Ärztl. Forsch. 26 (1971) 379–385

Gebert, G., Friedman, S. M.: An Implantable Glass Electrode Used for pH Measurement in Working Skeletal Muscle. Appl. Physiol. 34 (1973) 122–124

Hinke, J. A. M.: Cation-selective Microelectrodes for Intracellular Use. In: Glass Electrodes for Hydrogen and Other Cations, ed. by *Eisenman.* Dekker, New York 1967

Khuri, R. N., Agulian, S. K., Harik, R. I.: Internal Capillary Glass Microelectrodes with a Glass Seal for pH, Sodium and Potassium. Pflügers Arch. 301 (1968) 182–186

Lübbers, D. W., Baumgärtl, H., Saito, Y.: Isolation Properties and Ion-Sensitivity of Thin Layers of Dielectrica Deposited by RF Sputtering on Glass Surfaces this book, pp 110–115

Sager, O.: Ein Trioden-Hochfrequenz-Zerstäubungssystem zur Herstellung dünner Schichten im 10^{-4}-Torr-Bereich. Vakuumtechnik, 20 (1971) 225–231

v. Stackelberg, W. F.: A 0.1 Nanoliter Glass Capillary Electrode. Pflügers Arch. 321 (1970) 274–282

Thomas, R. C.: New Design for Sodium-Sensitive Glass Micro-electrode. J. Physiol. 210 (1970) 82–85

Uhlich, E., Baldamus, C. A., Ullrich, K. J.: Verhalten von CO_2-Drucken und Bicarbonat im Gegenstrom des Nierenmarks. Pflügers Arch. 303 (1968) 31–48

D. W. Lübbers, H. Baumgärtl and Y. Saito

Insulation Properties and Ion-sensitivity of Thin Layers of Dielectrics Deposited by RF Sputtering on Glass Surfaces

(2 Figures)

The technique of RF sputtering can be used to produce thin layers of relatively high insulation (*Hass* 1950; *Davidse, Maissel* 1966; *Pliskin, Kerr, Perri* 1967; *Wu, Formigoni* 1968; *Harrop, Campbell* 1970; *Mattox, Kominiak* 1973; *Maissel, Francombe* 1973). In the foregoing paper it could be shown that with a special technique it is possible to produce pH electrodes with a small-sized tip (*Saito, Baumgärtl, Lübbers* 1974). In the following we will report on details of the resistance and behavior of thin layers of dielectrics deposited by RF sputtering. Since the form and size of the glass has an important influence on the formation and behavior of the layer, we had to test the different materials by making the corresponding pH microelectrodes. In the ideal case the shaft should have an insulating resistance much higher than that of the pH tip and no pH sensitivity.

The measurements of the total electrode resistance, R_0, were carried out in H_2O by applying known DC voltages, E_x, in steps of 0.5 V from -1 V to $+1$ V during a period of 3 min and measuring the resulting voltage, E_m. Figure 1 shows a diagram of the measuring circuit.

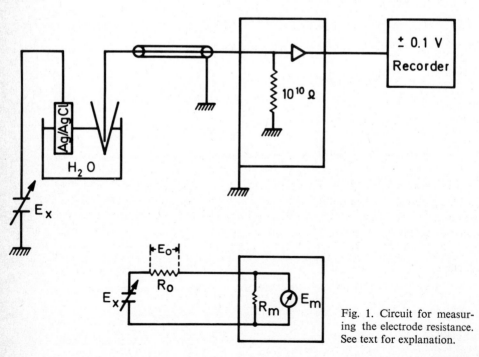

Fig. 1. Circuit for measuring the electrode resistance. See text for explanation.

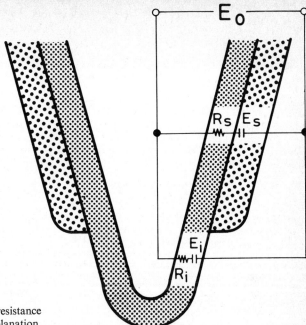

Fig. 2. Diagram of the electrode resistance and potentials. See text for explanation.

The resistance, R_0, can be calculated from E_m, if the input impedance, R_m, is known (*Schanne* et al. 1968, *Lee* and *Fozzard* 1974). The voltage decrease along R_0 is

$$E_0 = E_x - E_m. \tag{1}$$

With (1) we obtain for the current in R_0 and R_m

$$\frac{E_x - E_m}{R_0} = \frac{E_m}{R_m} \tag{2}$$

and from (2) for the total resistance of the electrode

$$R_0 = \left(\frac{E_x}{E_m} - 1\right) R_m. \tag{3}$$

Figure 2 shows that the total resistance, R_0, of the pH electrode is composed of two resistances: the tip resistance, R_i, and the shaft resistance, R_s. The pH sensitivity of the total electrode, α_0, also has to be divided in the pH sensitivity of the tip, α_i, and that of the shaft, α_s. Correspondingly we obtain two potentials, the potential of the tip, E_i, and the potential of the shaft, E_s.

We find

$$R_0 = \frac{R_i R_s}{R_i + R_s} \tag{4}$$

$$R_s = \frac{R_i R_0}{R_i - R_0} \tag{5}$$

The total current

$$i_0 = \frac{E_0}{R_0} \tag{6}$$

is divided into the two currents, $\frac{i_i}{R_i}$ and $\frac{i_s}{R_s}$.

$$i_0 = i_i + i_s = \frac{E_i}{R_i} + \frac{E_s}{R_s} = E_0 \frac{R_i + R_s}{R_i R_s} \tag{7}$$

$$E_0 = \frac{R_s}{R_i + R_s} E_i + \frac{R_i}{R_i + R_s} E_s \tag{8}$$

Using

$$k = \frac{R_s}{R_i} \tag{9}$$

we obtain

$$E_0 = E_i - (E_i - E_s) \frac{1}{1+k} \tag{10}$$

Assuming

$$\alpha_i = 1$$
$$E_s = \alpha_s E_i \tag{11}$$

we obtain

$$E_0 = \alpha_0 E_i \tag{12}$$

and with (10)

$$E_0 = 1 - (1 + \alpha_s) \frac{1}{1+k} E_i. \tag{13}$$

It follows from (12), (13), and (9) that

$$\alpha_0 = \frac{E_0}{E_i} = 1 - (1 + \alpha_s) \frac{1}{1+k} \tag{14}$$

and

$$R_s = \frac{1 + \alpha_s}{1 - \alpha_0} R_0 \tag{15}$$

Using (5) we find that

$$R_i = \frac{1 + \alpha_s}{\alpha_0 + \alpha_s} R_0 \tag{16}$$

From (9), (15) and (16) we obtain

$$k = \frac{1 + \alpha_s}{1 - \alpha_0} - 1 \tag{17}$$

and

$$R_s = \left(\frac{1 + \alpha_s}{1 - \alpha_0} - 1\right) R_i \tag{18}$$

Equations (5) to (8) are used to calculate the tip resistance, R_i, the shaft resistance, R_s, and its relationship R_s/R_i. For a rough estimation it is often possible to assume that $\alpha_s = 0$.

Table 1

	Electrical Insulating Resistance (R_0)			pH-Sensitivity (α_0)			
	mean value (Ohm)	min (Ohm)	max (Ohm)	min mV/pH	max mV/pH	mean value mV/pH	%
a) pH-glass tip	$9 \cdot 10^{10}$	$4 \cdot 10^8$	$3.5 \cdot 10^{11}$	49.3	59.5	57.5	97.5
b) SiO_2	$2.5 \cdot 10^{11}$	$1.5 \cdot 10^{10}$	$6.8 \cdot 10^{11}$	17.8	53.8	41.5	70.5
silicon + Al_2O_3	$5.8 \cdot 10^{11}$	$5.3 \cdot 10^{11}$	$6.4 \cdot 10^{11}$	17.2	28.0	22.7	38.4
silicon	$6.4 \cdot 10^{11}$					39.0	66.2
Pb glass	$1.1 \cdot 10^{12}$	$1.2 \cdot 10^{11}$	$4.1 \cdot 10^{12}$	20.0	59.3	48.5	82.3
Al_2O_3	$1.34 \cdot 10^{12}$	$2.9 \cdot 10^{11}$	$3.1 \cdot 10^{12}$	1.0	56.5	37.3	64.0
Al_2O_3 + Pyrex + Si_3N_4 + SiO_2	$4.6 \cdot 10^{12}$	$2.5 \cdot 10^{12}$	$6.6 \cdot 10^{12}$	0.4	39.0	16.7	28.3

Temperature: 25 °C

b) The material is deposited by RF sputtering so that it covers the whole pH-glass tip.

The resistance of the pH glass cone without any insulating layers amounts to $4 \cdot 10^8$ to $3.5 \cdot 10^{11}$ Ohm depending on the degree of hydration (Table 1 a). An average value of $9 \cdot 10^{10}$ Ohm is found after 10 days' watering. The sensitivity was 57.5 mV/pH at 25 °C, i.e., 97.5% of the theoretical value.

The effect of different insulating layers was tested by sputtering the total area of the pH glass tip with different materials (Table 1 b). SiO_2 layers of a thickness between 0.04–1 μm have an average resistance of $2.5 \cdot 10^{11}$ Ohm only. The pH sensitivity is 41.4 mV/pH, i.e., 70.5% of the theoretical value. The materials listed in the table are arranged from lower to higher resistances. Hydrophobic silicone layers deposited by glowing discharge do not yield adequate results. Al_2O_3 layers are much better, but the best results are obtained with multilayers, such as layers consisting of Al_2O_3 + Pyrex + Si_3N_4 + SiO_2.

The data demonstrate the well-known phenomenon that thin layers of many materials show a sensitivity to hydrogen or other ions. The resistance of thin layers is considerably influenced by small amounts of foreign atoms or ions (*Lark-Horovitz* 1931; *von Lengyel, Matrai* 1932; *von Lengyel* 1931; *Doremus* 1968, *Baucke* 1974). The multilayer electrode shows the lowest pH-sensitivity. Considerable variances in the resistance result, though the same materials and thicknesses are applied. We think that this depends on the sputtering conditions. Probably, we did not always obtain homogeneous layers. Possibly some molecules of the ground material are mixed with the sputtered molecules by the sputtering process itself, thus changing

Table 2

pH Electrode with Different Insulating Materials of the Shaft			Electrical Resistance (mean values)					
Insulating material	mV/pH α_0	%	R_0 (Ohm)	R_i (Ohm)	R_s (Ohm)	R_s/R_i	R_s/R_i min	R_s/R_i max
SiO_2	47.7	80.6	$2.3 \cdot 10^{11}$	$3.3 \cdot 10^{11}$	$8.1 \cdot 10^{11}$	5.2	2.4	8.5
Pb glass	56.0	95.0	$3.1 \cdot 10^{11}$	$3.3 \cdot 10^{11}$	$6.9 \cdot 10^{12}$	20.1	11.5	32.3
Al_2O_3	55.5	94.0	$4.9 \cdot 10^{11}$	$5.3 \cdot 10^{11}$	$1.4 \cdot 10^{13}$	32.9	6.1	70.5

Temperature 25 °C

Table 3

Date	El.	R_0 (G Ω)	R_i (G Ω)	R_s (G Ω)	R_s/R_i	Response-time (sec)	pH-Sensitivity mV/pH α_0
8. 3.	1	439	459	10 200	22.2	180	56.7
	2	485	493	30 300	61.5	150	58.3
30. 5.	1	14	14.4	480	33.4	25	57.4
	2	31	31.6	1980	62.5	35	58.1
23. 7.	1	7.7	8.6	71.3	8.3	15	55.5
	2	16.6	17.4	331.9	19	25	57.6

El = electrodes 1 and 2   Temperature 25 °C

the behavior of the sputtered layer. Further refinements of our technique are necessary and, as we hope, possible.

Table 2 shows the data for microelectrodes with a pH-sensitive tip of about 10 μm length as described above. For proper function the ratio of the resistances R_s/R_i should be as high as possible. As expected from the foregoing, the Al_2O_3-covered shaft yields the best results. On an average the shaft resistance is about 30 times that of the tip.

In the first few days after manufacture all the electrodes show a rather long response time of 3–5 min, but after a period of 25–100 days' watering the 90% response time decreases to 5–20 s. Table 3 shows, as an example, the improvement of the response time from 180 to 15 s. Stability also improves, for example, from a drift of 17 to 0.5 mV/h. The ratio of the resistances R_s/R_i increases from 22.2 to 61.5. It shows that, in our electrodes, the pH sensitivity is practically restricted to the tip.

In summary, with single layers of dielectric materials sputtered on pH- and shaft-glass it is difficult to reach the high electric resistance necessary for the insulation of the shaft of a pH electrode. Testing up to four layers shows that multilayers give far better results. In pH microelectrodes the ratio of the shaft to tip resistance improves during the watering process of the electrodes. It shows an optimum after which it changes for the worse.

References

Baucke, F. G. K.: Investigation of Surface Layers Formed on Glass Electrode Membranes in Aqueous Solutions by Means of an Ion Sputtering Method. J. Non-Crystallin Solids 14 (1974) 13–31

Davidse, P. D., Maissel, L. T.: Dielectric Thin Films through RF Sputtering. J. appl. Physics 37, (1966) 574–580

Doremus, R. H.: Membrane Potentials of Pyrex Glass Electrodes. J. Electrochem. Soc. 115 (1968) 924–927

Harrop, P., Campbell, D. S.: Dielectric Properties of Thin Films In: Handbook of Thin Film Technology, pp. 2–33, ed. by *L. I. Maissel* and *R. Glang, McGraw-Hill,* New York 1970

Hass, G.: Preparation, Structure and Applications of Thin Films of Silicone Monoxide and Titanium Dioxide. J. Amer. Ceramic Soc. 33 (1950) 353–360

Lark-Horovitz, K.: Die Phasengrenzkräfte an der Grenze Dielektrikum – wäßrige Lösung, Naturw. 19 (1931) 397

Lee, C. O., Fozzard, H. A.: Electrochemical Properties of Hydrated Cation-selective Glass Membrane. A Model of K^+ and Na^+ Transport. Biophys. J. 14 (1974) 46–68

v. Lengyel, B.: Über das Phasengrenzpotential Quarz/Elektrolytlösungen. Z. physikal. Chem. (A) 153, (1931) 1425

v. Lengyel, B., Matrai, T.: Beiträge zum Verhalten der Quarzelektroden. II. Z. physikal. Chem. (A) 159 (1932) 393

Maissel, L. I., Francombe, M. H.: Insulating Films. In: An Introduction to Thin Films, pp.

245–262. Gordon & Breach, New York – London – Paris 1973

Mattox, D. M., *Kominiak*, G. J.: Physical Properties of Thick Sputter-Deposited Glass Film. J. Electrochem. Soc.: Solid-state Science and Technol. 120 (1973) 1535–1539

Pliskin, W. A., *Kerr*, D. R., *Perri*, J. A.: Thin Glass Films. In: Physics of Thin Films, Vol. 4, pp. 257–324, ed. by *G. Hass* and *R. E. Thun*. Academic Press, New York 1967

Saito, Y., *Baumgärtl*, H., *Lübbers*, D. W.: The RF Sputtering Technique as a Method for Manufacturing Needle-shaped pH Micro-Electrodes. This book, pp 103–109

Schanne, O. F., *Lavallée*, M., *Laprade*, R., *Gagnes*, S.: Electrical Properties of Glass Microelectrodes. Proc. IEEE 56 (12968) 1072–1082

Wu, S. Y., *Formigoni*, N. P.: Charge Phenomena in DC Reactively Sputtered SiO_2 Films. J. appl. Physiol. 39 (1968) 5613–5618

Ion-selective Liquid Ion Exchanger Microelectrodes

(3 Figures)

To make ion-selective liquid ion exchanger electrodes that are small enough (0.5 μm diameter) to use for making intracellular measurements, it is necessary to use glass (Pyrex) as the structural material within which the liquid ion exchanger membrane is formed. The basic problem is that a clean glass surface is hydrophilic because it is covered with polar groups, primarily silanol (silicic acid) groups (*Hair* and *Hertl* 1973). In order to make the surface hydrophobic so a stable organic liquid membrane can be formed, it is necessary to mask the silanol groups. Furthermore, if not masked, the silanol groups behave as cation exchangers which, under some conditions, cause the pipettes to behave as cation electrodes. This is illustrated in Figure 1. The potentials of three different, untreated micropipettes filled with 3 M KCl have been plotted as logarithmic functions of KCl activity. From top to bottom they have cation slopes of approximately 3 mv, 16 mv and 32 mv respectively.

One way to mask the silanol groups is to chemically react them with chloro- or alkoxysilanes as shown in Figure 2. The silanes can be mono-, di- or trifunctional. The silanes are applied in

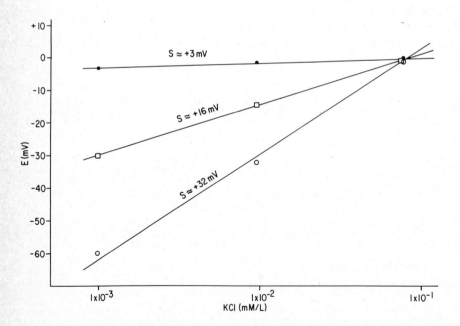

Fig. 1. The potentials of three different untreated, 3 M KCl filled micropipettes are plotted as logarithmic functions of KCl activity. The approximate slope of the response of each electrode is noted by the respective curve.

Ion-selective Liquid Ion Exchanger Microelectrodes

$$\geqslant Si-OH + R_3-Si-Cl \longrightarrow \geqslant Si-O-Si-R_3 + HCl$$

$$\geqslant Si-OH + R_3-Si-OR' \longrightarrow \geqslant Si-O-Si-R_3 + R'OH$$

$$\begin{matrix}\geqslant Si-OH \\ \geqslant Si-OH\end{matrix} + R_2-Si-Cl_2 \longrightarrow \begin{matrix}\geqslant Si-O \\ \geqslant Si-O\end{matrix} Si-R_2 + 2HCl$$

$$R-Si-Cl_3$$

$$\begin{matrix} & CH_3 & CH_3 & & CH_3 & CH_3 \\ & | & | & & | & | \\ H_3C-Si-O-Si-O- & \cdots & -O-Si-O-Si-CH_3 \\ & | & | & & | & | \\ & CH_3 & CH_3 & & CH_3 & CH_3 \end{matrix}$$

Fig. 2. The top three reactions show how mono- and di-functional chloro- and alkoxysilanes react with the silanol groups on a glass surface to render it hydrophobic. The third formula is the structural formula of dimethylpolysiloxane.

an organic solvent, followed by a heat treatment to fix them to the surface. A variety of chloro- and alkoxysilanes have been tried the best of which, according to *Blackman* and *Harrop* (1968), should be methyltrichlorosilane. Silanizing micropipettes, however, is difficult because the tips can be easily plugged and my experience with methyltrichlorosilane has been that I cannot apply enough to the surface to make it hydrophobic without plugging the tip. The most consistent results have been obtained with Siliclad (*Clay-Adams*) which is a trimethoxysilane. It is applied by dipping the pipettes into approximately a 2% solution of Siliclad in 1-chloronaphthalene for 15 seconds and then baking them for 1 hour at 300 °C. The exact concentration and dipping time depend on the shape of the pipette. Too much plugs them and too little is not effective.

Attempts to make Ca^{2+} microelectrodes with silanized micropipettes and using Orion and Corning Ca^{2+} exchangers have failed. Figure 3 shows that such an electrode has Nernstian

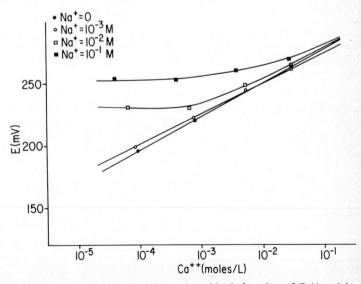

Fig. 3. The potential of a Ca^{++} microelectrode is plotted as a logarithmic function of Ca^{++} activity in the presence of varying concentrations of NaCl as indicated by the key in the upper left corner.

Table 1

	Intact	Broken
$1/k_{Na, K}$	26	76
	52	80
	59	83

slope in pure $CaCl_2$ but the Ca^{2+} : Na^+ selectivity is so poor that the electrode is useless in 0.1 m NaCl. In contrast, macroelectrodes using the same exchangers are very little affected by 0.1 M Na^+ down to 10^{-4} M Ca^{2+}. The loss of Ca^{2+} selectivity in the micro-application is consistent with the hypothesis that the glass surface is not sufficiently well covered to prevent it from behaving like a monovalent cation electrode.

Another way to treat the pipettes is to apply a dimethylpolysiloxane (DC 200 fluids, Dow Corining) which, according to *Blackman* and *Harrop* (1968), should be the best treatment available. The dimethylpolysiloxanes are applied in the same manner as silanes. Practically, it is more difficult to get consistent results with dimethylpolysiloxane than Siliclad because the concentration and dipping times are more critical and therefore less reproducible. It is possible to produce better electrodes using dimethylpolysiloxane, as judged by selectivity, but the yield of usable electrodes is substantially lower than with Siliclad.

This is illustrated by the results obtained with K^+ electrodes made with dimethylpolysiloxane-treated micropipettes and Corning K^+ exchanger. These pipettes were dipped for 60 seconds in an 0.1% solution of dimethylpolysiloxane (100 cs) in CCl_4 and then baked for 2.5 hours at 250 °C. The heat treatment schedule is provided by *Johannson* and *Torok* (1946). It is possible to obtain a low yield of K^+ microelectrodes that have K^+ : Na^+ selectivities close to that of the Corning macroelectrode, 80 : 1, but more typical results are shown in Table I. The K^+ : Na^+ selectivity is considerably lower than 80 : 1 for intact electrodes, but when the tips are broken to about 2 μ diameter the selectivity increases to the proper value. Again, this is consistent with the idea that the glass at the tip of the pipette is behaving as a monovalent cation electrode and contributing to the observed electrode potential.

One more example of the problems caused by using glass micropipettes to make ion-selective liquid ion exchanger microelectrodes is the observation that the Cl^- electrodes usually, but not always, do not have Nernstian slopes. This too can be attributed to the fact that the glass behaves to some extent as a cation electrode.

The conclusion drawn from this data is that, given a suitable ion-selective liquid ion exchanger, the major problem in using it in a glass micropipette is the treatment of the glass surface to minimize the cation electrode properties of the glass.

Acknowledgement:

This work was supported by a grant-in-aid from the American Heart Association.

References

Blackman, L. C. F., Harrop, R.: Hydrophilation of Glass Surfaces. I. Investigation of Possible Promoters of Filmwise Condensation. J. appl. Chem. 18 37–43 (1968)

Hair, M. L., Hertl, W.: Reactivity of Boria-silica Surface Hydroxyl Groups J. Phys. Chem. 77 1965–1969 (1973)

Johannson, O. K., Torok, J. J.: The Use of Dimethylsilicones to Produce Water-repellant Surfaces. Proc. I.R.E. 34 296–302 (1946)

I. A. Silver

Multi-Parameter Electrodes

(3 Figures)

During the last few years a number of microprobes have been developed which include micro oxygen electrodes, micro hydrogen clearance electrodes, micro pH electrodes and more recently other ion selective microelectrodes based on glass membranes or liquid ion exchangers. Although these probes in themselves are intrinsically very useful it has become obvious from work with them that much more critical information can be obtained when it is possible to measure several parameters at the same time in the same micro anatomical location rather than sequentially or simultaneously in different locations. This paper reports on the design and production of a multi parameter probe which includes its own reference electrode and has the capability of measuring three separate events within a tip diameter of approximately 1 micron. Currently three or four barrelled electrodes can be constructed according to the type of investigations being undertaken. Construction of probes is as follows:

(1) Three or four pieces of borosilicate glass tubing (Jencon Limited, Hemel Hempstead, Herts.) are stuck together with Araldite (Ciba, Duxford, U.K.) or are merely held together in the chuck of an electrode puller. For some kinds of multibarrelled probes the configuration of the glass is important, as is the wall thickness. We normally use a wall thickness either of 0.15 mm or 0.25 mm. The glass is heated in the coil of the electrode puller and twisted together as described by *Zeuthen, Hiam* and *Silver*, (1974). If liquid ion exchanger electrodes only are to be produced the glass tubes are arranged in a square and pulled so that each tube assumes a quarter segment shape in the electrode tip (Fig. 1). When an all-glass pH or sodium electrode is to be made in one of the barrels two thicker walled capillaries (OD 2.0 mm) are placed together so that they touch and two thinner walled capillaries (OD 1.8 mm) are placed on the outside, thus giving a rhomboidal arrangement of glass. When such a collection of glass is pulled the two thicker walled centre electrodes pull in an almost circular configuration or a very slight oval shape, whereas the two outer electrodes are pulled into the triangular section (see Fig. 2). The electrode barrels are treated as follows:

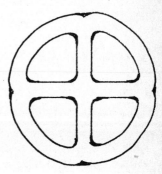

Fig. 1. Diagrammatic end view of 4 barrel microelectrode with barrels arranged in a square.

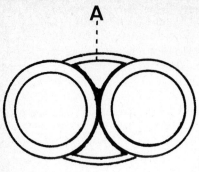

Fig. 2. Diagrammatic end view, 4 barrel electrode with 2 'central' and 2 'peripheral' barrels (one (A) metal-filled).

(2) Those which are to be filled with liquid ion exchanger are exposed from their back ends to the vapour of dichlorodimethylsilane for between 20 and 60 seconds according to the temperature and humidity of the laboratory. One barrel is left untreated to act as a reference, or in some cases two barrels are left untreated, one of which is to be the reference and the other is to contain a marker solution such as Procion Yellow (*Stretton* and *Kravitz* 1968), which can be injected iontophoretically to identify the site at which the electrode tips have been recording.

(3) Where other parameters are to be measured than ion activities, one barrel of the electrode may be metalised by filling it with an organic liquid compound of palladium and gold (bright platinum PBV 158, Blythe Colour Ltd., Stoke-on-Trent, England). The electrodes are centrifuged to force the liquid into the tip and when this has been accomplished the electrode is warmed until the metal is deposited from the solution. Excess solution is shaken from the back of the electrode after enough metal has been deposited on the inside of the barrel high enough up so that a good contact can be made from behind. The filled barrel should contain a solid plug of metal but if this has not been achieved, the partially filled barrel can be collapsed by local external heating with a 20 micron wire. It is necessary of course to use one of the thin walled capillaries for this treatment so that the other walls do not collapse during the heat treatment. Contact is made with metalised electrodes through the barrel by filling it with a conducting epoxy resin (Acheson DAG 1624) or high conductivity paint (DAG 962 Acheson Colour Co., Plymouth).

(4) If a pH glass or a sodium glass membrane is required in a multi-barrelled configuration, considerable difficulty is experienced in sealing the inner glass to the outer glass to form a "Thomas"-type of probe. It has been our experience that it is necessary to use a round or very slightly oval section capillary to achieve this purpose. The electrode is pulled in the rhomboidal configuration and one of the round central capillaries is measured and a suitable tip is prepared of the appropriate glass. It is dropped down into the electrode in the usual way and can be sealed by local heating and expanding the pH glass by pressure from the inside of the electrode applied through a hypodermic syringe. We have found it almost impossible to get effective sealing with sodium glass since it is usually necessary to heat the outer capillaries strongly enough to make them collapse onto the inner cone with this particular material. Sealing with wax or silicone rubber poured in through the back of the electrode (*Neild* and *Thomas* 1973) has not been very successful in our barrels.

The application of these electrodes to biological situations requires some care and it is frequently necessary to cut the tips on a diamond wheel to make sure that no silicone tracking has occurred between one electrode tip and another. If this happens either a liquid ion exchanger which is introduced into the hydrophobic barrel may track into another barrel, or the ion exchanger may escape into the tissue.

When using the metal filled barrel either as an oxygen electrode or as a hydrogen clearance probe it is necessary to use an external reference in the tissues. The reference electrode in the assembly is used in the usual way against the ion selective electrode. A single reference can be used with two or three ion selective probes.

The use of these electrodes can pose problems since interference between the competing barrels may be observed under certain circumstances. Two metal filled barrels, one used as a hydrogen detector and the other as an oxygen detector, frequently interfere. With regard to the liquid ion exchanger barrels no problem is encountered unless the siliconising runs from one barrel to the next, in which case the ion exchangers may mix near the tip producing equivocal results. The development of new ion selective liquid ion exchangers for sodium and for protons by *Simon* and his group has greatly increased the ease with which multi-barrel probes may be produced, since this will almost certainly eliminate the need for the all-glass type of detector in a multi-barrelled array. Attempts to make a Whalen-type electrode (*Whalen, Riley* and *Nair* 1967) have been unsuccessful since although the probe can be made the procedure for making it results in the contamination of the other electrode barrels with gold and with various cyanide residues which cannot be removed and which interfere with their subsequent performance.

Electron micrographs of the shafts of multi-barrelled electrodes which have been broken back from the tip show that there is a central air space between the electrodes (Fig. 3) but this does not extend into the extreme tip where the space is eliminated and the glass seals to form a solid plug.

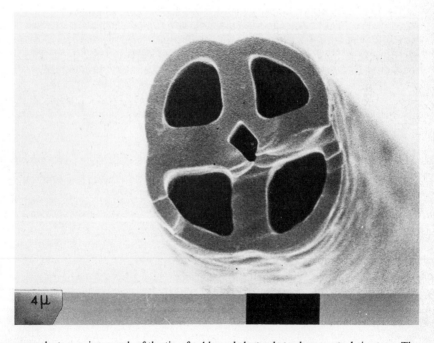

Fig. 3. Low power electron micrograph of the tip of a 4 barrel electrode to show central air space. The tip has been broken back 50 μ.

I. A. Silver

Summary

Multi-barrel probes containing both polarographic and ion exchanger electrodes within a tip size of 1.0 µ are described, and some of their characteristics discussed.

References

Neild, T. O., Thomas, R. C.: New Design for a Chloride Sensitive Microelectrode. J. Physiol. Lond. 231 (1973) 7P

Stretton, A. O. W., Kravitz, E. A.: Neural Geometry: Determination with a Technique of Intracellular Dye Injection. Science 162 (1968) 359

Whalen, W., Riley, J., Nair, P.: A Microelectrode for Measuring Intracellular PO_2. J. appl. Physiol 23 (1967) 798

Zeuthen, T., Hiam, R. C., Silver, I. A.: Microelectrode Recording of Ion Activity in Brain. In: Ion Selective Microelectrodes, ed. by *H. Berman* and *N. C. Hebert*, Plenum Press, New York 1974 p.p. 145—156

Raja N. Khuri

Microelectrodes Utilizing Glass and Liquid Ion-exchanger Sensors

(7 Figures)

Since 1960 we have been continuously involved in the construction and physiological applications of microelectrodes employing two general types of ion-exchanger electrochemical sensors: (1) glass and (2) liquid-liquid ion exchangers. Using these two types of sensors, a wide variety of ion-selective microelectrodes were developed and used in different biological systems under in-vivo and in-vitro conditions to measure ionic activities in extracellular and intracellular fluids.

I. Glass Microelectrodes

The cation-specific glasses, as ion sensors, possess some special advantages in direct analytical potentiometric analyses of biological systems. The glass membrane is insensitive to oxidation-reduction reactions, not much affected by proteins and rather indifferent to anions in general. Glass is a highly workable material which gradually and continuously softens with progressive heating. This allows the construction of a variety of microelectrodes using ion-selective glasses.

Two configurations of ion-selective glass microelectrodes were generally used in our work. These are the sealed micropipette (spear) type and the flow-through (internal capillary) type. These two general configurations of glass microelectrodes are represented diagrammatically in Figure 1.

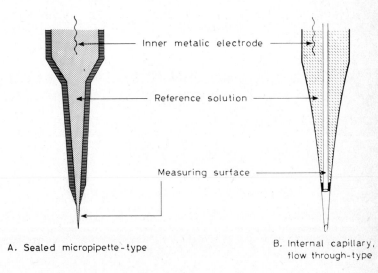

Fig. 1. Two configuration of indicator glass microelectrodes.

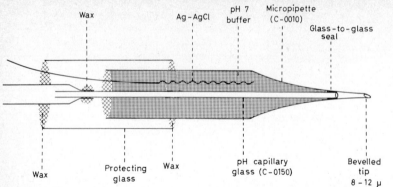

Fig. 2. A flow-through type of glass microelectrode (half-cell) with a glass-to-glass seal.

A. *Flow-Through Type*

The flow-through or internal-capillary-suction type of glass microelectrode offers some distinct advantages. With the measuring surface located internally, the test sample is protected from evaporation and gas exchange with the ambient air. The flow-through type has a much greater area-to-volume ratio than the spear type of microelectrode. They tend to have very small asymmetry potentials. The inner surface of a microcapillary which serves as the measuring surface tends to possess superior electrode function as compared to the outer surface (*Hamilton* and *Hubbard* 1941). The internal-capillary type has a short response time; perhaps because the turbulent inflow may serve as a substitute for stirring.

Figure 2 (*Khuri* et al 1968) represents a measuring halfcell (the reference being physically separate) of the flow-through type employing a glass-to-glass seal. The glass electrode capillary is fused at both ends to an appropriate inert glass selected on the basis of the similarity of their thermal characteristics. In Table 1 (*Hebert* 1969) the soft point and coefficient of thermal expansion of the pH, sodium and potassium electrode glasses are tabulated along with those of the inert glasses with which they fuse readily.

Figure 3 (*Khuri* et al. 1967) represents a single-unit pH glass microelectrode of the flow-through type. The measuring and reference half-cells were combined to form a single physical unit. However, this single-unit microelectrode employs wax seals. The glass-to-glass seal is superior to wax or cement seals, both for mechanical and electrical reasons. The use of a glass seal gives a higher yield of microelectrodes with ideal electrode function over a long time

Table 1. Some thermal properties of the electrode glass and "inert" glass pairs that account for their readiness to fuse together.

Glass		Soft point °C	Thermal expansion 10^{-7} in./in./°C
pH electrode	Corning 0150	655	110
	Corning 0120	630	89
Na$^+$ electrode	NAS$_{11-18}$	>970	53
	Corning 1720	915	42
K$^+$ electrode	KAS$_{11-18}$	796	102
	Corning 0080	695	92

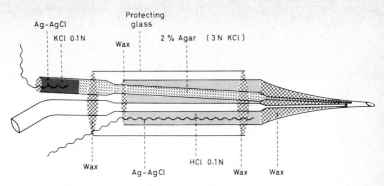

Fig. 3. Single unit pH glass microelectrode of the flow-through type.

span. This accounts for the replacement of the model represented in Figure 3 by the single-unit microelectrode with a glass seal and a microhole shown in Figure 4 (*Khuri, Agulian* and *Aklanjian* 1973). The microhole junction is a new feature in microelectrode construction. pH, sodium and potassium microelectrodes with a capacity as small as 10 nl can be constructed. These microelectrodes are ideally suited for in situ applications in extracellular systems both of the flowing and static varieties. The combination of the glass microelectrode and its reference into one physical unit is advantageous particularly for in situ ion analyses since only a single impalement of the tissue is made.

Based on the principles of the *Sanz* (1957) capillary electrode, *Siggard-Andersen* et al. (1960) developed the pH glass capillary which later became the commercially known Radiometer pH Microelectrode with a capacity of 25 µl. A significant jump in miniaturization of electrode capillaries with capacities of 10–15 nl represented in Figures 2 and 3 (*Khuri* et al. 1967, 1968). The process of miniaturization was carried one step further by *Uhlich, Baldamus* and *Ullrich* (1968) who used pH microcapillaries on samples as small as 2 nl in volume.

B. *Spear Type*

The sealed micropipette or spear type with its sharp tip and tapered shank is ideally suited for intracellular analyses. However, in our experience, three factors have thus far prevented glass

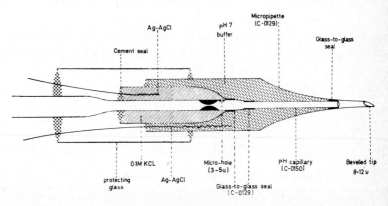

Fig. 4. Single unit pH glass microelectrode of the flow-through type with a glass seal and a microhole junction.

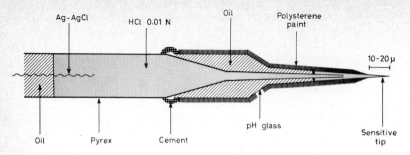

Fig. 5. Spear type pH glass microelectrode with external insulation.

microelectrodes of the spear configuration from becoming the intracellular microelectrodes of choice. Firstly, there is the uncertainty of the external insulation of all but the sensitive tip of the probe. To illustrate, Carter et al. 1967 claimed that they could insulate their pH microelectrodes down to a terminal sensitive portion of less than 20 μ. Using their original glaze insulation techniques *Paillard* 1972 concluded that the Carter pH microelectrode was inadequately insulated and its sensitive portion about 90–100 μ in length. Secondly, there is the degree of miniaturization. This is partly related to difficulties in controlling the insulation. Before the development of liquid ion-exchanger microelectrodes, the highest degrees of miniaturization were achieved with the spear-type glass microelectrode. One decade (1954 bis 1964) witnessed the reduction in the length of the cation-sensitive tip from 200 μ (*Caldwell* 1954) to 4 μ (*Lev* 1964), an effective miniaturization of two orders of magnitude. Thirdly, there is the difficulty in combining the measuring ion-selective glass half-cell (closed tip with insulation) with the reference half-cell (open tip) in a double-barrelled microelectrode configuration. This feat has been claimed by only one group (*Carter, Rector, Campion* and *Seldin* 1967). With single-barrel ion-selective glass microelectrodes one needs to correct for the value of the electrical gradient measured separately and on different cells. Since the highest degree of miniaturization we (*Khuri* 1968) could achieve with ion-selective glasses is a spear-type electrode with a sensitive tip of 10–20 μ in length (Fig. 5), we re-oriented our efforts in 1969 to the newborn field of liquid ion exchanger microelectrodes.

II. Liquid Ion-exchanger Microelectrodes

A liquid ion-exchanger is an organic electrolyte selected on the basis of its high selectivity for a given ionic species. The organic exchangers and the organic solvent in which it is dissolved form an organic membranous phase which is permeable to the counterion to which the coionic (organic exchanger) is highly selective. A selective ion-exchanger membrane develops a potential difference across itself which has a Nernstian relationship to the activity of the ion in question. Thus an ion-exchange process in the sensing interface is the electrode potential determining step in both electrode glasses and liquid-liquid ion exchangers.

The clear-cut advantage of using ion-selective micro-electrodes whose sensing element is a liquid ion-exchanger is the degree of miniaturization that can be achieved. Whereas with the spear-type glass electrode the dimensions of the sensing element were claimed to have been reduced to several micron (*Kostyuk* et al. 1969, *Lev* 1964, *Carter* et al. 1967), the sensitive

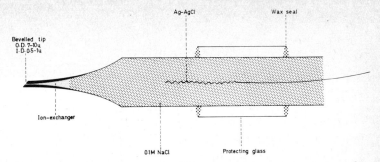

Fig. 6. Single barrel, ion-selective liquid ion-exchange microelectrode with a bevelled tip.

element of a liquid ion-exchanger microelectrode is the cross-sectional area of the exchanger filling the tip of a micropipette. This area is well below one square micron.

A. Single-Barrel Microelectrodes

With liquid ion-exchanger microelectrodes, as with glass microelectrodes, one can either use the chemical-sensing micropipette half-cell (single barrel) separately or in physical combination with the electrical reference micropipette in a double-barrel configuration.

The single-barrel liquid ion-exchanger microelectrode may be of two types depending on the size and shape of the sensing tip: the bevelled-tip and the conventional unbroken-tip micropipette. The bevelled-tip variety shown in Figure 6 (*Khuri* et al. 1971a) may be fabricated with an outer tip diameter of 5–10 μ and with a much smaller inner diameter. The combination of the thickening of the glass membrane of the tip portion of the shank and the bevelling itself renders the tip mechanically sturdy for tissue impalement. The bevelled-tip micropipette can be used in vivo for extracellular applications, as for intratubular ionic measurements (*Khuri* et al. 1971a), or for intracellular ion analyses in large cells. The unbroken-tip micropipette with its very fine tip (outside diameter less than one micron) is suited for intracellular measurements and was developed by *Walker* and *Brown* (*Walker* and *Brown* 1970; *Walker* 1971). However, the major limitation of the single-barrel liquid ion-exchanger microelectrode is that its potential, while its tip is within a cell, is the algebraic sum of the electrical potential across the cell membrane and the intracellular ionic potential. To correct for that would require either a separate determination of the membrane potential in the same tissue or preferably its simultaneous determination within the same cell. It would mean that the chemical and electrical sensing micropipettes would have to impale a single cell at two separate points. This is feasible only in large non-mammalian cells.

B. Double-Barrel Microelectrodes

One advantage of combining the chemical and the electrical micropipettes in a double-barrel configuration (Fig. 7), which was first developed by *Khuri* et al. (1971b; 1972a and b), is that such a unit impales a cell at a single point. This tool has been applied to single cells of renal tubules in situ (*Khuri* et al. 1972b; *Khuri* et al. 1972c; *Khuri* et al. 1974). The difficulties encountered in the fabrication of the double-barrelled microelectrode relate to the differential treatment of the chemical (organic exchanger) and the electrical (aqueous salt bridge) barrels.

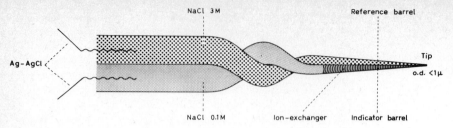

Fig. 7. Double-barrelled, ion-selective liquid ion-exchange microelectrode.

This involves rendering the chemical exchanger barrel hydrophobic-organophilic by unilateral siliconization. After filling the two barrels separately with the organic exchanger and the reference electrolyte salt solution, special care must be taken to prevent any interference between the organic and the electrolytic solutions in the region of the tip.

In addition to eliminating the uncertainty involved in the correction for the membrane electrical potential difference (PD) the combination of the electrical and the chemical sensing micropipettes into a single double-barrelled unit offers the added advantage of increasing our confidence in the intracellular localization of the microelectrode tip. The simultaneous technique of monitoring the electrical PD across a cell membrane and the intracellular ionic activity has been described and applied by our team using cation and anion-specific exchangers in different biological systems in situ. The leads of the two barrels were connected to one electrometer and this represented the ionic (K*) potential (Figs. 8 and 9; lower

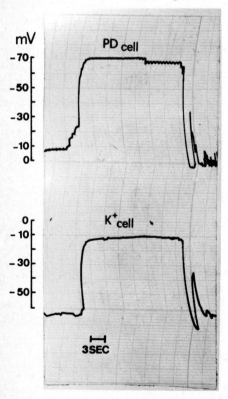

Figs. 8 and 9. Simultaneous in situ recordings of the peritubular membrane PDs and the potassium potentials in single proximal (Fig. 8) and distal (Fig. 9) tubular cells of the rat kidney.

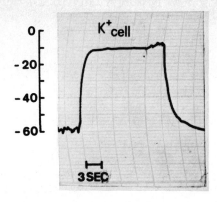 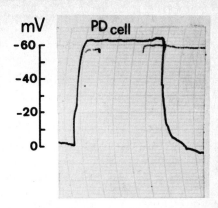

Fig. 9

tracings). The leads of the reference barrel and an external salt bridge reference electrode were connected to another electrometer and the electrical PD between them represented the membrane PD (upper tracings of Figs. 8 and 9). The tracings in Figs. 8 and 9 represent simultaneous in situ recordings of the peritubular membrane PDs and the potassium potentials in single rat renal tubular cells of the proximal and distal tubules respectively. As the double-barrelled microelectrode impales the peritubular membrane there is an abrupt rise in the membrane potential associated with a more gradual rise of the potassium potential. This is because the K*- sensor has a slower response time than the electrical cell. Thus as measured with the double-barrelled microelectrode, the electrical membrane PD and the intracellular ionic potential are both simultaneous electrometric determinations with the reference barrel half-cell (including its tip potential), which is common to both.

Needless to say these microelectrode techniques are applicable to the different liquid ion-exchangers that are currently available and to new ion-specific exchangers that may yet be developed.

Acknowledgement:

This work was supported in part by the Lebanese National Research Council.

References

Caldwell, P. C.: An Investigation of the Intracellular pH of Crab Muscle Fibers by Means of Micro-glass and Micro-tungsten Electrodes. J. Physiol. (London) 126 (1954) 169–180

Carter, N. W., Rector, F. C., Campion, D. S., Seldin, D. W.: Measurement of Intracellular pH of Skeletal Muscle with pH-sensitive Glass Microelectrodes. J. clin. Invest. 46 (1967) 920–933

Hamilton, E. H., Hubbard, D.: Effect of the Chemical Durability of Glass on the Asymmetry Potential and Reversibility of the Glass Electrode. J. Res. Nat. Bur. Std. 27 (1941) 27–33

Hebert, N. C.: Properties of Microelectrode Glasses. In: Glass Microelectrodes, pp. 25–31, ed. by *M. Lavallée, O. F. Schanne* and *N. C. Hebert.* Wiley, New York 1969

Hinke, J. A. M.: Glass Micro-electrodes for Measuring Intracellular Activities of Sodium and Potassium. Nature 184 (1959) 1257–1258

Khuri, R. N.: pH Glass Microelectrode for in Vivo Applications. Rev. Sci. Instrum. 39, (1968) 730–732

Khuri, R. N.: Intracellular Potassium and the Electro-chemical Properties of Striated Muscle Fibers. Proc. Int. Union Physiol. Sc. 9 (1971b) 301

Khuri, R. N., Agulian, S. K., Aklanjian, D. A.: A Single Unit Microelectrode with a Glass Seal and a Microhole. Pflügers Arch. 345 (1973) 265–270

Khuri, R. N., Agulian, S. K., Bogharian, K.: Electrochemical Potentials of Potassium in Proximal Renal Tubule of Rat. Pflügers Arch. 346 (1974) 319–326

Khuri, R. N., Agulian, S. K., Harik, R. I.: Internal Capillary Glass Microelectrodes with a Glass Seal for pH, Sodium and Potassium. Pflügers Arch. 301 (1968) 182–186

Khuri, R. N., Agulian, S. K., Kalloghlian, A.: Intracellular Potassium in Cells of the Distal Tubule. Pflügers Arch. 335 (1972b) 297–308

Khuri, R. N., Agulian, S. K., Oelert, H., Harik, R. I.: A Single Unit pH Glass Ultramicro Electrode. Pflügers Arch. 294 (1967) 291–294

Kjuri, R. N., Agulian, S. K., Wise, W. M.: Potassium in Rat Kidney Proximal Tubules in situ: Determination by K^+ Selective Liquid Ion-exchange Microelectrodes. Pflügers Arch. 322 (1971a) 39–46

Khuri, R. N., Goldstein, D. A., Maude, D. L., Edmonds, L., Solomon, A. K.: Single Proximal Tubulues of Necturus Kidney; VIII: Na and K Determination by Glass Electrodes. Amer. J. Physiol. 204 (1963) 743–748

Khuri, R. N., Hajjar, J. J., Agulian, S. K.: Measurement of Intracellular Potassium with Liquid Ion-exchange Microelectrodes. J. appl. Physiol. 32 (1972a) 419–422

Khuri, R. N., Hajjar, J. J., Agulian, S., Bogharian, K., Kalloghlian, A., Bizri, H.: Intracellular Potassium in Cells of the Proximal Tubule of Nectums Maculosus. Pflügers Arch. 338 (1972c) 73–80

Kostyuk, P. G., Sorokina, Z. A., Kholodova, Yu. D.: Measurement of Activity of Hydrogen, Potassium and Sodium Ions in Striated Muscle Fibres and Nerve Cells; In: Glass Microelectrodes, pp. 322–348, ed. by M. Lavallée, O. F. Schanne and N. C. Hebert, Wiley, New York 1969

Lev, A. A.: Determination of Activity Coefficients of Potassium and Sodium Ions in Frog Muscle Fibres. Nature 201 (1974) 1132–1134

Paillard, M.: Direct Intracellular pH Measurement in Rat and Crab Muscle. J. Physiol. (London) 223 (1972) 297–319

Sanz, M. C.: Ultramicro Methods and Standardization of Equipment. Clin. Chem. 3 (1957) 406–419

Siggaard-Andersen, O., Engel, K., Jorgensen, K., Astrup, P.: A Micro Method for Determination of pH, Carbon Dioxide Tension, Base Excess and Standard Bicarbonate in Capillary Blood. Scand. J. Clin. Lab. Invest. 12 (1960) 172–176

Uhlich, E., Baldamus, C. A., Ullrich, K. J.: Verhalten von CO_2-Druck und Bicarbonat im Gegenstrom des Nierenmarks. Pflügers Archiv 303 (1968) 31–48

Walker, J. L.: Ion Specific Liquid Ion Exchanger Microelectrodes. Analyt. Chem. 43 (1971) 89 A to 93 A

Walker, J. L., Brown, A. M.: Unified Account of the Variable Effects of Carbon Dioxide on Nerve Cells. Science 167 (1970) 1502–1504

N. C. Hebert and R. R. Deleault

The Art of Making Small Glass Electrodes

(6 Figures)

Three different categories of small glass electrodes are fabricated in our laboratory. These are miniature electrodes, microelectrodes, and intracellular electrodes.
The miniature electrodes require test solution volumes of about one microliter. The microelectrodes require smaller test solution volumes and can also be used in large single cells. The intracellular electrodes are glass micropipettes with tips filled with liquid ion-exchanger.
A variety of sensing and insulating glasses are used. Some of these along with their thermal expansion coefficients and softening points are listed in Table I.

Table 1. Properties of Sensing and Insulating Glasses. (*Hebert* 1969).

Glass	Type of Glass	Thermal Expansion Coefficient deg. C.	Softening Point deg. C.	Log. P Ohm-Cm. 25 C
Corning 0150	pH Sensing	110	655	10.8
NAS_{11-18}	Sodium Sensing	53	>970	9.45
NAS_{27-4}	Potassium Sensing	123	647	9.00
Corning 0120	Insulating	89	630	17+
Corning 7740	Insulating	33	820	15

Miniature Glass Electrodes

The miniature pH glass electrodes consist of a sensing glass hemisphere affixed to the end of a Code 0120 insulating glass tube.
The glass electrode contains an internal reference solution and a silver-silver chloride internal reference electrode.
Figure 1 depicts such an electrode sealed inside a 2 mm stainless steel tube. The pH sensing

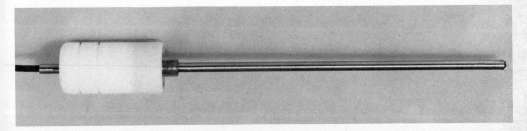

Fig. 1. A pH glass electrode protruding from the end of a 2.0 mm stainless steel tube. The electrode is equipped with a Luer lock for locking into a catheter.

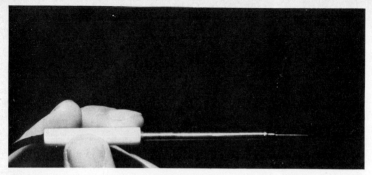

Fig. 2. A pH glass electrode sealed inside a 21 gauge stainless steel tube.

hemisphere protrudes from one end of the steel tube. The white plastic cap contains a Luer lock for fastening it into a placement catheter. Another glass electrode having a hemispherical diameter of 0.4 mm and sealed inside of a 21 gauge (0.75 mm in diameter) stainless steel tube is shown in Figure 2. The pH sensing hemisphere can either protrude from the blunt end (Fig. 3) or recessed in the bevel of a needle (Fig. 4). These electrodes are designed for tissue pH measurements and are used with research-type pH meters. Small glass electrodes for potassium and for sodium are also made using similar techniques.

A small combination pH electrode is fabricated by heat-shrinking a Code 0120 glass tube around a smaller tube which has a pH hemisphere affixed (Fig. 5). An asbestos fiber is inserted between the inner and the outer tube before they are sealed together. The asbestos fiber serves as the reference junction connecting the outer chamber which is filled from above with 3 M KCl. The electrode can measure in test solution volumes as small as 5 microliters with a depth of immersion of about 1.5 mm.

Miniature Reference Electrodes

A small reference electrode with liquid-liquid junction is fabricated by sealing an asbestos fiber inside a small diameter glass tube. The tube is scratched with a scribe and broken, leaving a 2 mm length of the fiber sealed in glass. The glass tube is filled with a 3 M KCl

Fig. 3. A pH-sensitive hemisphere protruding from the end of a 21 gauge stainless steel tube. The glass hemisphere has an outside diameter of 0.4 mm (18X).

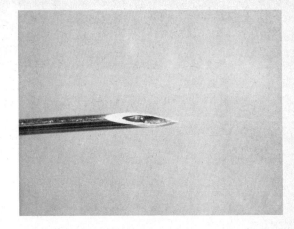

Fig. 4. A pH-sensitive hemisphere recessed in the bevel of a 21 gauge needle (9X).

solution, saturated with AgCl. A silver-silver chloride electrode is inserted into the glass tube and the reference electrode is ready for use. The flow rate of the junction is controlled by the size of the fiber and the length of the seal. A short length of tubing with a sealed-in fiber can be inserted into a polyethylene tube. The polyethylene tube serves as the 3 M KCl reservoir and is inserted into a holder containing a silver-silver chloride electrode. Such an electrode is shown in Figure 6. The white plastic cap holder also contains a Luer lock. The polyethylene tube can easily be pulled out of the cap for refilling.

Microelectrodes for pH, Sodium, Potassium, Chloride, and Calcium

Glass microelectrodes for pH and solium are fabricated according to the method of *Hinke* (*Hinke* 1959). A pH glass micropipette (Code 0150 tubing) is inserted through a larger insulating glass micropipette (Code 0120 tubing) having a tip diameter of about 40 microns. The outer glass is carefully ring-sealed around the pH pipette. The tip of the pH pipette is then sealed with a microforge. Our microforge consists of a left-hand micromanipulator which holds the 25 micron diameter platinum-iridium forge wire and a right-hand micromanipulator which holds the glass micropipette. The glass sealing is viewed through a horizontally mounted microscope having a total magnification of 150X. Electrodes made on this apparatus have sensing lengths of about 50 microns and tip diameters of 1–3 microns.

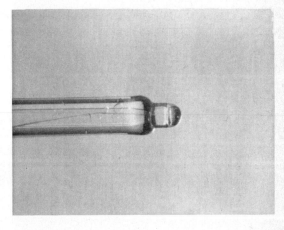

Fig. 5. The sensing portion of a combination pH electrode consisting of a pH-sensitive glass hemisphere of 1.2 mm diameter and of an asbestos fiber sealed in the glass 1 mm from the pH hemisphere.

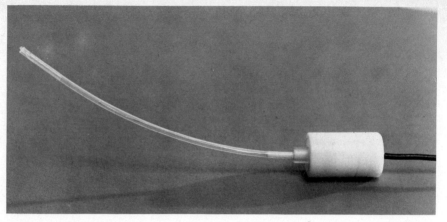

Fig. 6. A miniature reference electrode consisting of an asbestos fiber sealed in glass and attached to a polyethylene tube containing a 3 M KCl solution and a silver-silver chloride electrode. The white cap contains a Luer lock for fastening it in a catheter.

The critical step in making the pH electrode is to have the inner pH pipette fit tightly at the opening of the outer insulating pipette end when they are sealed together. A careful and slow sealing is necessary to prevent caving in of the glass. This step is not as critical when making the sodium microelectrodes since the outer insulating pipette (Code 7740 tubing) has a much lower softening point than the inner sensing, NAS_{11-18}, glass. After the outer insulating pipette is sealed (or heat-shrunk) around the inner sensing pipette, the electrode tip is reworked several times by pushing it horizontally into the hot microforge wire (glass loaded) and withdrawing it after the wire is cooled slightly. This generally takes a lot of practice to obtain one micron tips.

The pH microelectrodes are filled by boiling in distilled water for about 20 minutes. After cooling, the distilled water is displaced with 0.1 N HCl. Electrodes are tested approximately two hours later. This method of filling electrodes does not work well for sodium microelectrodes. When filled by boiling, these electrodes lose their selectivities in the presence of potassium ions. Sodium microelectrodes are filled according to the method of *Hinke* (*Hinke* 1967). A small glass fiber is inserted into the microelectrode and air bubbles which are left in the tip are removed by expanding them and moving them into the shank using a warm microforge wire. The sodium microelectrodes may still lose their ion selectivities after several weeks of storage. This seems to be a property of the NAS_{11-18} glass itself. The only method of restoring selectivity that we have found is by quickly dipping the electrode tip in a dilute hydrofluoric acid solution.

The pH microelectrodes, on the other hand, last for many months of storage. Both microelectrodes can be cleaned easily by dipping in warm chromic acid solution.

The microelectrodes for potassium, chloride, and calcium are fabricated from glass micropipettes (Code 7740 tubing) which have special tips formed on the microforge. A special tip is made by pushing a micropipette tip into the hot microforge which has been loaded with Code 0120 glass. The forge wire is slightly cooled and the micropipette is withdrawn from the forge pulling some of the Code 0120 glass with it. The forge is turned off and the micropipette is pulled away until it breaks. This produces sharp breaks and with proper timing of the various operations, will yield fattened tips with inside diameters of 1–3 microns and outside diameters from 3 to 40 microns. The sharp break occurs at the Code 0120 to Code 7740 seal which breaks easily because of the miss-match of thermal expansion coefficients. The glass micropipette tips are silanized by a one-second dip in a 0.5 to 1% solution of silanizing solution in xylene (Trifluoro propyl-trimethoxy silane or a Dow Corning 200 fluid). The solution is forced out of the tip with air pressure after about ten minutes and a small amount of liquid ion exchanger is placed in the shank of the micropipette with a fine needle. Air pressure is applied through the stem forcing the ion-exchanger to the tip. A 0.5 M KCl solution is added to make contact with the ion-exchanger. The electrode is inserted into a holder containing a silver-silver chloride electrode. Electrodes made this way have near theoretical responses and last for many months. All the liquid ion-exchangers used are commercially available form Corning Glass Works.

Intracellular Electrodes for Potassium, and Chloride

The intracellular electrodes are similar to those already described by *Walker* (1971). These consist of a silanized glass micropipette having a liquid organic sensing material in the tip. A glass micropipette (Code 7740 tubing) is silanized by dipping for one second in a 0.5 to 1% silane solution in xylene. After the xylene has evaporated, a droplet of exchanger is placed in the shank of the micropipette with a fine needle or polyethylene tube. A polyethylene tube which is attached to a 20 ml syringe is then inserted into the stem opening of the micropipette until it makes a tight seal at the shoulder of the micropipette. By applying pressure on the syringe, the droplet of exchanger moves into the shank and tip of the micropipette. The internal reference solution is then added. Intracellular elctrodes made this way have tip diameters of about 0.5 microns and D. C. electrical resistances in the order of 10^{10} ohms. The same technique of filling electrodes is used in our laboratory for filling 3 M KCl micropipettes. The D.C. electrical resistances of a batch of the electrodes were found to be around 2 Megohms.

Conclusion

We have found small glass electrodes when properly pre-conditioned and shielded to be superior to the larger laboratory electrodes with respect to stability, response time, and electrical noise. Also, they have found wide applications in the analysis of small fluid volumes and for in-vivo measurements.

References

Hebert, N. C.: Properties of Microelectrode Glasses, In: Glass Microelectrodes, pp. 25–31, ed. by *M. Lavallée, O. Schanne, N. C. Hebert,* Wiley, New York 1969

Hinke, J. A. M.: Glass Micro-electrodes for Measuring Intracellular Activities of Sodium and Potassium, Nature 184 (1959) 1257–1258

Hinke, J. A. M.: Cation-selective Microelectrodes for Intracellular Use, In: Glass Electrodes for pH and Other Cations, pp. 467–477, ed. by *G. Eisenman,* Dekker, New York 1967

Walker, J. L. Jr.: Ion Specific Liquid Ion Exchanger Microelectrodes. Anal. Chem. 43 (1971) 89A–93A

Kessler, M., Hájek, K., Simon, W.

Four-Barrelled Microelectrode for the Measurment of Potassium-, Sodium- and Calcium-Ion Activities

(5 Figures)

Introduction

Investigations performed using different mammalian organs clearly indicate that the direct and continuous monitoring of ion activities with ion-selective liquid membrane electrodes opens a wide range of new possibilities in physiological research and in clinical medicine. Our experiments showed that the multicomponent ionic systems can be interpreted only when the important cation activities are measured simultaneously within a small area. We therefore developed a new multiple-ion microelectrode.

Methods and Results

For making multiple ion-selective microelectrodes we use four-barrelled glass tubes. As shown by scanning electron microscopy, these tubes can be pulled in two steps to a tip diameter of less

 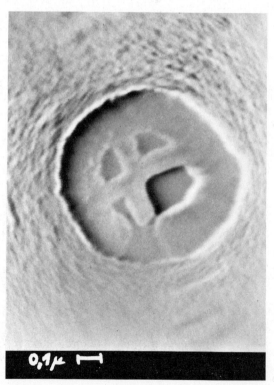

Fig. 1. Tip of four-barrelled microelectrode. The micrographs were taken by *K. Zierold* and *D. Schäfer* with the scanning electron microscope.

Fig. 2. Electromagnetic puller for production of multibarrelled microelectrodes.

than one micron (Fig. 1): In order to get tips with a reproducible size a puller was developed which has three new principles of construction (Fig. 2):
1. The axes are conducted very precisely by longitudinal ball bearings.
2. Since small movements of air cause distinct changes of temperature at the heating spiral the pulling process is performed within a closed chamber of lucite.
3. The temperature of the heating element is frequently controlled with a thermoelement.
The further procedure of filling the electrodes is done according to the method developed by *Neher* and *Lux* (1972).
Three barrels of the electrode are filled with ion-selective ligands. The fourth channel is used for the silver/silver-chloride reference electrode. As ion-selective carriers we use valinomycin for potassium and synthetic electrically neutral molecules for sodium and calcium. The neutral ligands developed recently (*Ammann* et al. 1973, *Pretsch* et al. 1974) show sufficient selectivity for measurements in blood and in tissue. The calibration curves (Fig. 3) give slopes of 50–55 mV per -pK and -pNa respectively 25–28 mV per -pCa.
Figure 4 shows the diagram of the amplifier circuit used for the electrode measurements of ion activities (*Hájek* and *Kessler* 1975).
In a first series of investigations extracellular measurements were performed with the multiple electrodes. In these studies the electrodes were inserted into the tissue (Fig. 5) with hydraulic

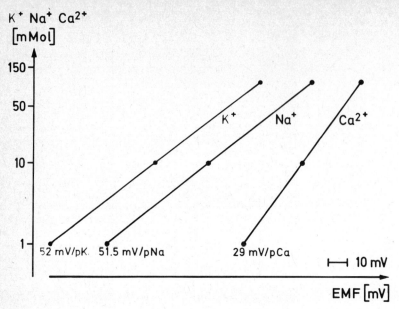

Fig. 3. Calibration curves of a four-barrelled microelectrode (tip diameter 2 microns).

micromanipulators (David Kopf, Los Angelos, USA). The electrodes had a tip diameter of 1 micron. We found rather uniform ion activities throughout the tissue when microcirculation was undisturbed.

In this context it should be emphasized that we were not able to distinguish between intracapillary ion activities and interstitial activities. For theoretical reasons it seems most unlikely that even with microelectrodes with a tip diameter of less than 0.1 micron it would be possible to get into the extremely small interstitial space without completely destroying the local environment.

We therefore believe that the interstitial cation activities can be monitored with sufficient accuracy only when ion-selective surface electrodes are used (*Kessler* et al. 1974).

First attempts were made to produce multiple electrodes for intracellular measurements with a tip diameter of 0.5 micron as used by *Walker* (1971). It became apparent that the filling procedure of multiple ion electrodes is much more difficult when microelectrodes of this size are used. At present we have best results when the filling of the very thin electrodes is induced

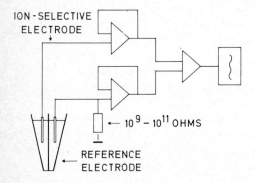

Fig. 4. Diagram of electrical circuit for microelectrode measurements in perfused liver.

Fig. 5. Arrangement with hydraulic micromanipulators for micropuncture under microscopic observation.

by a temperature gradient. The filling is controlled with a microscope at a magnification of 500. For illumination we use light pipes which are commercially available (Schott, Mainz). When ion measurements are performed with microelectrodes the organ can easily pick up parasitical potentials and therefore it should be isolated and shielded electrically from its surroundings. We try to avoid such noise potentials by carefully grounding the organ.

In experiments performed in the isolated perfused liver the preparation is protected against electrostatic charge by two grounded Ag/AgCl electrodes which are placed in the tubings of the influent and effluent perfusate.

Another source of disturbances in extracellular ion measurements with multibarrelled electrodes containing the reference electrode are streaming potentials which are in the range of 3–6 mV.

A possibility to avoid such streaming potentials could be the measurement of ionic ratios using a second ion carrier ligand as reference. We were able to perform such ratio measurements using a double-barrelled electrode filled with a potassium-sensitive ligand in one channel and a calcium ligand in the other (*Simon* et al. 1974). The EMF of the two electrodes was measured with two floating amplifiers (Burr Brown 3532L).

Measurements with two high-impedance electrodes and floating amplifiers are very susceptile to electric noise. New electronic techniques may eliminate these problems making tissue measurements with the ionic ratio method possible.

References

Ammann, D., Pretsch, E., Simon, W.: Darstellung von neutralen, lipophilen Liganden für Membranelektroden mit Selektivität für Erdalkali-Ionen. Helv. Chim. Acta 56 (1973) 1780

Pretsch, E., Ammann, D., Simon, W.: Design of Ion Carriers and their Application in Ion-selective Electrodes. Research/Development 25 (3) (1974) 20

Hájek, J., Kessler, M.: Amplifier System for the Simultaneous Measurement of pH, pK, pNa, pCa, pCl. 1976 in preparation

Kessler, M., Höper, J., Simon, W.: Methodology and Practical Application of a Multiple Ion-selective Electrode (pH, pK, pNa, pCa, pCl) for Tissure Measurements. Fed. Proc. 33 (1974) 279

Neher, E., Lux, H. D.: Messung extrazellulärer K^+-Anreicherung während eines Voltage-Clamp-Pulses. Pflügers Arch. 332 (1972) Suppl. R 176

Simon, W., Pretsch, E., Morf, W. E., Güggi, M., Kahr, G., Kessler, M.: New Ion-selective Electrodes Based on an Organic Ligands. The Society for Analytical Chemistry Centenary Celebrations, Paper C 55, London, July 1974

Walker, J. L., Jr.: Ion Specific Liquid Ion Exchanger Microelectrodes. Anal. Chem. 43 (1971) 89 A

R. C. Thomas

Construction and Properties of Recessed-Tip Micro-Electrodes for Sodium and Chloride Ions and pH

(8 Figures)

To enable an ion-sensitive microelectrode to penetrate a cell without causing excessive damage, the effective tip (that part of the electrode that must cross the cell membrane) needs to be as small as possible. With a protruding-tip design, as shown in Figure 1 A, the smaller the effective tip the higher the electrical resistance, and the greater the construction problems. The recessed-tip design (*Thomas* 1970), as shown in Figure 1 B, enables a relatively long length of ion-sensitive glass to be exposed to the cell interior, while only the extreme tip of the electrode needs to cross the cell membrane. Thus recessed-tip microelectrodes are much easier to insert into cells and have lower resistances than corresponding protruding-tip designs.

Of the three microelectrode designs shown in Figure 2, and to be described here; – the Na^+-sensitive microelectrodes (*Thomas* 1970, 1972) have resistances of about 10^{11} ohms and are best used soon after filling; the pH-sensitive microelectrodes (*Thomas* 1974) have a somewhat lower resistance (10^{10}–10^{11} ohms) and only work well after hydration; and the Cl^--sensitive electrodes (*Neild* and *Thomas* 1973) have low resistance and are best used soon after making.

Construction

General. A microelectrode puller and a microforge are needed. The puller can be of any design that allows wide control over the heat and pull; I use a simple vertical machine with a long, narrow heating coil of 22 gauge Kanthal A steel wire. The microforge setup consists of three Prior micromanipulators, one for the heating element and two to hold electrode components, arranged around a horizontally-mounted compound microscope. The heating element is a loop of 25 μ diameter Pt: Rh wire.

Na^+-sensitive microelectrodes. The method, illustrated in Figure 3, depends on the fact that the

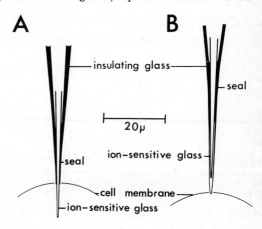

Fig. 1. Comparison of construction of and depth of penetration required by two designs of ion-sensitive microelectrode. A: protruding-tip design. B: recessed-tip.

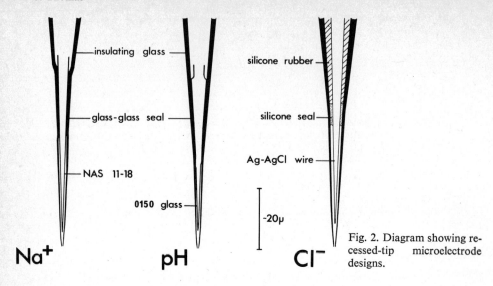

Fig. 2. Diagram showing recessed-tip microelectrode designs.

softening point of NAS 11–18 is higher than that of the insulating glass. (A): Using high heat and strong pull, make micropipettes out of NAS 11–18 glass (Corning Glass Works, N.Y.). Measure the external diameters at about 100 μ and 200 μ from the tip: suitable values are 6 and 12 μ respectively. (B): Make a batch of insulating glass (Corning code 7740 or 1720) micropipettes and measure their apparent internal diameters, which should match those of the NAS 11–18 micropipettes fairly well. One way is to heat and elongate the glass, in the puller, by about 1 cm. Then recentre and complete the pull at a relatively low heat. (C): Seal the NAS 11–18 pipettes by touching their tips briefly, two or three times, to the hot microforge element; adjusting the heat so that the glass melts only on contact with the element. (It is better to over rather than under-seal.) Mount an insulating glass micropipette vertically in the microforge. (D): Scratch near the neck of a matching NAS 11–18 micropipette with a diamond and break off the end. (E): Carefully drop it down inside the insulating glass micropipette. With vibration if necessary, it should fall into the correct position. (F): Position the microforge element close to the micropipettes and about 100 μ from their tips. Apply heat

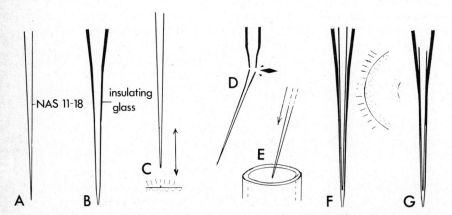

Fig. 3. Method of construction of recessed-tip Na⁺-sensitive microelectrode. Details in text.

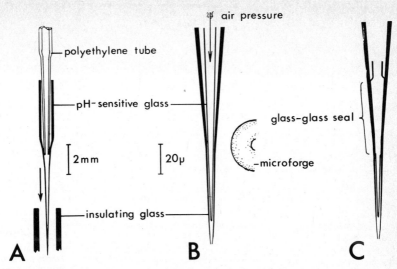

Fig. 4. Construction of recessed-tip pH-sensitive microelectrode. Description given in text.

as briefly as possible to melt the insulating glass so that it collapses onto the NAS 11–18, as shown in (G): Even though heat is applied only on one side, the glass will melt sufficiently to form a seal all round. The length of NAS 11–18 exposed inside the recess should be 40–100 μ, and the distance between the tips should be as small as possible to minimize the volume of the recess.

Until use, finished microelectrodes can be stored dry. Before filling, remove the surplus NAS 11–18 by tapping the inverted electrode. The filling solution can then be injected with a syringe, using a cat's whisker and the microforge to remove remaining air bubbles. The recommended filling solution is 1 M Na EDTA, pH 9, and 0.1 M NaCl.

pH-sensitive microelectrodes. The pH-sensitive glass, code 0150 (NCaS 22-6), has a rather low softening point. The method is illustrated in Figure 4. Insulating glass and pH glass micropipettes are made, and the latter sealed, as described above. Then (A): the pH micropipette is mounted on the end of a fine polyethylene tube connected to a 2.5 ml syringe full of air. The micropipette and tube are then lowered into a vertically mounted, matching, insulating glass micropipette and positioned as far down as possible. (B): Heat is applied and the syringe squeezed until the pH glass near the microforge element is blown outwards to seal to the inside of the insulating glass, as shown in Figure 4(C). Withdrawing the polyethylene tube usually pulls out the surplus pH glass. Electrode dimensions to aim at are: an exposed length of pH glass 10–50 μ, and a minimum recess volume.

Finished pH-sensitive microelectrodes can be stored dry. They are best filled by boiling in distilled water for about 2 hours, thus also hydrating the pH glass. The water is then replaced with 0.1 M NaCl buffered to pH 6 with 0.1 M Na citrate.

Cl-sensitive microelectrodes. Since glass cannot be sealed to silver, this method involves the formation of a "silicone" seal between silver wire and the insulating glass. Lengths of pure Ag wire are etched in aqueous 5% NaOH, 10% NaCN solution, to form tapering points. (A potential of about 5 volts is applied with the wire positive until the current falls to 40% of its peak value. Then 2 volts is applied for a few seconds to polish the tip.) The tip is then chlorided electrolytically, and the wire glued in position inside an insulating glass micropipette. The latter is then dipped in glycerol until it rises inside the glass to cover the tip of the Ag wire. Air is then bubbled through a 3:1 mixture of trichloromethylsilane and dichlorodimethylsilane and into the top of the micropipete. This forms a rather weak "silicone" seal at the glycerol-air interface. The body of the electrode is then carefully filled with silicone rubber. (Since the seal is delicate, it is advisable before forming it to glue fine polyethylene tubes into the top of the pipette and use them for applying the silane and injecting the silicone rubber.) The electrodes are ready for use when the rubber has set.

Protruding-tip ion-sensitive microelectrodes. It is quite possible to make these using the same techniques as described above, except that the insulating glass tips are broken back to the required diameter before the ion-sensitive material is inserted.

Properties

Those that arise from the recessed-tip configuration will be considered first, and then those which depend on the ion-sensitive material.

The recess is initially filled with air, which dissolves and disappears soon after an electrode is immersed in saline. When the electrode is inserted into a cell, it thus carries in a small volume of the extracellular medium. The volume can be calculated for a typical electrode by considering the recess to be the space between two cones, as shown in Figure 5A. Thus the volume of a cone of base radius 3 μ and height 50 μ (representing the ion-sensitive glass) can be subtracted from the volume of a cone of the same radius but height 75 μ. This gives a recess volume of 0.24×10^{-12} litres, or about 2000 times less than the volume of a 100 μ diameter sphere. It will only be with spheres of diameter less than 35 μ that this recess volume will be more than 1% of the sphere volume, and it is certainly possible, if necessary, to make microelectrodes with recess volumes much smaller than that calculated.

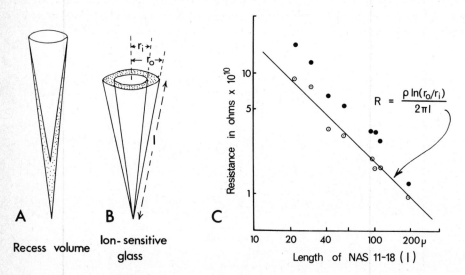

Fig. 5. A: Diagram showing the recess-volume as the space between two cones with the same base. B: Diagram of the tip of an ion-sensitive micropipette, with the ratio of the internal and external radii constant. C: Resistance of a batch of protruding-tip Na$^+$-sensitive microelectrodes plotted against length of NAS 11–18 exposed. The line is drawn according to the calculated resistances. Open circles: resistance measured in 0.1 M NaCl, filled circles: resistance measured in 0.1 M KCl.

The response times of recessed-tip electrodes are inevitably slower than for protruding-tip or liquid ion-exchanger (*Walker* 1971) microelectrodes. The ions must diffuse into the recess and equilibrate with the recess contents before a complete response can occur, so that half-response times are in the range 1–5 seconds.

Tip blockage only becomes a problem at the end of a long experiment, when presumably, intracellular material which has entered the tip may solidify on contact with the external medium. Since in use the recess is normally full of water, it otherwise rarely becomes blocked. It is important to avoid having an oil film on the saline surface, as this leads to oil blocking the initially dry recess. *Na^+-sensitive microelectrodes.* It is interesting to compare the theoretical resistance of Na^+-sensitive microelectrodes with that obtained in practice. Since the conduction of heat and electricity is analogous, the formulae used in heat conduction calculations can be used to calculate electrical resistances. The thermal resistance between the inside and outside of a hollow cylinder is given by *Kreith* (1965) as:

$$\text{Thermal resistance} = \frac{\ln(r_0/r_i)}{2\pi k l}$$

where r_0 and r_i are the external and internal radii, k is the thermal conductivity, and l is the length in cm. By analogy, the electrical resistance R will be the same. Substituting resistivity for conductivity, the equation becomes:

$$R = \frac{\rho \ln(r_0/r_i)}{2\pi l}$$

where ρ is the specific resistivity of the glass, in ohm/cm. Assuming that the ratio of the radii near the tip of a microelectrode is reasonably constant, as shown in Figure 5B, this equation can be applied to sealed-tip microelectrodes. Figure 5C shows that for a batch of NAS 11–18 microelectrodes with different exposed lengths there is good agreement between calculated and measured resistances of recently-filled electrodes. These resistances also agree well with published resistances of previous NAS 11–18 microelectrodes. For example; *Hinke* (1959)

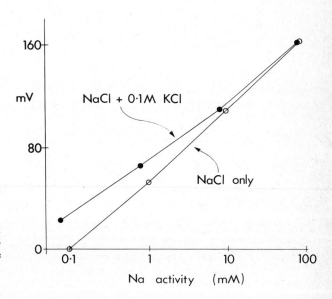

Fig. 6. Calibration curves of a recessed-tip Na^+-sensitive microelectrode measured in pure NaCl solutions and in NaCl + .1M KCl.

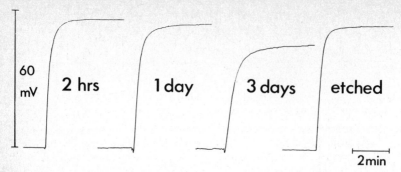

Fig. 7. Pen-recordings of the responses of a recessed-tip Na^+-sensitive microelectrode to ten-fold Na^+ activity changes at various times after filling. Each response is to a change from 10 mM NaCl, 100 mM KCl to 100 mM KCl, 10 mM KCl.

gives resistances of 10^{10}–10^{11} ohms with exposed lengths of about 150 μ, and *Kostyuk, Sorokina* and *Kholodova* (1969) resistances 10^{10}–10^{12} ohms with lengths of 10–20 μ.

The properties of NAS 11–18 as a Na^+-sensitive glass are well-established, and Figure 6 confirms that recessed-tip microelectrodes made from this glass give a near-perfect response to changes in Na activity, and are little affected by expected intracellular concentrations of K^+. As well as varying with the electrode dimensions, the response times and resistances of recessed-tip Na^+-sensitive electrodes change with time. Figure 7 shows that the responses of a microelectrode (exposed length: 60 μ, tip diameter ca. 1 μ, distance between tips 55 μ) to a ten-fold increase in Na^+ activity deteriorate with time; over 3 days both the size and the speed of the response clearly decrease. The electrode resistance was initially 7.2×10^{10} ohms, but after 3 days it had risen to 26×10^{10} ohms. Both response and resistance can be restored by etching the electrode for an hour or two in a solution containing 4 M NaOH and 0.1 M EDTA, tetrasodium salt, as shown in the last record of Figure 7. This treatment, preceded by a period in chromic acid if necessary, can restore an electrode many times, so that one electrode can last for many days.

It seems clear that for the best response, NAS 11–18 glass should not be hydrated. With time the glass surface becomes insensitive and develops a high electrical resistance. Removal of this surface with the etching treatment then restores the electrode's sensitivity, resistance and response time. *pH-sensitive microelectrodes.* In many ways these electrodes behave very differently from the Na^+-sensitive ones. Resistance decreases and sensitivity improves with time. The response time depends not only on the electrode dimensions, but also on the buffering power of the external medium. This last is illustrated in Figure 8, where the responses to a unity

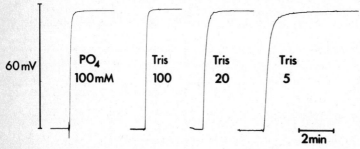

Fig. 8. Pen-recordings of the responses of a recessed-tip pH-sensitive microelectrode to unit pH changes (from pH 7.5 to pH 6.5) in four differently buffered solutions.

pH change using differently buffered solutions are shown. The electrode had an exposed length of 25 µ, tip diameter ca. 1 µ, and a distance between the tips of the insulating glass and pH glass of only 5 µ. Thus the responses are for an electrode with a minimal recess volume. The strange aspect of these responses is of course the dependance of response speed on buffering power. The higher the buffering power, (at near neutral pH) the faster the response.

For microelectrodes of similar dimensions, pH electrodes have lower resistances than Na^+-sensitive ones, in spite of the much higher specific resistivity of pH glass. Fresh, cold-filled pH electrodes *do* have the expected high resistance, coupled with a poor response to pH. Three weeks soaking at room temperature, or about two hours boiling, reduces the resistance and perfects the pH response, suggesting that hydration of the glass is essential. Once working, pH microelectrodes can be used for several months before losing sensitivity, and they can be stored and cleaned in chromic acid cleaning solution without harm.

Typical resistances of hydrated pH electrodes are in the range 10^{10}–10^{11} ohms. Although this is much lower than the resistance of about 10^{12} ohms calculated from the specific resistivity, it is still much higher than resistances quoted by other workers for their protruding-tip pH microelectrodes. For example *Kostyuk* et al. (1969), with exposed lengths of 10–20 µ, quote resistances of 0.5–5×10^8 ohms, and *Carter, Rector, Campion* and *Seldin* (1967) exposed lengths of 10–50 µ and resistances of around 10^9 ohms. Presumably these low resistance pH electrodes have, in effect, open tips.

Cl⁻-sensitive microelectrodes. These electrodes have relatively low resistances and respond very well to chloride changes in test solutions. In snail neurones, however, (*Neild* and *Thomas* 1974) they give erroneously high values for internal chloride, but there is some evidence that this problem may not occur in muscle.

Conclusion

Recessed-tip Na^+ and Cl^--sensitive microelectrodes are best when first used, but pH-microelectrodes improve with time. The disadvantages of the recessed-tip design are slow response time, the volume of the recess, and the possibility of tip blockage. The main advantage is that only the extreme tip of the electrode, typically 1 µ in diameter, needs to cross the cell membrane, so that the difficulties and uncertainties of penetration are minimized.

References

Carter, N. W., Rector, F. C., Campion, D. S., Seldin, D. W.: Measurement of Intracellular pH of Skeletal Muscle with pH-sensitive Glass Microelectrodes. J. Clin. Invest. 46 (1967) 920–933

Hinke, J. A. M.: Glass Microelectrodes for Measuring Intracellular Activities of Sodium and Potassium. Nature (London) 184 (1959) 1257–1258

Kostyuk, P. G., Sorokina, Z. A., Kholodova, Y. D.: Measurement of Activity of Hydrogen, Potassium and Sodium Ions in Striated Muscle Fibres and Nerve Cells, In: Glass Microelec-trodes, pp. 322–348, ed. by *M. Lavallée, O. F. Schanne* and *N. C. Hebert*, Wiley, New York 1969

Kreith, F.: Principles of Heat Transfer. Int. Textbook Co., Scranton, Pa. (1965).

Neild, T. O., Thomas, R. C.: New Design for a Chloride-Sensitive Microelectrode. J. Physiol. (London) 231 (1973) 7–8 P

Neild, T. O., Thomas, R. C.: Intracellular Chloride Activity and the Effects of Acetylcholine in Snail Neurones. J. Physiol. (London) 242 (1974) 453–470

Thomas, R. C.: New Design for Sodium-Sensitive Glass Microelectrode. J. Physiol. (London) 210 (1970) 82–83 P

Thomas, R. C.: Intracellular Sodium Activity and the Sodium Pump in Snail Neurones. J. Physiol. (London) 220 (1972) 55–71

Thomas, R. C.: Intracellular pH of Snail Neurones Measured with a New pH-Sensitive Glass Microelectrode. J. Physiol. (London) 238 (1974) 159–180

Walker, J. L.: Ion Specific Liquid Ion Exchanger Microelectrodes. Anal. Chem. 43 (1971) 89 A–93 A

Discussion of Session II

Chairman: *I. A. Silver*

(1 Figure)

Discussion of Paper by Saito, Baumgärtl, Lübbers: The RF Sputtering Technique as a Method for Manufacturing Needle-Shaped pH Microelectrodes*)

Hinke: Can you tell me how you insulate the tip when you were sputtering?

Lübbers: Under the microscope we put a photoresistor droplet on the very tip of the electrode. The trick is that this droplet has a tail. Then the whole electrode tip together with the droplet is sputtered. After the sputtering is finished the electrode is put in an ultrasonic bath. This procedure breaks off the tail. After dissolving the photoresistor the excess glass breaks off and the tip is free. This is a rather easy method without any micromanipulation.

Hinke: May I ask you two other very short questions? What in the durability of these special layers that you have used as an isolator?

Lübbers: We have not enough experience to answer this in general. Some electrodes work properly for 50 to 100 days and more. I think at the moment our technique is not so good that we get the best results.

Hinke: My final question is: Have you put them inside the cell?

Lübbers: No, we have not. We are just in the procedure of testing our method. We will use the electrode to measure in the perivascular space.

Gebert: Do you have an explanation for the short response time for metal-connected electrodes?

Lübbers: We think that the membrane thickness in front of these very small 1 μm-tip platinum electrodes is extremely slight. This may partly explain the short response time.

Gebert: Just a short comment to that explanation. We make metal-connected electrodes with a membrane about 200 μm-thick. We also get short response times with these electrode.

Lübbers: You are working with silver as reference electrode. It is known that especially silver atoms diffuse easily into the glass layer. Thus the silver atmos may be situated close to the H^+ exchange layer. But I cannot say if this really is a reason for the short response time.

---

*) and: *Lübbers, Baumgärtl, Saito:* Insulation Properties and Ion-sensitivity of thin Layers of Dielectrics Deposited by RF Sputtering in Glass Surfaces.

Discussion of Session II

Discussion of Paper by Walker: Ion-selective Liquid Ion-exchanger Microelectrodes

Eisenman: I wondered if you have ever tried coating with fixed positive charges such as a long chain hydrocarbon with quaternary ammoniums at both ends or using a hydroxyapatite-like glass composition.

Walker: No, I have not. I have tried exchanging quaternary ammonium compounds into the surface of the glass to dry and cover it up and also to get the alkalide cations, in this case primarily sodium, out of the glass, but the problem that you run into repeatedly is that working with a half micron tip you plug them up so easily with almost anything you do to them.
You just have to accept a compromise although Drs. *Lux* and *Hebert* have both apparently circumvented this problem.

Zeuthen: I have two points. The first is that we are using dichlorodimethylsilane vapour evaporated into one barrel of a double-barrelled electrode, which we then fill with ion exchanger. This gives a yield of 90% usable electrodes and you don't get any plugging of tips of the size of 0.3 μm. The second point is about the tip potential. There has always been the concern that you don't know whether the tip potential changes when you enter into tissue. We have made measurements of the resting level in the brain cortex of the rat and by choosing electrodes with a reference barrel the potential of which varied less than 2 mV in a range of isotonic solutions of sodium and potassium chloride. If we assume that tip potential does not change when you enter into the brain tissue, then we get the correct value for the potassium about 3.0 mM. Furthermore if we go on to measure in the cerebral ventricles we get the correct potassium value, which indicates that the tissue matrix as such does not influence the tip potential. I think that a sufficient criterion for selecting electrodes in terms of tip potentials is to select them so that they vary less than about 1 mV depending on the accuracy you are interested in.

Walker: But each electrode must be examined for tip potential otherwise you can find yourself in the same situation.

Zeuthen: Yes, but I am sure that even the worst of ours could have been used if you worked in a tissue of constant isotonicity.

Walker: In fact if you run them through constant ionic strength mixtures of sodium and potassium you still get a slope, because they do discriminate between sodium and potassium.

Hinke: A very short question: When you dope the surfaces beforehand, do you do both, and have you done just the inside and not the outside and is there any difference?

Walker: I have not done just the inside and not the outside because I always treat them by dipping the pipette into an organic solution of silane or the dimethylpolysiloxane.

Baucke: Everyone is concerned with coating the surface with silicones. We are using dimethylpolysiloxane dissolved in concentrated sulphuric acid. By that method you get a condensation between SiOH groups of the organic substrate and the glass with very good adhesion to the surface. After applying the solution, dissolve the excess with water. The procedure yields good hydrophobic monolayers.

Zeuthen: We have made these double-barrelled electrodes with one reference barrel and the other with ion exchanger. You can also siliconize them on the outside by putting them into vapor of dichlorodimethylsilane. This will give you the advantage that, if you want to go inside one cell only, then the cell membrane probably seals better to the electrode. On the other hand, if you want to push this electrode through a lot of tissue, siliconizing the outside may cause the tissue to stick to the electrode and you are then pushing the tissue when advancing the electrode.

Discussion of Paper by Silver: Multiparameter Microelectrodes

Silver: A lot of this work has been done by *Thomas Zeuthen*, by *Rachel Yerbury* and *Charles Drown* in my laboratory.

Walker: I just want to make a brief comment on relative humidity. We find with 'Siliclad' that when the relative humidity goes down we have to increase the concentration and when the relative humidity goes up we decrease it, but it is a problem.

Silver: In Philadelphia just recently we have had a problem with humidity, in fact I think about 70–80% of the tips were either not covered or they clogged.

Khuri: Whenever we deal with tips of the double-barrelled electrodes which are more than 2 μm in diameter, the micro-electrode is susceptible to interference between the reference and the exchanger barrels. Only when we are dealing with total tip diameters of below 1 μm, do we obtain satisfactory separation. I am really concerned about multi-barrelled electrodes where one may have a large cross sectional area.

Silver: These electrodes have a maximum diameter of 0.6 μm.

Lehmenkühler: Have you ever seen artefacts from the density of the oxygen reduction current on the potential of the ion-sensitive electrodes?

Silver: Yes, I have, but they are absolutely unpredictable. Sometimes it interferes, and sometimes not, but I do not know why this happens. One thing which I think may happen is that there is a leakage of this metal outside the electrode with smearing across the surface. This is something that we have only been playing with recently. It works, but we certainly have not got all the bugs out of the system yet.

Lehmenkühler: This effect might be dependent on the oxygen reduction current and it might falsify the measurement of ion activity.

Silver: The other thing which you may measure, which was of course mentioned, is that sometimes the insulating glass acts as an ion-selective membrane especially to calcium and sodium.

Lehmenkühler: If we measured with an electrode about 10 μm approximately from the tip of a platinum oxygen electrode we found no interference. However, if we bring the O_2-electrode nearer to the tip than 10 μm we have seen falsifying effects.

Silver: The size of the metal spot is somewhat less than 0.1 μm in diameter which is rather smaller than most oxygen electrodes. Our major problem is an electronic one.

Lehmenkühler: Do you have a calibration curve for this electrode?

Discussion of Session II

Silver: It is like that of other microelectrodes. Sometimes it is good and sometimes it is extremely bad. It is useful only as an indicator, and I do not think one can possibly use it for an absolute measurement.

General Discussion

Kessler: I would like to comment on the presentation by Dr. *Walker*. We use the Lux-type double-barrelled and multi-barrel electrode and we did not see this problem you described. I wonder what is the reason for that. There is one point which seems to be important. I think we should not speak of the tip Diameter of the electrode but rather of the pore diameter of the electrode. I would like to show a slide which was made last year. Figure 1 shows a scanning electron micrograph of a Thomas-type pH electrode and you can see that the size of the pore is much smaller than the outer diameter of the tip. In general microelectrodes can become ion-sensitive, e. g., pH-sensitive. I wonder what the reason could be for the potassium sensitivity caused in your experiments.

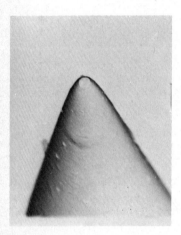

Fig. 1.

Walker: You are worried about the tip potential on the electrodes filled with 3 molar KCl. I think that generally what happens in a pipette of that sort is that there is some sort of plug in it, e.g., a salt crystal or a piece of dust or something of that sort. *Walter Woodberry* observed some years ago, that if you reduce the pH of the filling solution to about 3 this obviates the tip potential problem.

Lux: Cleaning the capillaries well is of some importance. We adapted a method of Křίž et al. (1972) who used a mixture of 30% H_2O_2 acid and concentrated sulphuric acid. The glass was left for 2 hours in the mixture and then cleaned in the usual way with water and acetone before pulling. After pulling and after breaking the tips down to 1 to 3 μm apparent diameter to reduce tip potentials, the Theta capillaries are first filled with the standard solutions 0.1 M NaCl and 0.1 KCl. During siliconization we keep the later reference barrel under pressure and like to observe a steady flow out of that barrel. Simultaneously, by suction a solution of 1% dimethyldichlorosilane in chloroform is brought into the tip of the other, later K-sensitive barrel. This solution is pressed out then and the same treatment done with trimethylchlorosilane which is more hydrophoric. This is done under the microscope and is of advantage for preventing clogging of the electrode tips.

Zeuthen: There are also old reports that if you have thorium ions in your electrode this reduces the tip potential.

Discussion of Paper by Khuri: Microelectrodes Utilizing Glass and Liquid Ion-exchanger Sensors

Zeuthen: What do you use as a bicarbonate ion exchanger?

Khuri: The bicarbonate-selective liquid ion exchanger consists of a 3:1:6 mixture of tri-n-octylpropylammonium chloride: octanol solvent.

Baum: There was a paper by *Rechnitz* in Science during this year, which described the chemistry of the bicarbonate ion exchange system and the performance characteristics in a dip-type microelectrode. The exchanger was dissolved, I think, in decanol.

Simon: I think the solvent cannot be decanol because that would not prefer bicarbonate, it must be a fluorinated hydrocarbon.

Khuri: Dr. *Simon* is correct; the solvent is trifluoroacetyl-p-butyl benzene.

Discussion of Paper by Kessler, Hajek, Simon: Four-barrelled Micro-electrode for the Measurement of Sodium, Potassium and Calcium Activity

Zeuthen: I want to make two points, and the first is that maybe at a workshop like this we could agree to a nomenclature for the logarithm of potassium activity or sodium activity, because I think the term of pK could be misleading in some context where you are dealing with ionic strengths as well. The second point is on the principal difference between measurements on surface of tissues and in the tissue with microelectrodes. There is some evidence at least in the brain that the extracellular activities deduced from surface measurements are not the same as those you deduce from interstitial measurement with a microelectrode during a dynamic change in tissue. It is a point I will return to on Wednesday, because I have compared some measurements with a microelectrode in the extracellular space to measurements from the surface either in the ventricle or in the subarachnoid space during an anoxic incidence. I wonder if you have some direct comparisons between surface measurements and tissue measurements with your electrodes.

Kessler: The ion-selective electrodes monitor ion activity. When you look at the literature you will find that ion activity is given in pX values as well as in m-values. The original meaning of pH was the negative logarithm of hydrogen-ion concentration. This definition, which was introduced by *Sörensen*, has changed in the meantime, since now pH is usually measured with glass electrodes. When we apply one of the two terms in our measurements, we should not forget to mention that their meaning is "activity". A special definition that may replace the mole value, e.g., Nernst or milli-Nernst (mN) could be advantageous.
Coming to your second point, I would like to repeat my opinion that, for theoretical reasons, it is impossible to monitor the true interstitial ion activity with needle electrodes. As compared to the diameter of the interstitial space, the tip of the ion electrode is too big. It

would, therefore, completely destroy the interstitial environment when, by chance, it entered this small space between plasma membrane and capillary wall containing polyelectrolyte gel.

Silver: This is a major point and varies with tissues. In brain the electrode tip seems to make an artificial extracellular space which rapidly equilibrates with the E.C.F. It is quite difficult to enter a capillary lumen in the liver.

Brown: What is the response time of the various ion-exchanger microelectrodes in your four-barrelled assembly?

Kessler: We did not try to get so fast electrodes as Dr. *Lux* uses, because, at present, we are mainly interested in relatively slow changes in ion activities. Our microelectrodes have response times of about two to five seconds. With the surface electrode we find response times between three and ten seconds (95%).

Gebert: I cannot understand a special feature of your design. You have a single unit and then you put two big reference electrodes in the outflow and in the inflow and connect them altogether to the ground. I think the resistance of the two reference electrodes in the circuit, i.e., in the flow-through tubing, might be in the range of kOhms and the other one in the range of MOhms and therefore you do not measure any reference potential with your reference electrode of the single unit. You should disconnect it and measure its potentials separately. Then you would get the real D.C. value, but with your kind of design you do not need the reference in the single unit.

Kessler: Normally we use symmetric amplifiers with an input impedance of 10^{13} ohms. In some experiments we wanted to test if this input impedance was high enough and therefore used electrometer amplifiers. At first we had some problems, because the reference channel was connected to the case of the amplifier. Therefore, we disconnected the case of the electrometer and grounded the reference electrode by placing a resistor of 10^{10} ohms between the reference electrode of the double-barrelled sensor (the impedance of the reference electrode was 10^7 ohms) and the other two grounding electrodes. This was a sort of compromise which caused electrical noise. However, recordings made in this way showed that, in our measurements, the impedance of the ion-selective electrodes did not increase after they were inserted into tissue.

Lehmenkühler: I have some comments on this problem. From many organs, such as brain and heart, it is known that DC potential variations occur at different tissue boundaries. I suggest there are such potentials between the surface of the liver and the extra-cellular space. The problem is that you have the reference micropipette connected with electrodes in a blood vessel and a potential may exist over the vessel wall. This potential is a component of the signal attributed to K activity. The true K potential therefore is different.

Kessler: Usually, we measure the potentials between the four-barrelled microelectrodes and the reference electrodes used for grounding the perfused liver preparation. During normal perfusion these potentials are in the range of 1 to 8 mV. After perfusion stop they show strong changes. When we used the electrometer in the way described, the preparation was maintained in such a condition that the DC potentials generated by microcirculation and by the tissue did not change.

Lehmenkühler: Have you measured any potential difference between the internal side of the vessel and the liver?

Kessler: We have not measured these potentials as yet.

Discussion of Paper by Thomas: Construction and Properties of Recessed-tip Micro-Electrodes for Sodium and Chloride Ions and pH

Hinke: Did the response time you spoke of refer to standard aqueous solutions. Also, did you notice after you had made an impalement and came out whether you had altered your response time?

Thomas: Firstly, with the sodium-sensitive electrode all the responses have been measured in what were basically Ringer solutions modified to make them suitable for testing, so they were all about 100 mM sodium chloride, potassium chloride or whatever. Secondly, with sodium-sensitive electrodes certainly, and to some extent with pH electrodes, during a long experiment I imagine that some protein gets inside the recess. When you pull the electrode out into what is a fairly high calcium medium, you probably get some precipitation inside the tip, particularly with the sodium-sensitive electrodes; it is more of a nuisance with them anyhow. The tip does get partially blocked, rarely completely blocked, so that the response time is doubled or trebled after an experiment of say 3 or 4 hours. It can be a nuisance, but of course the pH electrodes can be easily put in chromic acid which cleans them quite rapidly. If you put sodium-sensitive electrodes in chromic acid you have to give them an etching treatment afterwards to revive them. So, yes, the tips do block after a long experiment, but not in "in and out" use.

Eisenman: I have two comments, one that might be pertinent to your comment on the higher resistance in potassium than sodium and a related one on what looks like a very nice calcium-free cleaning solution. In glass, potassium mobility is generally considerably lower than sodium mobility, so your observations might suggest that you are actually getting enough of K^+ and Na^+ superficial ion exchange to contribute to the overall resistance of your electrode. What intrigues me about your cleaning solution is that I've found by tracer studies on hydrated NAS_{27-4} glass that Calcium shuts down Na^{24} efflux. If calcium could raise the resistance of your thin glass films, this might account for why your Ca^{++}-free EDTA cleaning solution helps.

Thomas: It is certainly possible.

Eisenman: How did you happen to hit on the EDTA?

Thomas: Well, that was not so much to chelate calcium. I started off with sodium hydroxide as a cleaning agent because I read somewhere that it etched glass. I tried that with a sodium-sensitive electrode and it worked the first time but the next time it really did not work too well. The electrode remained noisy; so I figured that there might be heavy metal ions somewhere. I had EDTA – so I put some of that in, and it made all the difference!

Kessler: We have used your design for pH electrodes since last year. The problem we have in intracellular measurements with electrodes having a tip diameter of 1 μm is to pick up real intracellular pH values in mammalian cells. However, in extracellular measurements the electrode works very well and I think it is a most beautiful design.

Discussion of Session II

Thomas: Thank you. I feel it must be possible to make them a bit sharper, but I have not myself tried because I do not need to. I feel sure that by changing the microelectrode puller I could perhaps make them sharper, but of course it would increase the response time.

General Discussion

Zeuthen: How large is the volume you have underneath your surface electrode? If you have to press them onto the tissue, you might cause trauma, and if the volume is too large you will see changes different from those really going on in the extracellular space. For example, during anoxic depolarization in brain cortex, samples taken from the subarachnoid fluid give you a potassium change which corresponds to the interstitially measured values, but the samples do not reflect the sodium changes during this depolarization. That could be explained if, in a depolarized brain, the potassium diffuses easily *through* the cell or *between* the cells, but sodium still has to diffuse *between* the cells only. Therefore, for slowly diffusing ions the surface electrode would measure slower changes than those actually happening within the tissue.

Kessler: When we use ion-selective surface electrodes for tissue measurements, the layer of liquid between the electrode and the tissue must be small enough to enable precise monitoring of ion activity. The thickness of this layer depends upon the weight of the electrode. When the surface electrodes are used which we have developed, it is 1 to 3 µm. Concerning the use of tissue surface electrodes we have some experience, since we have carried out hundreds of thousands of measurements with Po_2 surface electrodes. In these measurements the limiting factor is the response time of the electrodes and not the thickness of the liquid layer between the tissue and the surface of the electrode. The tremendous advantage of surface electrodes is that the trauma which they cause is minimal as compared to that effected by needle electrodes (see paper by *Olfers-Weber* et. al. pp. 205 and *Schäfer* et. al. pp. 217). Concerning your second question, namely, extracellular ionic alterations in the brain tissue during cycles of anoxia, I would like to mention that we didn't find significant changes in potassium and sodium activity in the brain as long as low-flow or no-flow anoxia was avoided. However, when there was a brain edema, characteristic alterations of ion activity were found. Lassen in Copenhagen observed similar results during anoxia (personal communication).

As to your third question, I would like to draw your attention to the fact that the diffusion times of the cations we are talking about are fairly short. Thus, for the response of the surface electrodes the diffusion distance ist much more important than are the differences between the diffusion coefficients (see Tab. of *R. Strehlau*). Nevertheless, we cannot completely neglect the different diffusion coefficients as they can cause diffusion potentials. We all know these diffusion potentials from our reference electrodes.

Table 1. Diffusion time of sodium and potassium in water $\left(Na^+: 1.334 \cdot 10^{-5} \frac{cm^2}{sec}, K^+: 1.957 \cdot 10^{-5} \frac{cm^2}{sec}\right)$ time in seconds for 90% of the final value (*Strehlau*).

Distance	K^+	Na^+
3 µ	0.15	0.21
5 µ	0.40	0.59
10 µ	1.62	2.37

Zeuthen: Oxygen is probably a molecule you can measure satisfactorily on the surface, because it is easily diffusible, but sodium may not be an easily diffusible ion in tissue.

Kessler: The net diffusion time of oxygen and that of monovalent cations, such as potassium and sodium, does not differ very much, and in fact, when we induce simultaneous slug injections of oxygen and cations, the response times as measured with different microeletrodes as well as with different surface electrodes do not show great differences.

Zeuthen: In a steady-state situation, ECF and CSF in the brain are usually considered to be about the same, but that does necessarily apply for a dynamic situation. For example, in the ventricle we recorded during anoxia a decrease in potassium before the brain depolarized, whereas with microelectrodes inserted in the cortex you get increase in potassium.

Kessler: In the brain which has such a complex structure and function, it may be dangerous to draw general conclusions from one local measurement. The only possibility to analyse such a multicomponent system is to perform multiple measurements at the different points you are interested in.

Session III

Enzyme Electrodes

Chairman: *L. C. Clark Jr.*

L. C. Clark and C. R. Emory

The Measurement of Cholesterol and its Esters Using a Polarographic Anodic Enzyme Electrode

(11 Figures)

Introduction

There is a rapid expansion of knowledge in the area of enzyme-coupled electrochemical sensors, an increasing knowledge of means of stabilizing enzymes and an increase in the use of enzymes for analytical purposes. Most enzyme electrochemical sensors are based either on polarographic principles where current flow is measured or potentiometric principles where voltage is measured.

The first report of an enzyme-coupled electrode was in 1962 (*Clark* and *Lyons* 1962). I have published a number of papers where oxygen oxidoreductases (oxidases) have been used with electrodes (*Clark* 1970, 1971, 1973a, b; *Kessler* et al. 1973). There are several reviews relating enzymes to electrodes (*Christian* 1971; *Durst* 1971; *Guilbault* 1970). The Corning Bibliography (1972) is a valuable compilation of data on immobilized enzymes.

In this paper I will report on the use of two enzymes in combination with a platinum anode sensitive to hydrogen peroxide. Oxygen measurements are not involved. However, oxygen is required by the enzymatic reaction which converts cholesterol to cholestenone.

Cholesterol and its esters have been of interest for many decades (*Bloor* 1943). The *Liebermann-Burchard* and the *Zak* reactions are common colorimetric analytical methods. Most involve a saponification step, a step which has long been suspected of causing a loss of cholesterol. The enzymatic methods do not involve saponification. This and other reasons why enzymatic methods are superior will be reported in this article.

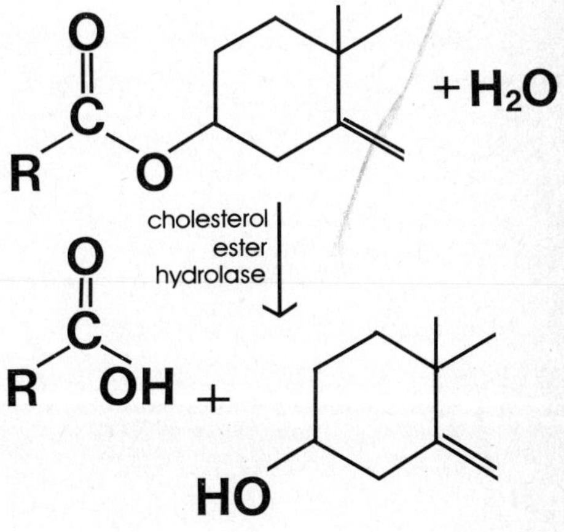

Fig. 1.

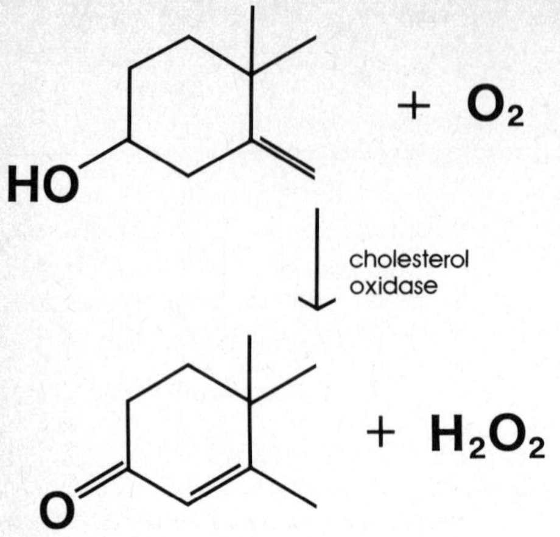

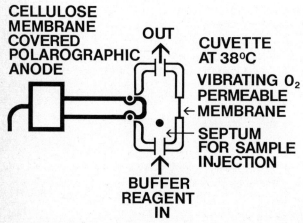

+ O_2

↓ cholesterol oxidase

+ H_2O_2

Fig. 2.

It is the purpose of this article to report preliminary findings which support our belief that a cholesterol electrode based on the principles developed here will make possible new rapid means of measuring cholesterol in blood, tissue and other substances.

The chemical reactions involved are shown in Figures 1 and 2.

The publications of *Richmond* (*Buckland* and *Richmond* 1973) and *Tarbutton* and *Gunter* (1974) stimulated the research reported here. Other pertinent references are those of *Allain* et al. (1974), *Klein* et al. (1974), and *Flegg* (1973).

Materials and methods

Polarograph. Most measurements were made in a modified Model 23 Yellow Springs Instrument Company glucose analyzer. The instrument we used here had a cuvette volume of 0.40 ml and was thermostated at 38 °C. Samples of from 5 to 100 µl were injected. A cellophane-covered ($1^1/_8$ in. dialysis tubing, A. H. Thomas Co., Cat. No. 4465-A2) platinum polarographic anode was mounted as shown in Figure 3 with an adjacent silver reference electrode. The anode had a diameter 0.5 mm and the reference silver cathode had an area of about 0.5 sq. cm. Full scale output is of the order of magnitude of nanoamperes. A G2500

Fig. 3. Diagram of cuvette and electrodes.

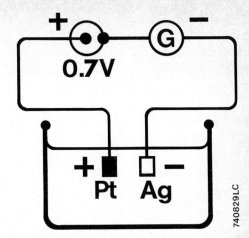

Fig. 4. Schematic diagram of the electrode system.

Varian strip chart recorder was used. The polarographic principle involved is shown in Figure 4; the platinum is attached to the positive pole of the battery using an applied voltage of 0.70.

Cuvette reagent. The composition of the cuvette reagent is shown in Table I. If the pH is not 6.6, adjustment is made using the two phosphate salts. Although the azide acts as a preservative, the buffer should be made fresh weekly and stored at 4 °C.

Cholesterol oxidase (EC 1.1.3.6). This enzyme, derived from *Nocardia*, was obtained from Abbott Scientific Products, Miles Research Products and Boehringer Mannheim Corporation. The enzyme was also obtained from the Beckman Instrument Co.; their preparation is derived from *Brevi bacterium* sp. The enzyme was reconstituted by dissolving 9.4 U of enzyme in 1 ml of 1% Triton X-100. In most of the research reported here the Miles preparation was used. Enzyme activity is reported as International Units (U).

Cholesterol ester hydrolase (EC 3.1.1.13). This enzyme was obtained from each of the above companies. It was reconstituted by dissolving 9.4 U in 1 ml of 1% Triton X-100. All of the work reported here was performed with enzyme prepared from bovine pancreas by Miles Laboratory.

Cholesterol standards. Cholesterol was purified by crystallization from acetone and dried over silica gel. It was emulsified in the surfactant solution using a Branson Model J17A sonicator. The preferred cholesterol emulsion standard at present is 200 mg% in 1% sodium stearate with the pH adjusted to 7.1. Standards of the American Society of Clinical Pathology cannot be used. The standards used for most of this work were human plasma (LCC) having a mean total cholesterol content of 232 mg% and dog plasma having a total cholesterol content of 154 mg%.

Cholesterol esters were emulsified by sonication in 1% Triton X-100 and were the equivalent of 200 mg% free cholesterol. The esters were used as received from Eastman Kodak, Aldrich, or Sigma.

Other. The 0.01% hydrogen peroxide was made by diluting a 30% solution with water. It was stored in a glass bottle in the dark. The alcohols were reagent grade and were not purified further. Sodium taurocholate was obtained from ICN-K&K Labs, New York. Human albumin was of clinical grade and was obtained from Parke, Davis & Company, Detroit. Soy lecithin was purchased from Nutritional Biochemical Inc.

Surfactants. Pluronic F68 (Wyandotte), Triton WR 1339 and Triton X-100 (Rohm & Haas), Surfal (P & L Biochemical, Inc.), sodium stearate (Fisher Scientific Co.), Monflor 51 and 71 (Imperial Chemical, Industries, Ltd.) Zonyl FSK (duPont), and polyoxyethylated octyl phosphate (Chem. Service) were used as received.

Plasma and sample preparation. Blood was collected by venipuncture in heparin-containing tubes, centrifuged and the plasma harvested and stored at 4 °C. Blood from small animals was collected by orbital bleeding or heart puncture. Solid samples, such as brain, were homogenized in 1% Pluronic F68 using an all-glass rotating pestle and tube.

Sample analysis. The cuvette was arranged so that it was readily rinsed between analyses with 0.5 M phosphate buffer at 38 °C and pH 6.6. It was then drained and filled with cuvette reagent having the composition shown in Table 1. Then 50 µl of plasma were injected using a microliter syringe followed by 100 µl (0.94 U) of cholesterol oxidase. After a stable current reading was recorded, usually in about one minute, 20 µl (0.2 U) of cholesterol ester hydrolase were injected and, after a second stable reading was obtained, also in about 1 minute, the cuvette was rinsed with phosphate buffer and the instrument was ready for a new test. Standards were run periodically. The procedure was varied as indicated in Results.

Animals. Beagles and other animals were treated in accordance with the Animal Welfare Act of 1970 and the guideline established by the U.S. Department of Health, Education and Welfare as set forth in publication (NIH) 73–23 revised 1972. They were anesthetized with intravenous sodium pentobarbital (Nembutal) at a dose of 20 mg/kg of body weight.

Results

Methanol, n-butanol, n-pentanol, ethanol, iso-butanol, 2-propanol, n-propanol, tert-butanol, and octanol at a concentration of 10.3 molar (the concentration of a 400 mg/dl suspension of cholesterol) caused virtually no polarographic current alone or in combination with cholesterol oxidase. Larger concentrations of these alcohols caused current flow as shown in Table 2.

Cholesterol, isoandrosterone, 5-pregnen-3-ol-20-one, dehydroisoandrosterone, pregnenolone, isodehydroandrosterone acetate gave a distinct color, desoxycorticosterone acetate, $\Delta 5$-androstenediol-3 (β), 17 (α) a slight color, and dihydrotestosterone, pregnandiol, androsterone, estrone cholestanone-3-enol acetate, androstanediol-3 (α), 17 (α), $\Delta 4$-androstene-3, 17 dione, etiocholanol-3 (α) dione 11, 17 acetate, methyl testosterone, dehydro-cis-testosterone no color when incubated at room temperature overnight with *Tarbutton-Gunter* reagent.

Table 1. Cuvette Reagent. The pH of the reagent is 6.6. Adjustment can be made using the two phosphate salts. It should be stored at 4 °C or made in batches to last a week or so.

Substance	Concentration gm/100 ml	Function
KH_2PO_4 (0.5 M)	13.0	Buffer
K_2HPO_4 (0.5 M)	8.7	Buffer
Sodium azide	0.07	Inhibit catalase
Sodium taurocholate	0.30	Hydrolase co-factor
Triton X-100	0.05	"Solubilize" cholesterol
Potassium chloride	1.0	Stabilize Ag Cathode

Table 2. Polarographic current from various alcohols 10 µl samples were injected in the cuvette which contained 0.5 M phosphate buffer at pH 6.6 at 38 °C.

Alcohol	Peak Meter Reading	Time to Peak
		minutes
methanol	90	13.2
ethanol	31	0.5
iso-propanol	2	3.7
n-propanol	28	1.0
n-butanol	15	1.5
iso-butanol	12	1.1
tert-butanol	0	0.0
pentanol	8	1.2
cyclohexanol	25	7.0
1-octanol	0	0.0

It was found that sodium azide, used as a buffer preservative and a catalase inhibitor, is also effective in stabilizing hydrogen peroxide injected in the cuvette. The current found after repeated injections of 35 µl of 0.01 % hydrogen peroxide is shown in Figure 5 together with the stabilizing effect of azide.

It was not possible to inhibit the catalase-type of activity sufficiently to analyze whole blood. In fact the concentration of azide we selected (see Table 1) is sufficient to allow only the use of plasma having the color of pale Rosé wine.

Increasing concentrations of enzyme were injected into the cuvette containing cuvette reagent and 50 µl of dog plasma with the results shown in Figure 6. From this we concluded that about 1 unit of activity was sufficient for a fast response and an adequate current. A similar procedure was used to determine the optimum cholesterol ester hydrolase activity as shown in Table 3.

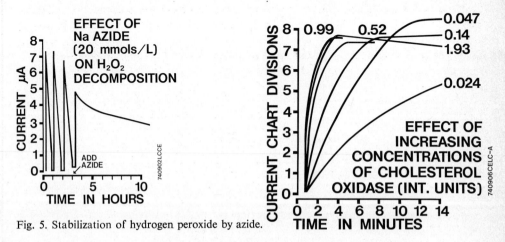

Fig. 5. Stabilization of hydrogen peroxide by azide.

Fig. 6. Response time of the anode to various concentrations of the oxidase enzyme.

Table 3. The effect of the concentration of cholesterol ester hydrolase on electrode response to dog plasma samples.

Volume of CEH	International Units of CEH	Peak meter Reading	Time to Peak Reading
μL			minutes
1	0.0094	315	11.0
5	0.047	240	3.0
10	0.094	385	1.5
20	0.188	343	1.5
50	0.47	370	1.0
100	0.94	343	0.9

Having found the optimum activity for the two enzymes, we studied the effect of temperature as indicated in Figure 7. Although it is apparent that the enzymes increased in activity, or more precisely a greater current was obtained as the temperature was increased to 57 °C, we selected 38 °C for routine work until more can be learned about all the effects of the high temperatures.

Surprisingly, changing the pH had but little effect on the peak current for both enzymes. Furthermore, the reading of the ratio of free to total cholesterol did not change. (Table 4). In order to determine if cholesterol suspensions could be used as standards we compared them with plasma. As shown in Figure 8, the curve for cholesterol in 1% Triton X-100 is almost identical to that for plasma following the injection of the oxidase. Cholesterol suspended in 5% Pluronic F68 shows a much flatter curve, due presumably to the state of aggregation of the cholesterol. When plasma was added, a typical curve for plasma was obtained thus showing that the enzyme was behaving normally in the presence of the Pluronic.

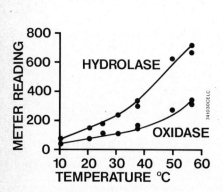
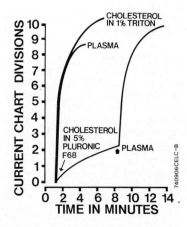

Fig. 7. The effect of temperature on current output of the anode. The current increases due both to increased enzyme activity and faster oxidation of the peroxide by the anode.

Fig. 8. Comparative response curves to cholesterol in Triton X-100, Pluronic F68, and plasma.

Table 4. Effect of pH on electrode current in the presence of dog plasma and cholesterol oxidase and added cholesterol oxidase plus cholesterol ester hydrolase.
The pH of the cuvette reagent was adjusted with phosphoric acid or a concentrated potassium hydroxide solution. After filling the cuvette, 50 µl of dog plasma was injected followed by 100 µL (0.94 U) of cholesterol oxidase. After reaching a peak in current, 20 µl (0.188 U) of cholesterol ester hydrolase was injected and the current was recorded after it had peaked and stabilized.

pH	Peak Meter Reading		Ratio Free/Total	
	Oxidase	Hydrolase		
3.9	154	322	1	2.1
5.0	153	346	1	2.3
6.1	172	336	1	1.95
6.8	153	336	1	2.2
8.1	174	360	1	2.1
8.9	182	375	1	2.1
10.7	193	403		

We are studying various detergents to find one that can be used to make cholesterol emulsions which resemble plasma and some of the first results are shown in Table 5. Sodium stearate emulsion (pH 7.1) appears to be the closest to plasma of those tested so far.

Table 5. The Effect of Surfactants on Solubilizing Cholesterol as a substrate for Cholesterol Oxidase. 50 µl of 200 mg % cholesterol suspensions in 1% surfactant in water were injected into the cuvette (0.40 ml volume) containing our standard cuvette reagent and 0.94 U of cholesterol oxidase (Abbott) 50 ml batches of emulsion were sonicated for 2 to 5 minutes.

Surfactant	Type	Time to Peak	Peak Reading
		minutes	meter reading
Synthetic *Hydrocarbon*			
Pluronic F 68	non-ionic	3.5	75
Triton X-100	non-ionic	6.0	116
Polyoxyethylated-Octyl phosphate	cationic	4.5	149
Hydroxyethyl-starch	non-ionic	3.0	120
Triton WR 1339	non-ionic	12.0	145
Surfal	non-ionic	7.0*)	169
Na stearate	anionic	0.8	283
Synthetic *Fluorocarbon*			
Monflor 51	non-ionic	3.0	82
Monflor 71	cationic	4.0	178
Zonyl FSK	anionic	9.0	132
Other			
Human albumin	polyionic	11.0	144
Na taurocholate	anionic	3.5	144
Refined soy lecithin	non-ionic	12.0	138
Dog plasma	polyionic	0.7	268

*) 5% Surfal = 1 minute, peak 405.

Table 6. Hydrolysis of Cholesterol Esters. 50 μl of cholesterol ester emulsions equivalent to 200 mg % cholesterol were injected. pH 6.7. Temp 38 °C. Both oxidase and hydrolase were present.

Cholesterol ester emulsion	Peak Reading	Time to peak
	division	minutes
Palmitate	81	7.0
hydrogen succinate	209	18.0
oleyl carbonate	256	9.0
pelargonate	172	25.5
2-(2-butoxyethoxy)ethyl carbonate	443	5.5
linoleate	368	12.5
hydrogen phthalate	51	46.0

In order to determine if cholesterol ester suspensions can be used as standards, a number of esters are being examined with the result shown in Table 6. The 2-(2-butoxyethoxy)ethyl carbonate shows the most promise so far since it forms a stable emulsion and yields a high current with a prompt response.

Figure 9 shows the kind of curves one gets when the oxidase is added and this is followed by the hydrolase. Full scale response for both the free and esterified cholesterol is obtained in less than four minutes. Analysis of blood of other animals is shown in Table 7.

The linearity of the response is given in Figure 10, Analysis of a variety of liquids is given in Table 8. Where data are available it is more or less in agreement with the literature.

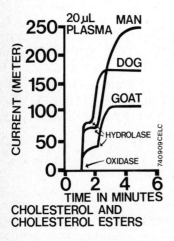
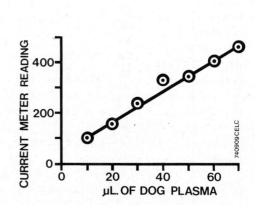

Fig. 9. Anodic response curves for free and esterfied cholesterol in various plasmas.

Fig. 10. Relationship between anodic current and sample size. Both enzymes were present at the same time, giving total cholesterol.

Table 7. Analysis of Animal Plasmas for Cholesterol and Cholesterol Esters.

Animal Plasma	Free Cholesterol		Cholesterol Esters	
	Peak Reading	Minutes to Peak	Peak Reading	Minutes to Peak
Mouse	112	0.5	320	2.0
Rat	55	0.5	288	2.0
Gerbil	99	0.8	314	3.0
Rabbit	90	0.8	289	1.3

Table 8. Analysis for Cholesterol and Cholesterol Esters in a Variety of Liquids.

Sample	Peak meter Readings	
	Free Cholesterol	Cholesterol Esters
Egg yolk ¼ with H_2O	643	0
Mouse brain homogenate 1/6	454	474
Whipping cream	340	0
3.5% butterfat milk	102	141
Non-dairy creamer	50	94
Human semen	215	0
Lobster tail homogenate 1/6	121	0
Shrimp homogenate 1/8	111	0
Juice from shrimp	122	0
Oyster homogenate 1/3	89	0

Discussion

The new polarographic enzyme electrode system may make possible the rapid determination of free and esterified cholesterol. Our present report describes the progress made to date and is designed to emphasize the future possibilities of enzymes for the determination of these steroids rather than to be a detailed report of the best method of procedure.

We found that both the oxidase and hydrolase react with cholesterol and its esters in plasma and serum. These steroids are not bound in lipid micelles so as to be unavailable as substrates even though they are difficult to extract with some cholesterol solvents, such as ligroin.

Our method depends upon the consecutive or simultaneous reaction of the oxidase and the esterase to give free and esterified cholesterol or total cholesterol.

Our method is based on a hydrogen peroxide measuring platinum polarographic anode covered with a cellophane-like membrane. No membrane has yet been found which will allow cholesterol to pass through without allowing the enzymes to escape or catalase to penetrate.

Our cholesterol electrode cannot be used in whole blood because the very high catalase activity of whole blood destroys the hydrogen peroxide as rapidly as it is formed by the oxidase. The catalase activity of slightly hemolyzed plasma can, however, be overcome by the use of azide. We have found that azide also inhibits the spontaneous decomposition of hydrogen peroxide. By virtue of its bacteriocidal properties it also preserves the buffer. In the concentration used here it does not contribute to the polarographic background current. Prohibitively large concentrations of azide would be required to inhibit the catalase activity of whole blood, if in fact it can be done.

The ionic composition and pH of the cuvette buffer reported here is based largely on the optimum conditions for the enzymes reported in the literature, largely that of *Tarbutton* and *Gunter* (1974). We have used taurocholate although the less expensive cholate will serve as well as a cofactor. The azide concentration is based upon that required to counteract the catalase like activity present in plasma obtained from slightly hemolyzed blood. Surfactants other than Triton X-100 will be needed since we understand that its manufacture will soon be discontinued. Chloride is added to the buffer to stabilize the silver reference electrode. Turbidity of the cuvette reagent, due to a salting out of the surfactants, sometimes occurs and can be prevented by dilution.

In this enzymatic procedure for the measurement of cholesterol the physical state of these particles affects both the rate and magnitude of the peak current. Therefore plasma or serum at present is preferred as the standard although emulsions offer promise. Morpholine hemisuccinate of cholesterol which forms a clear aqueous solution has been proposed by *Klein* et al. (1974) as a standard.

In the pilot experiments reported here we did not make corrections for volume changes in the cuvette caused by enzyme or sample injections. Such corrections can be minimized by the proper sequence of injection, by the use of concentrated enzyme preparations, or by adding enzyme or enzymes to the cuvette reagent itself.

There is a dearth of information regarding normative data for free and esterified cholesterol in plasma.

Cholesterol oxidase is a stable robust enzyme but the dry hydrolase tends to lose activity on exposure to moist air. There may be different kinds of these enzymes.

At present the price of enzymes is such that we estimate the cost for both tests to be about $ 3.50. The oxidase is at least ten times the cost of the hydrolase. Costs can be reduced, obviously, by reducing the cuvette volume or increasing the reaction time and temperature.

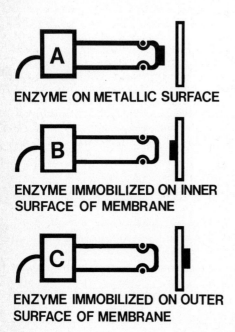

Fig. 11. Various means of entrapping enzymes on the anode and its covering membrane.

Most enzyme electrode work is done with the enzyme between the membrane and the electrode surface but here because there is no practical membrane permeable to cholesterol the enzyme must be in the cuvette reagent or mounted on the outside of the membrane as shown at the bottom of Figure 11. In that configuration the cholesterol is acted on by the enzyme and the water soluble hydrogen peroxide diffuses through the membrane to the anode. Peroxide also diffuses into the cuvette reagent and will accumulate causing an upward drift of the electrode current unless destroyed by catalase which is possibly immobilized in the cuvette.

Summary

The cholesterol electrode reported here is based upon the polarographic measurement of hydrogen peroxide liberated when cholesterol oxidase acts upon cholesterol. The cholesterol esters are measured by converting them to free cholesterol using cholesterol ester hydrolase immediately before or during treatment with cholesterol oxidase. At present the electrode is calibrated using serum analyzed by other methods. Cholesterol emulsions show promise as standards. A 10 to 50 μl serum sample can be analyzed for free and esterified cholesterol in less than four minutes. Optimum concentration of the enzymes, sodium azide, and the sample size were determined. We report on the effects of temperature and pH. This preliminary report indicates that our method has great promise for the rapid measurement of cholesterol in blood and tissues. Some suggestions are made concerning possible future designs of electrode – enzyme systems. The electrode also may prove useful for determining the physical state of cholesterol aggregation.

Acknowledgements

This work is supported by the Children's Hospital Research Foundation, Elland and Bethesda Avenues, Cincinnati, Ohio 45229. The authors are also indebted to the Deutsche Forschungsgemeinschaft. The polarographic instruments were provided by the Yellow Springs Instrument Co., Yellow Springs, Ohio 45387. The help of *Alan Brunsman, Jeff Huntington, Jay Johnson* and *Frank Williams* of the Yellow Springs Instrument Company is acknowledged. Abbott Scientific Products, 820 Mission Street, South Pasadena, California 91030, Miles Research Products, 1127 Myrthle Street, Elkhart, Indiana 60901 supplied preparations of cholesterol oxidase and cholesterol ester hydrolase. Dr. *Claude Gunter, Peter Cooper*, Dr. *Thomas Kreiser, Pete Tarbutton* and *Steve Peach* provided technical information concerning prior research with these two enzymes. The help of Dr. *Eugene P. Wesseler, Sara Devine, Frank Knapke, Eleanor Brinkmoeller, Mary E. Schneider* and *Sandy K. Hoffmann* is acknowledged as well as the donation of blood by Dr. *Marian Miller* and *Esther Tombragel*.

Addendum

Since this work was completed clear water-based cholesterol standards suitable for use with enzymes have become available.

    (Tekit Cholesterol Standard)
    Searle Diagnostic Inc.
    Subsidiary of G. D. Searle & Co.
    Box 2440
    Columbus, Ohio 43216 USA

    (BM-Standard Preciset)
    Boehringer Mannheim Corporation
    Mannheim, W. Germany

References

Allain, C. C., Poon, L. S., Chan, C. S. G., Richmond, W., Fu, P. C.: Enzymatic Determination of Total Serum Cholesterol. Clin. Chem. 20, (1974) 470–475

Bloor, W. R.: Biochemistry of the Fatty Acids, pp, 1. Reinhold, New York 1943

Buckland, B. C., Richmond, W.: The Large-Scale Isolation of Intracellular Microbiological Enzymes: Cholesterol Oxidase from Nocardia. Proc. FEBS Industr. Aspects Biochem. (Dublin) 1973

Clark, L. C., Jr.: Membrane Polarographic Electrode System and Method with Electrochemical Compensation. U. S. Patent No. 3,539,455, Issued November 10, 1970

Clark, L. C., Jr.: A New Family of Polarographic Electrodes for the Measurement of Glucose, Ethanol, Amino Acids and other Oxidase Substrates. Proc, Intern. Union Physiol. Sci. IX (1971)

Clark, L. C., Jr.: A Polarographic Enzyme Electrode for the Measurement of Oxidase Substrates. In: Oxygen Supply, pp. 120–128 ,ed. by *M. Kessler, D. F. Bruley, L. C. Clark, Jr., D. W. Lubbers, I. A. Silver, J. Strauss.* Urban & Schwarzenberg, München 1973a

Clark, L. C., Jr., Clark, E.: Differential Anodic Enzyme Polarography for the Measurement of Glucose. In: Oxygen Transport to Tissue s. Ms. 12 37 A, pp. 127–133, ed by *H. I. Bicher, D. F. Bruley:* Plenum Press, New York 1973b

Clark, L. C., Jr., Lyons, C.: Electrode Systems for Continuous Monitoring in Cardiovascular Surgery. Ann. N. Y. Acad. Sci. 102 (1962) 29–45

Christian, G. D.: Electrochemical Methods for Analysis of Enzyme Systems. In: Advance in Biomedical Engineering and Medical Physics, Vol. IV. pp. 95–161, ed. by S. N. Levine. Interscience, New York 1971

Corning Glass Works: Immobilized Enzymes, a Compendium of References from the Recent Literature, prepared for Corning Glass Works, Corning, New York 14830 by the New England Research Application Center, University of Connecticut, Storrs, Connecticut 06268, 1972

Durst, R. A.: Ion-selective Electrodes in Science, Medicine, and Technology Amer. Sci. 59 (1971) 353–361

Flegg, H. M.: An Investigation of the Determination of Serum Cholesterol by an Enzymatic Method. Ann. Clin. Biochem. 10 (1973) 79

Guilbault, C. G. (Ed.): Enzymatic Methods of Analysis, pp. 1. Pergamon Press, Oxford 1970

Kessler, M., Bruley, D. F., Clark, L. C. Jr., Lübbers, D. W., Silver, I., Strauss, J. (Eds.): Oxygen Supply, pp. 1–312, Urban & Schwarzenberg, München 1973

Klein, B., Kleinman, N. B., Foreman, J. A.: Preparation and Evaluation of a Water-Soluble Cholesterol Standard. Clin. Chem. 20 (1974) 482–485

Tarbutton, P. N., Gunter, C. R.: Enzymatic Determination of Total Cholesterol in Serum. Clin. Chem. 20 (1974) 724–725

G. D. Christian

Amperometric Methods of Enzyme Assay

(4 Figures)

I. Introduction

Amperometric methods have been used in analytical chemistry for decades, but their use in routine clinical analyses is more recent. The primary uses have been the monitoring of oxygen in enzymatic glucose analysis (*Kadish, Little* and *Sternberg* 1968) and other oxidase reactions, and in respiration of microbiological samples (*Estabrook* 1967), using the Clark oxygen electrode. In recent years, there has been an interest in amperometric enzyme electrodes for monitoring oxidase reactions, whereby the production of hydrogen peroxide is measured with a platinum electrode on which an enzyme is immobilized (*Clark* 1972; *Clark* and *Clark* 1973; *Guilbault* and *Lubrano* 1972, 1974a, 1974b).

In this paper, methods are described for the amperometric monitoring of reduced nicotinamide adenine dinucleotide (NADH) in enzyme reactions, and for measuring oxidase reactions via oxygen depletion. Two instrumental techniques are described for the measurements, one a derivative technique and the other an integration technique.

II. Voltammetry of NADH

Although the electrochemical reduction of NAD and NADP has been well characterized (*Elving, O'Reilley* and *Schmakel* 1973, *Thevenot* and *Buvet* 1972a, 1972b), the electrochemical oxidation of their reduced form is less well understood. NADH is reported to give a poorly-defined oxidation wave on platinum, but it can be quantitatively oxidized to NAD at an applied potential of + 1.05 volts vs. S.C.E. (*Burnett* and *Underwood* 1965). In addition, NADPH has been oxidized to NADP at + 0.62 volts vs. S.C.E. (*Cunningham* and *Underwood* 1966). *Blaedel* and *Haas* (1970) have proposed a mechanism for the electrochemical oxidation of reduced pyridine nucleotide analogs. In basic medium, the oxidation apparently goes in two steps, with a free radical intermediate that exchanges a proton with a solvent base

$$NADH \xrightarrow{-e^-} NADH^{\cdot+} \qquad \varepsilon \sim 0.4 \text{ V} \qquad (1)$$

$$B + NADH^{\cdot+} \rightleftharpoons NAD^{\cdot} + HB^+ \qquad (2)$$

$$NAD^{\cdot} \xrightarrow{-e^-} NAD^+ \qquad \varepsilon \sim -1.0 \text{ V} \qquad (3)$$

with the overall reaction

$$\underset{\underset{R}{|}}{\underset{N}{\bigcirc}}\!\!\!\!\!\!\!\!\!\!\!\!\overset{H\ H}{}\!\!\!\!\overset{CONH_2}{} + B \rightarrow \underset{\underset{R^+}{|}}{\underset{N}{\bigcirc}}\!\!\!\!\!\!\!\overset{CONH_2}{} + HB^+ + 2e^- \qquad (4)$$

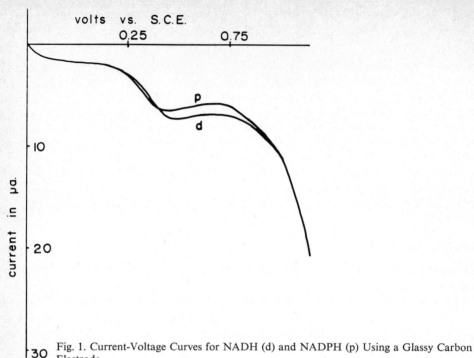

Fig. 1. Current-Voltage Curves for NADH (d) and NADPH (p) Using a Glassy Carbon Electrode.

Blaedel and *Jenkins* (1974) have recently reported on the oxidation of NADH by steady-state voltammetry.

The reduced forms of NADH and NADPH exhibit well-defined anodic voltammetric waves using either a carbon paste electrode (*Christian* and *Thomas* 1973) or a glassy carbon electrode (*Thomas* and *Christian* 1974) (Fig. 1). The residual current on the carbon electrodes is significantly lower than on platinum at potentials corresponding to the oxidation of NADH and NADPH. On glassy carbon, the half-peak potentials are approximately $+0.32$ volts vs. S.C.E. for NADH and NADPH. The potentials are independent of pH between 5.5 and 8.3, but are dependent on the concentration, shifting anodically with increasing concentration. A plot of the peak current vs. concentration of NADH or NADPH results in a straight line for 0 to 0.5 μmole/ml.

It should be possible to amperometrically monitor changes in NADH or NADPH concentrations during enzyme reactions. Such measurements would not be influenced by solution turbidity and would be simple to make. Studies utilizing this system are reported below.

III. Direct Amperometry of NADH in Enzyme Reactions

The direct monitoring of enzyme reactions by amperometric measurement of NADH has been investigated for serum lactic dehydrogenase (L-lactate: NAD oxidoreductase; LDH) (*Thomas* and *Christian* 1974). The amperometric current at $+0.75$ volts vs. S.C.E. was recorded at a glassy carbon electrode over a 5-minute time interval during the reaction of

NAD with D-L lactate in the presence of LDH. A linear relationship between LDH concentration in serum and a fixed-time current was achieved for physiological concentrations of LDH (20–1000 I.U./*l*); serum samples were diluted sixty-fold for the measurements.

Although the direct amperometric measurement of NADH in enzyme reactions is feasible, certain difficulties arise. There is an appreciable residual current in serum at the applied potential, possibly due to endogenous amines (*Porterfield* 1972; *Smith* 1972), which contributes to the noise level. Also, protein adsorption could affect the response of the electrode. In spite of this, *Park*, *Adams* and *White* (1972) were able to directly determine uric acid in blood amperometrically at a carbon paste electrode at + 0.64 volts vs. Ag/AgCl. Also, *Smith* (1972) performed measurements in blood at a bare carbon electrode at an applied potential of + 0.08 volts vs. S.C.E. (see below). The carbon paste electrode must not be exposed to air while the high positive potential is applied, or high residual currents result (*Christian* and *Thomas* 1973). The glassy carbon electrode is more stable in this respect.

IV. Coupling of NADH to Oxygen

Because of the potential difficulties of a bare electrode in serum samples, it was felt that if a membrane-covered electrode could be used, greater stability would result. For this reason, means were investigated for reacting NADH with oxygen; under such conditions, oxygen depletion could be monitored during a dehydrogenase enzyme reaction using the *Clark* oxygen electrode. This electrode is very stable and responds rapidly, continuously, and linearly to oxygen activity. It may be used under a variety of conditions with respect to pH, gas content and substrate. The membrane protects it from protein solutions.

Mitochondria have been used in biochemical studies for coupling reduced nucleotide to oxygen and measurement with the *Clark* electrode (*Chappell* 1964). Oxidation-reduction dyes have been used as electron transfer agents for such coupling in manometric monitoring of biochemical reactions since the 1930's (*Green* and *Brosteaux* 1936); for example, LDH has been analyzed by coupling NADH to oxygen using methylene blue. In recent years, the trend has been to use redox dyes to couple NADH to other dyes for spectrophotometric clinical analysis. *Smith* (1972) used either PMS or diaphorase enzyme to couple NADH to *Bindschedler's Green* (BG) and then measured the reduced BG amperometrically at + 0.08 volts vs. S.C.E. using a tubular carbon electrode. All solutions were de-aerated.

The redox dye, 5-methyl-5-phenazinium methyl sulfate (PMS) was first used in 1938 to couple NADH with oxygen, but it was virtually ignored until 1954 when *Singer* (see *Nachlas*, *Margulies* and *Seligman* 1960) used it for monometric measurements. The following reactions occur in dehydrogenase enzymatic determinations:

$$S_R + NAD^+ \xrightleftharpoons{E} S_0 + NADH + H^+ \tag{5}$$

$$NADH + PMS^+ \rightleftharpoons NAD^+ + PMSH \tag{6}$$

$$PMSH + O_2 + H^+ \rightleftharpoons PMS^+ + H_2O_2 \tag{7}$$

and the overall reaction is

$$S_R + O_2 \rightleftharpoons S_0 + H_2O_2 \tag{8}$$

S_R and S_0 are the reduced and oxidized forms of the substrate, respectively, and E is the enzyme.

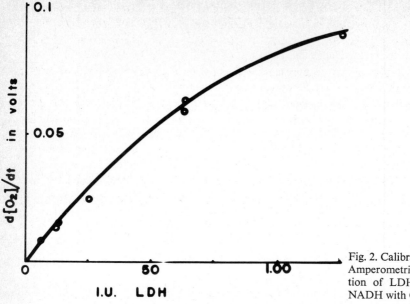

Fig. 2. Calibration Curve for Amperometric Determination of LDH by Reacting NADH with O_2 Using PMS.

We have investigated this procedure for the possible amperometric monitoring of enzyme reactions using the *Clark* electrode (*Christian* and *Thomas* 1973). Some of the important parameters are: (1) The half-time of the PMS-NADH reaction in air-saturated solution may be slower than the rate of NADH production, which can decrease sensitivity and introduce nonlinearity in calibration curves. By increasing the PMS concentration to 0.05 mg/ml, the rate is maximized. (2) PMS will inhibit some enzymes. This may by quite slow in some cases. For example, one hour is required for complete inhibition of yeast alcohol dehydrogenase. (3) PMS is reduced by many substances, causing depletion of oxygen in the solution and resulting in appreciable blank readings.

In spite of the limitations noted above, we have been able to demonstrate the determination of LDH levels corresponding to 100 µl of normal serum by measuring the rate of oxygen consumption (dO_2/dt) (Fig. 2); the instrumental basis of this measurement is described below. A second promising method for coupling oxygen to NADH involves the use of horseradish peroxidase enzyme along with Mn^{++} and certain phenols as catalysts (*Akazawa* and *Conn* 1958; *Yamazaki* and *Yokota* 1966; *Yokota* and *Yamazaki* 1965). In preliminary amperometric measurements, we have not detected appreciable blanks or side reactions (*Cheng* and *Christian* 1974).

V. Integration Methods of Measurement

A. *Theory*. Increased precision and sensitivity, particularly for noisy signals, can be obtained by integrating a transducer signal due to changing substrate or product concentration over a period of time much longer than the period of noise fluctuations. Integration techniques described (*Cordes, Brouch* and *Malmstadt* 1968; *Hicks, Eggart* and *Toren* 1970; *Ingle* and *Crouch* 1970) generally measure the rate of change of a linear transducer signal, using a time-averaged two-point differential-rate technique. The concentration-dependent signal is integrated over two different time intervals and the difference between these integrals is related to the rate of the reaction. Either two sequential integrals are subtracted or the integration intervals are separated by an "idle interval" (*Cordes, Crouch* and *Malmstadt* 1968).

We have developed equations describing the dependence of continuous integrals on enzyme and substrate concentrations for first- and second-order reactions (*Thomas* and *Christian* 1974). First- and second-order reactions may be kinetically modeled by the following simple enzymatic mechanism:

$$E + S \underset{k_{-1}}{\overset{k_1}{\rightleftarrows}} ES \underset{k_{-2}}{\overset{k_2}{\rightleftarrows}} E + P \tag{9}$$

where E is the enzyme, S the substrate, P the product, and ES the enzyme-substrate complex. $K_m = \frac{k_{-1} + k_2}{k_1}$. For the case where $K_m \ll [S]_t$, a pseudo first-order reaction results and the following equation is shown to apply if $[ES]_t$ is neglected:

$$\int_0^t ([P]_t - [P]_0) \, dt = \int_0^t \int_0^t k_2 [E]_0 \, dt^2 = \int_0^t ([S]_0 - [S]_t) \, dt \tag{10}$$

where $[S]_0$ represents initial concentrations and $[S]_t$ represents time dependent concentrations. Hence, for a fixed time interval, t_f,

$$\int_0^{t_f} [P]_i \, dt - \int_0^{t_f} [P]_0 \, dt = [E]_0 \cdot \text{constant} \tag{11}$$

That is, the net integral of a linear transducer signal due to product (or substrate) concentration change is linearly related to the enzyme concentration. If $[E]_0$ is fixed, then the reaction is zero-order and an integral that is independent of substrate concentration results. For an irreversible pseudo first-order reaction such as $\frac{d[P]_t}{dt} = k[S]_t$, where k is the rate constant for S → P, then $\int_0^{t_f}([P]_t - [P]_0) dt = [S]_0 \cdot \text{constant}$ for a fixed time, t_f, and the difference integral is directly proportional to the substrate concentration. For the case $K_m \approx [S]_t$, a relationship similar to Equation 11 holds.

When $K_m \gg [S]$, the reaction is second-order and

$$[P]_t - [P]_0 = \int_0^t \frac{k_2 [E]_0 [S]_t}{K_m} \, dt = [S]_0 - [S]_t \tag{12}$$

if $[ES]_t$ is again neglected. $[S]_t$ is, in general, a complicated time-dependent function. Using a method of iteration, it can be shown that the integral is approximated by

$$\int_0^{t_f} ([P]_t - [P]_0) \, dt \approx [S]_0 \cdot \text{constant if } [E]_0 \text{ is fixed} \tag{13}$$

$$\int_0^{t_f} ([P]_t - [P]_0) \, dt \approx \sum_{j=1}^n a_j [E]_0^j \text{ if } [S]_0 \text{ is fixed} \tag{14}$$

where n is the number of intervals used for approximation of the integral by the method of rectangles. a_j is a constant that includes the rate constants from Equation 12.

To summarize, a linear relationship between the difference integral for a fixed-time interval and the substrate or enzyme concentration exists for all conditions except those of Equation 14. In this last case, a non-linear relationship may exist with the enzyme concentration.

This difference (continuous) integration method should allow measurement of more complicated reactions than two-point rate methods. For example, the set of time vs. concentration curves described by Figure 3 may be distinguished by this method, whereas two-point rate methods with delay times greater than $\Delta\tau$ would yield identical results in all cases, even though the rates of reaction approaching equilibrium are different. There are other more complex examples.

In order to utilize the integration technique in the clinical laboratory, an electronic integrating instrument was constructed which is capable of measuring and integrating nanoampere currents with high precision (*Thomas, Christian* and *Danielson* 1974). The instrument will integrate a signal for a preselected time interval, store the result, and then either add or subtract a subsequent integral, allowing difference integral measurements; the net result is displayed digitally and can be adjusted to read directly in concentration units. The capabilities of the difference integral method were demonstrated by analyzing glucose samples using the Clark electrode and comparing results with those obtained with the *Beckman* Glucose Analyzer. For the integral measurements, the background oxygen tension was measured by integrating the amperometric current in the absence of added glucose for 16 seconds, and this result was read out and stored. The glucose sample was then added and the 16-second integration initiated 2 seconds later. The final readout represented the difference between the two integrals and, therefore, the area between the background current level and the decreasing current due to oxygen depletion during the enzyme reaction. A correlation

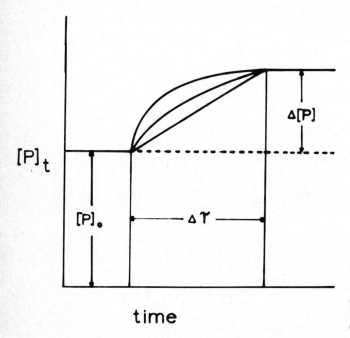

Fig. 3. Illustration of Differential Rate Methods in Which Equilibrium is Achieved within Delay Interval, $\Delta\tau$.

coefficient of 0.998 was obtained between the Beckmann Glucose Analyzer result and the integration results, with a least squares line of $y = 0.987 \times + 2.34$ for the range 0 to 150 mg percent glucose.

VI. Derivative Rate Measurements

The second instrumental method investigated for the amperometric measurement of enzyme reactions is based on measurement of the derivative, di/dt, of the amperometric current during the initial course of the reaction. The principle of the measurement is the same as described by *Kadish, Little* and *Sternberg* (1968) for glucose determination, which serves as the basis for the Beckmann Glucose Analyzer. Under proper conditions, the rates of enzyme reactions are most rapid very early in the reactions. By taking the derivative of the amperometric current with time, a maximum is generally obtained in 5–20 seconds. The magnitude of the maximum is related to the enzyme activity or substrate concentration.

We have employed the Beckman Glucose Analyzer for the measurement of other enzyme reactions using the derivative technique. Using this instrument with the Clark electrode, we have developed an extremely rapid method for the determination of tyrosine in serum (*Kumar* and *Christian* 1974). The method is based on the rate of oxygen consumption in the presence of the enzyme, polyphenol oxidase (tyrosinase):

$$\text{L-tyrosine} + O_2 \xrightleftharpoons{\text{p.o.}} \text{L-dopaquinone} + H_2O \qquad (15)$$

The only reagent required is the enzyme, tyrosinase, in a phosphate buffer. The enzyme is dissolved in a 0.1 M phosphate buffer, pH 7.4, to give an activity of 2 to 6 I.U./ml. One milliliter of this reagent is placed in the cell. After allowing a few seconds to allow the dissolved oxygen to come to equilibrium with atmospheric oxygen, a 10 to 50 µl serum sample or tyrosine standard is added. The peak (maximum rate) is obtained within 20–30 seconds and is directly proportional to the tyrosine concentration. Typical recordings are shown in Figure 4. A relative standard deviation of $\pm 2.7\%$ is obtained for a 250 ug/ml tyrosine sample.

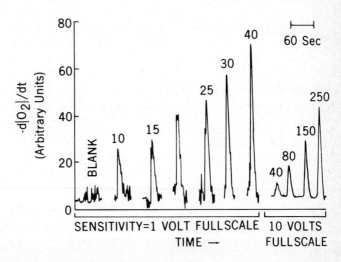

Fig. 4. Signals Recorded for Aqueous Tyrosine Standards Using a 50 µl Sample. Numbers on peaks represent µg/ml of tyrosine.

Results for the analysis of serum samples containing 3.5 ug/ml to 115 ug/ml tyrosine agree closely with those obtained using a fluorometric (*Williams* and *Besser* 1971) and an amino acid analyzer method, with correlation coefficients of 0.999 and 0.997, respectively.

The described method is applicable to other oxidase reactions as well as other enzyme reactions that can be measured amperometrically using other electrodes.

References

Akazawa, T., Conn, E. E.: The Oxidation of Reduced Pyridine Nucleotides by Peroxidase. J. Biol. Chem. 232 (1958) 403

Blaedel, W. J., Haas, R. G.: Electrochemical Oxidation of NADH Analogs. Anal. Chem. 42 (1970) 918

Blaedel, W. J., Jenkins, R. A.: Anodic Oxidation of Nicotinamide Adenine Dinucleotide (NADH) by Steady-State Voltammetry. 8th Great Lakes Regional Meeting of the Amer. Chem. Soc., West Lafayette/Indiana, June 3–5 (1974)

Burnett, J. N., Underwood, A. L.: Electrochemical Reduction of Diphosphopyridine Nucleotide. Biochem. J., 4 (1965) 2060.

Chappell, J. B.: The Oxidation of Citrate, Isocitrate and cis-Aconate by Isolated Mitochondria. Biochem. J. 90 (1964) 225

Cheng, F. S., Christian, G. D.: Unpublished work (1974)

Christian, G. D., Thomas, L. C.: Rapid Amperometric and Potentiostatic Measurement of Enzyme Reactions and Biological Components. Symposium on Electrochemical Techniques in Bioanalytical Chemistry, 166th National Meeting of the American Chemical Society, Chicago, Illinois, August 26–31 (1973) Anal. Chim. Acta 78 (1975) 271

Clark, L. C., Jr.: A Family of Polarographic Enzyme Electrodes and the Measurement of Alcohol. In: Biotechnol. and Bioeng. Symp. No. 3, pp. 377–394, ed. by *C. B. Wingard.* Wiley, New York 1972

Clark, L. C., Jr., Clark, E. W.: Differential Anodic Enzyme Polarography for the Measurement of Glucose. In: Oxygen Transport to Tissue. Instrumentation, Methods, and Physiology, pp. 127–133, ed by. *H. I. Bicher* and *D. F. Bruley.* Plenum Press, New York 1973

Cordes, E. M., Crouch, S. R., Malmstadt, H. V.: An Automatic Digital Readout System for Reaction-Rate Methods. Anal. Chem. 40 (1968) 1812

Cunningham, A. J., Underwood, A. L.: Electrochemical Reduction of Triphosphopyridine Nucleotide. Arch. Biochem. Biophys. 117 (1966) 88

Elving, P. J., O'Reilley, J. E., Schmakel, C. O.: Polarography and Voltammetry of Nucleosides and Nucleotides and Their Parent Bases as an Analytical and Investigative Tool. In: Methods of Biochemical Analysis, Vol. 21, pp. 287–465, ed. by *D. Glick.* Interscience, New York 1973

Estabrook, R. W.: Mitochondrial Respiratory Control and the Polarographic Measurement of ADP : 0 Ratios. In: Methods in Enzymology, Vol. 10, pp. 41–47, ed. by *R. W. Estabrook* and *M. E. Pullman.* Academic Press, New York 1967

Green, D. E., Brosteaux, J.: The Lactic Dehydrogenase of Animal Tissues. Biochem. J. 30 (1936) 1489

Guilbault, G. G., Lubrano, G. J.: An Enzyme Electrode for the Amperometric Determination of Glucose. Anal. Chim. Acta 64 (1973) 439

Guilbault, G. G., Lubrano, G. J.: Amperometric Enzyme Electrodes. Part II. Amino Acid Oxidase. Anal. Chim. Acta 69 (1974a) 183

Guilbault, G. G., Lubrano, G. J.: Amperometric Enzyme Electrodes. Part III. Alcohol Oxidase. Anal. Chim. Acta 69 (1974b) 189

Hicks, G. P., Eggert, A. A., Toren, E. C., Jr.: Application of On-Line Computers to the Automation of Analytical Experiments. Anal. Chem. 42 (1970) 729

Ingle, J. D., Jr., Crouch, S. R.: Fixed-Time Digital Counting System for Reaction Rate Methods. Anal. Chem. 42 (1970) 1055

Kadish, A. H., Little, R. L., Sternberg, J. C.: A New and Rapid Method for the Determination of Glucose by Measurement of Rate of Oxygen Consumption. Clin. Chem. 14 (1968) 116

Kumar, A., Christian, G. D.: A Rapid Enzymatic Assay of L-Tyrosine in Serum by Amperometric Measurement of Oxygen Consumption. Clin. Chem. 21 (1975) 325

Nachlas, M. M., Margulies, S. I., Seligman, A. M.: A Colorimetric Method for Estimation of Succinic Dehydrogenase Activity. J. Biol. Chem. 235 (1960) 499

Park, G., Adams, R. N., White, W. R.: A Rapid Accurate Electrochemical Method for Serum Uric Acid. Anal. Lett. 5 (1972) 887

Porterfield, R. I.: Redox Potentiometry and Application to Kinetic Analysis at the Carbon Electrode. Ph. D. Thesis, Ohio State Univ. 1972

Smith, M. D.: The Development of an Amperometric Method of Measuring NADH-coupled Enzyme Concentration Using Bindschedler's Green. Ph. D. Thesis, Ohio State Univ. 1972

Thevenot, D., Buvet, R.: Propriétés Électrochimiques de la Nicotinamide et de ses Dérivés en Solution Aqueuse. I: Analyse Critique de l'Ensemble des Données Disponibles Relatives Aux Propriétés Electrochimiques des Dérivés de la Pyridine en Solution Aqueuse. J. Electroanal. Chem. 39 (1972a) 429

Thevenot, D., Buvet, R.: Propriétés Électrochimiques de la Nicotinamide et de ses Dérivés en Solution Aqueuse. III: Influence du pH de la Solution et de la Concentration de Nicotinamide sur la Réduction Polarographique de ce Composé. J. Electroanal. Chem. 40 (1972b) 197

Thomas, L. C., Christian, G. D.: Amperometric Integration Methods of Enzyme Assay. 29th Annual Northwest Regional Meeting of the Amer. Chem. Soc., Cheney/Wash., June 13–14 (1974). Anal. Chim. Acta 77 (1975) 153

Thomas, L. C., Christian, G. D., Danielson, J. D. S.: A Versatile Amperometric Integrator. (1974) Anal. Chim. Acta 77 (1975) 163

Williams, T., Besser, G. M.: Effects of Treatment on Tyrosine Tolerance in Thyroid Diseease, a Modified Tyrosine Assay. Clin. Chem. 17 (1971) 148

Yamazaki, I., Yokota, K.: Analysis of the Conditions Causing the Oscillatory Oxidation of Reduced Nicotinamide-Adenine Dinucleotide by Horseradish Peroxidase. Biochim. Biophys. Acta 132 (1967) 310

Yokota, K., Yamazaki, I.: Reaction of Peroxidase with Reduced Nictonamide-Adenine Dinucleotide and Reduced Nicotinamide-Adenine Dinucleotide Phosphate. Biochim. Biophys. Acta 105 (1965) 301

W. W.-C. Chan, P. Davies and K. Mosbach

Immobilized Enzymes and Ligands in Biological Research

(5 Figures)

This presentation is divided into two sections. The major section consists of a discussion of certain special applications of immobilized enzymes. This is followed by a short section on the application of an immobilized coenzyme in enzyme electrodes.

Enzymes immobilized on insoluble carriers offer a number of practical advantages which have been frequently made use of in industry as well as in pure research (for review see *Goldman, Goldstein* and *Katchalsky* 1971). The one property of immobilized enzyme which has been generally found to be most useful lies in the insoluble nature of these derivatives. With immobilized enzymes, separation of soluble products at the end of the desired reaction is accomplished easily permitting the recovery of the enzyme derivative and the use of continuous-flow techniques. Such applications are often also facilitated by the increased stability of immobilized enzymes compared to the correspondling free enzymes in solution. These applications are quite straightforward in principle and will not be discussed further.

We wish now to consider one property of immobilized enzymes which may not be completely obvious: namely, the fact that each bound enzyme molecule is to a large extent fixed in its location relative to other bound enzyme molecules in the carrier. Several years ago one of us decided to make use of this property in a general method to investigate the activity of subunit form of oligomeric enzymes (*Chan* 1970). The purpose behind this kind of work is to find out what effects the quaternary structure of an enzyme has on activity. The experimental difficulty is that a number of oligomeric enzymes do not dissociate unless they are treated with protein denaturants such as urea or guanidinium chloride, which also unfolds the individual polypeptides. If the denaturant is subsequently removed, the refolding of the polypeptides is accompanied by reassociation. Therefore it is not possible to study isolated subunits under non-denaturing conditions unless some special technique is employed. The special technique in this case is to have the subunits immobilized on an insoluble matrix such as Sepharose to prevent their reassociation.

The experimental procedure consists of selecting a carrier (such as Sepharose 4B) which contains pores sufficiently large that the enzyme to be studied can diffuse readily into the interior. A relatively small number of "anchoring points" are then introduced onto the carrier (e.g., by activation of the hydroxyl groups of the Sepharose with low amounts of CNBr) so that the "anchoring points" are sufficiently far removed from each other. An oligomeric enzyme is then attached to the carrier and the low density of "anchoring points" makes it probable that each bound enzyme molecule will be attached covalently via only one subunit (Fig. 1). Dissociation with protein denaturant and exhaustive washing then removes all polypeptides which are not covalently bound. Finally by removing the denaturant the polypeptides remaining attached to the carrier are allowed to refold but are prevented by the rigidity of the carrier from reassociating with other bound monomers. This monomeric form of the enzyme can be compared to the original derivative containing bound oligomeric

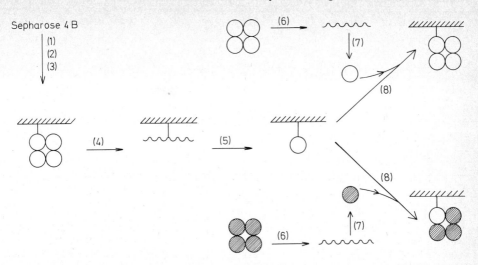

Fig. 1. Preparation of Immobilized Subunits of an Oligomeric Protein and Tests for the Presence of Monomers in the Product. The various operations are: (1) Activation with CNBr, (2) Coupling of the oligomer, (3) Washing and blocking of unused activated groups, (4) Washing with 6M guanidinium chloride, (5) Removal of guanidinium chloride, (6) Denaturation of soluble enzyme in guanidinium chloride, (7) Dilution into buffer to initiate renaturation, (8) Reassociation of soluble subunits with immobilized monomers. Shaded circles denote subunits of inactivated enzyme.

enzyme in order to gain insight into the effects of quaternary structure on enzyme activity. One may well ask why we have decided to bind the oligomeric enzyme and then to dissociate it rather than to bind the denatured polypeptide directly. The answer is that only by binding the folded native enzyme can one be sure that covalent attachment is (1) to a group on the protein surface which will not hinder the refolding process and (2) at a point on the carrier where there is sufficient space for refolding. To our knowledge, every experiment in which denatured polypeptides are attached has failed to yield refolded active enzyme.

How can one demonstrate that the bound enzyme is now in a monomeric form? Here the insoluble nature of the carrier becomes an obstacle since one can not perform molecular weight measurements. However, a number of indirect methods have been developed which provide fairly convincing evidence that the activity (if any) is due to the monomeric form of the enzyme. The most important of these tests are:

(1) Demonstration that the amount of bound protein on the carrier decreases upon treatment with denaturant by the expected quantity (i.e. 50% for a dimeric enzyme, 75% for a tetrameric enzyme etc.) This would confirm the hypothesis that each oligomer is bound via only one subunit.

(2) Demonstration that the activity of the putative monomer has different properties (e.g. heat or urea sensitivity or pH-activity profile) with control experiments to ensure that the difference is not due to irreversible damage by the denaturant.

(3) Demonstration that the bound monomer can pick up specifically an amount of soluble subunits approaching the amount which has been removed by the treatment with denaturant. Furthermore it might be shown that accompanying the reassociation of the subunits, the bound enzyme regains the properties of the oligomer. A valuable control experiment is to allow the bound monomer to reassociate with subunits which have previously been inactivated

(e.g., by modification at the active site) (Fig. 1). Since the subunit interactions are now reestablished, the properties of the oligomeric enzyme should re-appear. In the case of subunits which are inactive, this control experiment is crucial since it would show that addition of inactive subunits to inactive bound monomers generates bound activity. Such a result would provide clear-cut evidence for the necessity of the quarternary structure for enzyme activity. In the case of rabbit muscle aldolase (*Chan* 1970, *Chan* and *Mawer* 1972, *Chan* 1973), and yeast transaldolase (*Chan, Schutt* and *Brand* 1973) these tests have all led to the conclusion that the monomeric form is active. Although the evidence from these studies of immobilized enzymes is quite strong in our opinion, some enzymologists continue to have reservations regarding these results in view of possible interactions with the carrier itself. We have therefore turned to studies in free solution to seek confirmation of the activity of monomers. A general approach is to study the renaturation of the denatured enzyme at extremely low concentration so that reassociation is retarded (*Chan, Chong, Mort* and *Macdonald* 1973). With this approach we have observed that denatured aldolase refolds in a first-order reaction to give active monomers which then reassociate in a concentration-dependent process to yield the renatured tetramer. Of particular importance was the finding that the active monomer formed as an intermediate during the initial phases of renaturation was sensitive to inactivation by 2.3 M urea while both the renatured and the native tetrameric enzyme were stable at this concentration of urea. This property of the monomers in solution correlates precisely with properties of the immobilized subunit and provides strong support for the validity of the approach using immobilized enzymes. Further work which supports the conclusion that aldolase monomers are active comes from the preparation of stable active monomers by treatment with iodoacetate during renaturation (*Chan, Kaiser, Salvo* and *Lawford*, 1974). Presumably residues at the intersubunit region were exposed in the refolded monomer and were modified before reassociation could take place. In any case conclusive evidence was obtained that some material with the expected Stoke's radius of the monomer was enzymically active. Again the sensitivity in 2.3 M urea characteristic of immobilized aldolase subunits was also observed in this material.

Since its introduction in 1970, this approach has been applied by several other groups to various oligomeric proteins (Table 1). In several cases the subunits were found to be active but with distinct properties; in other cases, inactive monomers were obtained. It should be pointed out that the tests for monomers discussed in an earlier section have not always been vigorously applied in these studies and extensive confirmatory evidence from studies in solution is so far available only in the case of aldolase.

What are the limitations of such an approach? What improvements and developments are likely to be made in the near future? As with all enzymes covalently bound to carriers so far, the material is heterogeneous to a considerable extent. The "anchoring points" on the carrier are located in a variety of environments and the individual protein molecules are also attached at different points on the protein surface. This heterogeneity may be unimportant to workers who are interested in using immobilized enzymes only as a preparative tool. For studies of biological problems such as those discussed here, this property limits the conclusions to a semi-qualitative nature. Thus any measurable parameter (e.g., K_m or pH-activity profile) of the covalently-bound enzyme (subunit or oligomer) represents the mean value for a variety of molecules. The ideal carrier, consisting of perfectly regular polymers with "anchoring points" in identical environments is not yet available. It would also be desirable to attach proteins specifically at only one point on the protein surface.

Table 1. Subunits of proteins studied with immobilized derivatives.

Protein	Results	Reference
Muscle aldolase	Monomer is active but more sensitive to heat, urea, or high pH.	Chan 1970 Chan and Mawer 1972 Chan 1973
Muscle lactate dehydrogenase*)	Monomer is active	Cho and Swaisgood 1972 Cho and Swaisgood 1974
Glycogen phosphorylase	Monomer is inactive	Feldman, Zeisel and Helmreich 1972
Avidin	Monomer binds biotin with diminished affinity	Green and Toms 1973
Liver fructose-diphosphatase	Monomer is inactive, dimer is active but desensitized towards AMP	Grazi, Magri and Traniello 1973
Yeast transaldolase	Monomer is active but more sensitive towards urea, pH-activity profile is also changed	Chan, Schutt and Brand 1973
E. coli aspartate transcarbamylase	c_3r_6 complex lacks homotropic and heterotropic interactions	Chan 1974
Bacterial luciferase	Isolated α and β subunits are inactive	Chan, Meighen and Hastings unpublished results

*) The carrier in this work was porous glass; in all other cases Sepharose 4B was used.

A problem which is more amenable to solution in the near future is the rigidity of the carrier which is required to prevent reassociation of bound subunits. Thus over extended periods of time or at moderately elevated temperatures, the matrix of Sepharose 4B shows considerable mobility (Green and Toms 1973, Chan, unpublished results). The use of glass-beads (Cho and Swaisgood, 1972) may overcome this problem although the ability of glass in preventing the interaction of bound subunits over extended periods has yet to be tested.

Some recent work (Chan, in preparation) has been directed toward a combination of immobilized derivatives with studies in solution. If the protein is attached to the carrier via a

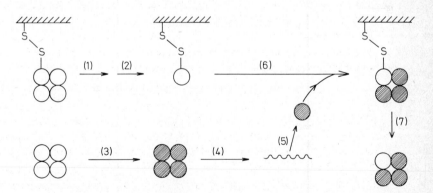

Fig. 2. Demonstration of the Presence of Immobilized Monomers. The various operations are: (1) Washing with 6M guanidinium chloride, (2) Removal of guanidinium chloride, (3) Succinylation of soluble enzyme, (4) Denaturation of succinylated enzyme, (5) Renaturation of succinylated enzyme, (6) Reassociation of succinylated subunits with immobilized monomers, (7) Detachment of product from carrier and identification by electrophoresis.

linkage which can be subsequently cleaved under mild conditions (e.g. a disulfide bond), then the bound protein could be recovered in solution and subsequently characterized. Three potential applications can be anticipated. First the bound monomer may be allowed to reassociate with added subunits of modified electro-phoretic mobility (e.g., succinylated subunits) and the product is then detached from the carrier (Fig. 2). The resulting hybrid can then be distinguished from other possible hybrids or from the native tetramer. This would provide a direct confirmation of the existence of carrier-bound monomers and estimation o the degree of any contamination by oligomers. Secondly, one could introduce bulky groups onto the intersubunit surface of the bound monomer by chemical modification. When these modified subunits are now detached from the carrier they might remain dissociated on account of the steric hindrance towards association (Fig. 3). This should provide a general method of preparing stabilized soluble subunits. Thirdly, irrespective of whether stabilized subunits are formed, this approach would allow the modification of residues at the intersubunit surface. The detachment of the modified protein from the carrier then allows the location of the modified residue in the amino acid sequence. Thus structural information can be obtained concerning the intersubunit surface. The development of these approaches should extend the general applicability of immobilized enzymes in biological research.

Two other kinds of work are also based on the fixed location of bound enzyme molecules relative to other parts of the carrier. Thus by introducing charged or hydrophobic groups onto the carrier, one could observe their effects on bound enzymes (for summary of this work see *Katchalsky*, *Silman* and *Goldman* 1971; *Srere* and *Mosbach*, 1974). These provide important models for enzymes found in a particulate state in the cell. Equally important are

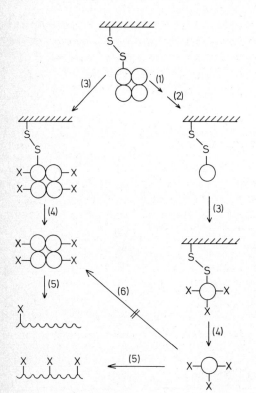

Fig. 3. Modification of Residues at the Intersubunit Surface using Immobilized Subunits. The various operations are: (1) Washing with 6M guanidinium chloride, (2) Removal of guanidinium chloride, (3) Treatment with a protein reagent, (4) Detachment from the carrier, (5) Identification of the location of the modified residue in the amino acid sequence, (6) Prevention of reassociation by the groups introduced onto the intersubunit surface.

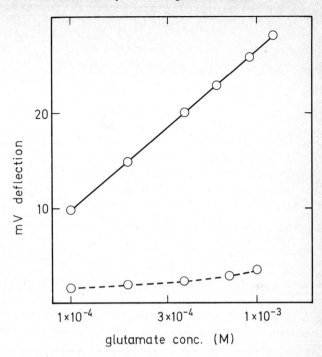

Fig. 4. Coupled Reactions used in the Enzyme Electrode.

studies in which two or more enzymes which act in a metabolic sequence are bound onto the same carrier (*Mosbach* and *Mattiasson* 1970), *Mattiasson* and *Mosbach* 1971). The proximity of the enzyme allows the study of conditions approximating those *in vivo*. The observed increased efficiency of such systems may help explain certain discrepancies between the properties of isolated enzymes and their presumed function *in vivo* (*Srere, Mattiasson* and *Mosbach*, 1973).

Application of Immobilized Coenzyme in an Enzyme Electrode

Recently we have used dextran-bound NAD^+ in conjunction with glutamate dehydrogenase to generate ammonium ions which are detected by an ion-selective electrode (*Davies* and *Mosbach*, 1974). The conversion of glutamate to α-ketoglutarate is coupled to the lactate-dehydrogenase-catalyzed reduction of pyruvate (Fig. 4). This system serves to regenerate the immobilized coenzyme. Linearity of response for glutamate was obtained in the 10^4 M to

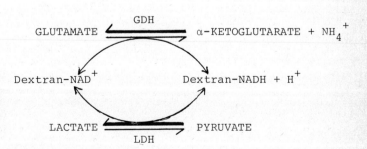

Fig. 5. Response of the Electrode as a Function of Glutamate Concentration. The broken curve represents control experiments in 8M urea.

10^{-3} M range and was shown to be dependent on the presence of active enzyme (Fig. 5). Similarly determination of pyruvate concentration was also possible. Fairly rapid response (3 to 4 minutes) and reasonable stability (more than two weeks) were found. Enzyme electrodes incorporating immobilized coenzymes should have significant applications in clinical and analytical chemistry.

References

Chan, W. W.-C.: Matrix-Bound Protein Subunits. Biochem. Biophys. Res. Commun. 41 (1970) 1198–1204

Chan, W. W.-C.: Studies of Protein Subunits, V: Specific Interaction between Matrix-Bound Subunits of Aldolase and Soluble Aldolase Subunits. Canad. J. Biochem. 51 (1973) 1240–1247

Chan, W. W.-C.: Subunit Interaction in Aspartate Transcarbamylase from *Echerichia coli* Studied Using Matrix-Bound Derivates. FEBS Letters 44 (1974) 178–181

Chan, W. W.-C., Chong, D. K. K., Mort, J. S., Macdonald, P. D. M.: Studies of Protein Subunits, III: Kinetic Evidence for the Presence of Active Subunits During the Renaturation of Aldolase. J. Biol. Chem. 248 (1973) 2778–2784

Chan, W. W.-C., Kaiser, C., Salvo, J. M., Lawford, G. R.: Formation of Dissociated Enzyme Subunits by Chemical Treatment During Renaturation. J. Mol. Biol. 87 (1974) 847

Chan, W. W.-C., Mawer, H. M.: Studies on Protein Subunits, II: Preparation and Properties of Active Subunits of Aldolase Bound to a Matrix, Arch. Biochem. Biophys. 149 (1972) 136–145

Chan, W. W.-C., Schutt, H., Brand, K.: Active Subunits of Transaldolase bound to Sepharose. Eur. J. Biochem. 40 (1973) 533–541

Cho, I. C., Swaisgood, H. E.: The Reactivitation of an Unfolded Subunit Enzyme covalently linked to a Solid Surface. Biochim. Biophys. Acta 258 (1972) 675–679

Cho, I. C., Swaisgood, H. E.: Surface-bound Lactate Dehydrogenase: Preparation and Study of the Effect of Matrix Microviroment on Kinetic and Structural Properties. Biochim. Biophys. Acta 334 (1974) 243–256

Davies, P., Mosbach, K.: The Application of Immobilized NAD* in an Enzyme Electrode and in Model Enzyme Reactors. Biochim. Biophys. Acta (1974)

Feldman, K., Zeisel, H., Helmreich, E.: Interaction between Native and Chemically Modified Subunits of Matrix-bound Glycogen Phosphorylase. Proc. Nat. Acad. Sci. USA 69 (1972) 2278–2282

Goldman, R., Goldstein, L., Katchalski, E.: Biochemical Aspects of Reactions on Solid Supports. In: Water-Insoluble Enzyme Derivatives and Artificial Enzyme Membranes, pp. 1–78, ed. by G. R. Stark. Academic Press, New York 1971

Green, N. M., Toms, E. J.: The Properties of Subunits of Avidin Coupled to Sepharose. Biochem. J. 133 (1973) 687–700

Grazi, E., Magri, E., Traniello, S.: Active Subunits of Rabbit Liver Fructose Diphosphatase. Biochem. Biophys. Res. Commun. 54 (1973) 1321–1325

Katchalski, E., Silman, I., Goldman, R.: Effect of the Microenvironment on the Mode of Action of Immobilized Enzymes. Adv. Enzymol. 34 (1971) 445–536

Mattiasson, B., Mosbach, K.: Studies on a Matrix-bound Three-enzyme System. Biophys. Acta 235 (1971) 253–257

Mosbach, K., Mattiasson, B.: Matrix-bound Enzymes II: Studies on a Matrix-bound Two-enzyme System. Acta Chem. Scand. 24 (1970) 2093–2100

Srere, P. A., Mattiasson, B., Mosbach, K.: An Immobilized Three-enzyme System: A Model for Microenvironmental Compartmentation in Mitochondria. Proc. Nat. Acad. Sci. USA 70 (1973) 2534–2538

Srere, P. A., Mosbach, K.: Metabolic Compartmentation: Symbiotic, Organellar, Multienzymic and Microenvironmental. Ann. Rev. Microbiol. 28 (1974) 61

I. A. Silver

An Ultra Micro Glucose Electrode

(2 Figures)

Currently available micro enzyme electrodes have been developed from the ideas of *Clark* (1973) who showed that a variety of oxygen oxydoreductase enzymes can be attached to the surface of platinum. The enzyme reaction releases hydrogen peroxide which is destroyed at the surface of the platinum polarized as an anode. The anodic degradation of hydrogen peroxide was first shown by *Chance* in 1949. The destruction of the hydrogen peroxide gives rise to a current in the circuit which is proportional to the amount of hydrogen peroxide released by the enzyme reaction. While the microelectrodes described by *Clark* (1973) and *Clark* and *Clark* (1973) have proved invaluable in investigation of glucose concentrations in solutions and at tissue surfaces, they are unsuitable for insertion into tissue. It seemed therefore appropriate to design a microelectrode which could be inserted into tissues and which would give information at the cellular level about local glucose concentrations.

Materials and methods

Glucose microelectrodes consist of a glass-insulated, platinum surface onto which glucose oxidase is adsorbed. The practical details of the electrodes are as follows:
(1) Micro-needle electrodes are prepared by electropolishing in sodium nitrite (*Silver* 1965). They are covered with Normalglas III (Jena, Leipzig) to give an uninsulated area of approximately 3 μ^2. The surface is coated with glucose oxidase by dipping the tip into a 1% solution of glucose oxidase in water and drying in air. This type of electrode can be used in normal well-perfused tissues or in tissue culture. The electrode is polarized at 0.6 V as an anode with reference to a silver/silver chloride electrode.
(2) The second type of electrode is based on the standard hollow glass micropipette, with a tip diameter of approximately 0.1 μm. The glass pipette is covered with platinum by evaporation in the electron microscope shadowing device to give a platinum layer of approximately 100 μm thickness. The tip of the electrode is insulated by drawing a capillary of glass similar to that of the main capillary but with a very slightly larger diameter. The tip of this second capillary is measured in a comparator microscope and then cut off and slipped over the tip of the electrode. When the two glass surfaces lie in close juxtaposition the space between the two is eliminated by melting the outer glass in a microforge so that it adheres to the platinum layer on the inner glass. The main electrode is kept patent by a stream of nitrogen which is passed through it during this process. The electrode is then dipped into an epoxy or vinyl insulating resin which covers the platinum on the body and shaft of the electrode and also incidentally covers the glass sheathed tip. Since the last two processes usually result in blocking of the tip of the electrode this is then cut to an oblique point on a rapidly rotating wheel covered with diamond dust. Figure 1 shows the electrode after this treatment. The probe will store in this condition in a dust-free situation almost indefinitely. Before use the probe is dipped into a 1%

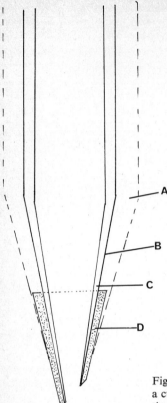

Fig. 1. Construction of a pipette-type glucose microelectrode with a cut tip. A, Resin; B, Platinum layer; C, Pipette; D, Glass insulation.

solution of glucose oxidase in water and dried. The enzyme attaches itself to the platinum ring in the electrode tip, from which it is very difficult to displace it. When the electrode is to be used it is inserted into the tissue and, if the situation is likely to be anoxic, the barrel of the electrode is filled with oxygen on which a pressure of approximately 10–20 cm of water is maintained through a hypodermic syringe. The oxygen within the lumen of the electrode diffuses out from the tip to the area of the platinum ring and provides an adequate oxygen tension for full effectiveness of the glucose oxidase. Excessive amounts of oxygen will of course diffuse into the tissue if the electrode tip is too large. This will convert the local situation from being hypoxic to being normoxic, or even hyperoxic. The supply of oxygen is therefore somewhat critical and is related to the size of the lumen in the tip of the electrode. The electrode is polarized as an anode at 0.6 V against a silver/silver chloride reference.

The response of both types of electrode to changes in glucose concentration is linear within the normal physiological range up to about 250 mg%. There is a tendency for the response to fall off above level. Similar findings have been reported by *Clark* and *Clark* (1973) in microelectrodes. The surface of the electrode may be protected with a membrane by dipping the probe into a solution of nitrocellulose in ether and alcohol. This membrane appears to have a beneficial effect only when the electrode is first used since under normal circumstances a bare electrode gains a protein covering shortly after its insertion into the tissues.

Ascorbic acid may interfere with the performance of the electrode but since the current obtained from ascorbic acid is very much less than that obtained from glucose it is only likely to be of significance where very low levels of glucose are to be measured. Since this electrode is by no means entirely reliable at very low levels, one can for practical purposes ignore the contribution of ascorbic acid.

Application

The application of the electrodes in biological situations has been carried out in two tissues, in brain and growing connective tissue. In the brain we have seen that glucose concentration remains extremely constant unless severe changes are imposed on the metabolism of the organ. In spreading depression (*Leão* 1947) there is a very marked change of blood flow, of cellular activity, of extracellular potassium concentration, and also of glucose concentration. This response is biphasic; there is first an increase in glucose followed by a decrease and return to normal (Fig. 2). Successive waves of spreading depression produce progressively less response in the glucose electrode, and the general level of glucose in the tissue may increase. It seems likely that this change is brought about firstly by the increase in blood flow which occurs during spreading depression, and secondly by blockage on the uptake of glucose which probably occurs during the time when the pyridine nucleotide fluorescence is minimal (*Mayevesky, Zeuthen* and *Chance* 1974).

In connective tissue during healing micro glucose electrodes have been used to follow the changes which occur at the edge of the healing tissue. In some normal situations it is possible to detect the slight arterio-venous difference in glucose between the arteriolar and venular ends of capillaries when perfusion is relatively slow. When the tissue is made hypoxic, and particularly when blood flow is reduced, our initial findings were that glucose concentration was dramatically reduced. This finding was obtained with solid electrodes and we assumed this was due to increased glucose uptake. When, however, we used hollow microelectrodes, each with its own internal oxygen supply, it became obvious that part of the apparent

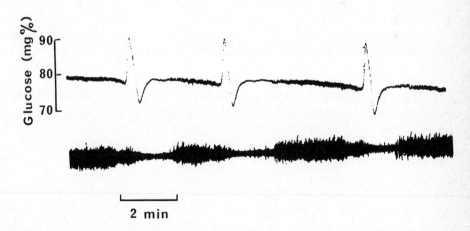

Fig. 2. Recording of tissue glucose changes and ECoG in rat cerebral cortex during 3 waves of spreading depression.
The drift in the glucose trace was due to protein deposition on the electrode immediately after insertion.

reduction in glucose concentration was illusory and due to the fact that the glucose oxidase had insufficient oxygen for full oxidation of the glucose and production of H_2O_2.

In under-perfused tissue the delivery of glucose at the arteriolar end of the capillary is normal but the fall off in glucose concentration along the length of the capillary is very dramatic. This might be expected when it is realized that the amount of energy obtained from a given amount of glucose is twelve times less under anaerobic than aerobic conditions. The efficiency of glucose extraction in hypoxia is extremely high and presumably compensates to some extent for this. Coincidentally with this increase in glucose consumption there is a fall in pH which is also a feature of under-perfusion, especially in connective tissue.

The reponse time of the electrodes is of the order of 2–3 seconds for 97% response in a bare electrode but if a membrane has been placed over the end the response time may increase to as much as 30 seconds.

Summary

A microelectrode is described which performs satisfactorily in soft tissues and in tissue cultures for the detection of glucose in physiological concentrations at the cellular level.

References

Chance, B.: The Properties of the Enzyme-Substrate Compounds of Horseradish Peroxidase and Peroxides. IV: The Effect of pH upon the Rate of Reaction Complex II with Several Acceptors and its Relation to their Oxidation-Reduction Potential. Arch. Biochem. 24 (1949) 410

Clark, L. C.: A Polarographic Enzyme Electrode for the Measurement of Oxidase Substrates. In: Oxygen Supply, p. 120, ed. by M. Kessler et al. Urban & Schwarzenberg, Munich 1973

Clark, L. C., *Clark*, E. W.: Differential Anodic Enzyme Polarography for the Measurement of Glucose. Adv. exp. Med. Biol. 37 A (1973) 127

Leão, A. A. P.: Further Observations on the Spreading Depression of Activity in the Cerebral Cortex. J. Neurophysiol. 10 (1947) 409

Mayevesky, A., *Zeuthen*, T., *Chance*, B.: Measurements of Extracellular Potassium, ECoG and Pyridine Nucleotide Levels during Cortical Spreading Depression in Rats. Brain Research 76 (1974) 347

Silver, I. A.: Some Observations on the Cerebral Cortex with an Ultramicromembrane Covered Oxygen Electrode. Med. Electron. Biol. Engng. 3 (1965) 377

G. Baum

An Automated Kinetic Analysis of Cholinesterase Activity by a Substrate-selective Ion-exchange Electrode

(8 Figures)

A fascinating application for ion-selective electrodes is the determination of enzymic activity. Several examples of published work in this area are given in Figure 1. Urease activity has been measured with a cation selective glass electrode (*Katz* and *Rechnitz* 1963). The activity of β-cyanoalanine synthetase was determined with an Ag_2S pellet type sulfide electrode (*Guilbault* et al. 1972). We have measured acetylcholinesterase activity using a liquid membrane acetylcholine selective electrode (*Baum*, 1972). A 1971 review of this area was given by *Christian* (1971).

Enzyme assay by electrodes

$$NH_2CONH_2 \xrightarrow{\text{Urease}} 2\,NH_4^+ + CO_2$$
Katz Rechnitz, 1963

$$CN^- + \text{Cysteine} \xrightarrow{\text{Synthetase}} HS^- + \beta\text{-Cyanoalanine}$$
Guilbault, 1972

$$CH_3CO_2CH_2CH_2\overset{+}{N}(CH_3)_3 \xrightarrow{\text{AChE}} CH_3CO_2H + HOCH_2CH_2\overset{+}{N}(CH_3)_3$$
Baum, 1972

Fig. 1. Enzyme Assay for Electrodes.

The acetylcholine selective electrode initially consisted of the potassium salt of tetraarylboron dissolved in an alkylated nitrobenzene. The electrode responds towards choline and its esters as shown in Figure 2. The sensitivity of the electrode towards choline esters increases as the hydrocarbon content of the acid increases. We have shown that the selectivity of the electrode can be attributed directly to hydrophobic interactions between the ester and the organic membrane (*Baum* 1972). I will briefly mention some of the studies conducted with this electrode relative to the assay of cholinesterase type enzymes.

These later studies were not conducted with the liquid membrane electrode, but with a plasticized PVC membrane electrode. The same ion-exchange salt was used; however, the solvent was changed in order to satisfy the requirements of the PVC for a satisfactory plasticizer. As shown in Figure 3, the plasticizer has a large influence on the selectivity ratio of the electrode for the choline ester relative to choline. The construction of the PVC membrane electrode, shown in Figure 4, is quite simple and a functioning electrode can be assembled in a few minutes.

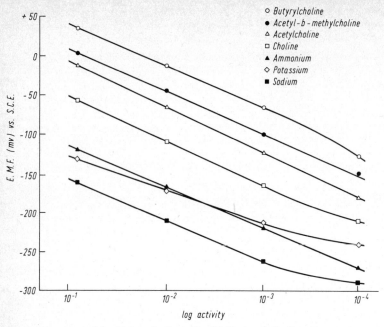

Fig. 2. Response of the Electrode Towards Choline and Choline Esters.

Since the electrode, after conditioning will respond to other choline esters than acetylcholine, it is possible to use the electrode to discriminate between different forms of cholinesterase. As indicated in Figure 5, acetylcholinesterase, the form bound to erythrocytes, will not hydrolyse butyrylcholine whereas cholinesterase, the form freely circulating in serum, will not hydrolyse acetyl-β-methylcholine. Thus by the use of these different substrates we have been able to determine the activity of both enzymes in whole blood (*Baum* et al. 1972).

The electrochemical output of the electrode is related to the log of the activity of the substrate ion being measured. As shown in Figure 6, by combining the first derivative of the Nernst equation with the pseudo first order kinetic equation for enzymic activity, the enzymic activity is seen to be linearly related to the first derivative of the potential with respect to time. The enzymic rate equation used is valid when substrate concentration is considerably greater than K_m and that condition is realized in the assay for cholinesterases.

Selectivity of membrane electrodes

Plasticizer	R ACh/Ch
Dioctylphthalate	5.4
p-Hexyl-Nitrobenzene	12
p-Hexyl-Nitrophenol	20
3-Nitro-O-xylene*	38

*) Liquid membrane.

Fig. 3. Effect of PVC Plasticizer on Selectivity.

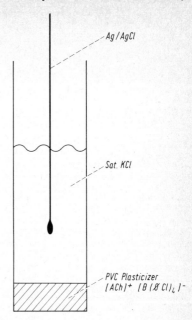

Fig. 4. Construction of the PVC Membrane Electrode.

Analog differentiators have been built for low impedance electrodes such as the oxygen electrode (*Sargent* and *Taylor* 1971), the PVC acetylcholine electrode has a resistance of about 10^6 ohms. A low cost stabilized differentiator has been designed by Mr. *D. Johnson* of our laboratories that can be interfaced either directly to the electrode or to the electrometer.

Fig. 5. Substrate Specificity of Cholinesterases.

Acetylcholinesterase-cholinesterase specificities		
	AChE	ChE
Acetylcholine	+	+
Butyrylcholine	−	+
Acetyl-β-methylcholine	+	−

Fig. 6. Relationship Between Rate of Change of Potential and Enzymic Activity.

Differentiator output is linear with enzymic activity

$$V = V' + \frac{RT}{F} \ln \gamma C$$

$$\frac{dV}{dt} = \frac{B \, d\gamma}{\gamma_t \, dt} + \frac{B \, dc}{C_t \, dt}$$

$$-\frac{ds}{dt} = k_1^1 (E)$$

$$k_1^1 [E] = \frac{0.95 \, Co}{B} \frac{dV}{dt}$$

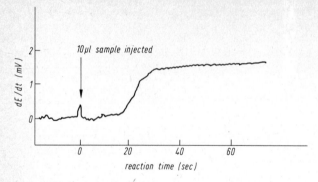

Fig. 7. Typical Experimental Run with the Differentiator.

The differentiator can accurately measure the rate of change of potential from 0.1 mV/min to 1 V/min. With the use of a flow-through reaction cell the cholinesterase activity of a sample can be determined within 30 sec. A typical experimental run is shown in Figure 7.

A major concern for serum cholinesterases is related to the presence, in a limited portion of the population, of an atypical cholinesterase which has greatly reduced esterolytic activity toward succinylcholine. Succinylcholine is often administered as a muscle relaxant in many forms of surgery. Succinylcholine functions as a blocking agent in the neuromuscular transmission system. In persons possessing atypical cholinesterase, an autosomal, recessive trait, the muscle also develops tension (*Kalow* 1972). The patient's enzyme is unable to hydrolyze succinylcholine and must be placed on a respirator. In extreme cases, prolonged apnea and death results.

The present methods for detecting the atypical phenotype are indirect. Correlations exist between the degree of inhibition cholinesterase activity towards a particular substrate by inhibitors such as Dibucaine or fluoride (*Garry* 1971). Approximately 4% of the male caucasian population are heterozygotes for the atypical enzyme. The electrode may be used to determine the estrolytic activity of serum or blood against succinylcholine directly. Dr. G. Resendes at the Terre Haute Medical Laboratory has instituted the method to guide the anesthesiologist for most cases of surgery. When an atypical phenotype is identified, the quantity of succinylcholine administered is either reduced to the level that is hydrolyzed in a few hours or an alternative blocking agent is used.

In the final Figure 8, a summary is given of the various research programs in which we have participated with investigators outside of our company. Although it would be inappropriate

---

Research interest in the ACh electrode

1. Esterolytic activity of blood fractions, cells, serum, platelets
2. Esterolytic activity of plant extracts
3. Esterolytic activity of synthetic polymers
4. Microcapillary electrodes
5. Detection of anticholinesterase agents

Fig. 8. Research Activities Employing the Acetylcholine Electrode.

to discuss their particular findings, we have received considerable confirmation of the utility and effectiveness of this research tool.

I want to acknowledge the experimental assistance of Mr. *F. B. Ward*; Dr. *M. Lynn* participated in the preparation of PVC membranes; and Mr. *D. Jonson* designed and constructed the analog differentiator.

References

Baum, G.: Determination of Acetylcholinesterase by an Organic Substrate Selective Electrode. Analyt. Biochem. 39 (1971) 65–72

Baum, G.: J. Phys. Chem. 76 (1972) 1872

Baum, G., Ward, F. B., Yaverbaum, S.: Kinetic Analysis of Cholinesterases Using a Cholinic Ester Selective Electrode. Clin. Chim. Acta 36 (1972) 405–409

Christian, G. D.: Adv. Biomed. Eng. Med. Phys. 4 (1971) 95

Garry, P. J.: Clin. Chem. 17 (1971) 183

Guilbault, G. G., Gutknecht, W. F., Kuan, S. S., Cochran, R.: Measurement of β-Cyanoalanine Synthase Activity Using a Sulfide Ion Selective Electrode. Anal. Biochem. 46 (1972) 200–203

Kalow, W.: Succinylcholine and Malignant Hyperthermia. Fed. Proc. 31 (1972) 1270–1271

Katz, S. A., Rechnitz, W.: Direct Potentiometric Determination of Urea after Urease Hydrolysis. Z. anal. Chem. 196 (1963) 248–252

Sargent, D. F., Taylor, C. P. S.: A Stable Long-Term Differentiator and its Use in the Automatic Recording of Enzyme Kinetics. Anal. Biochem. 42 (1971) 446–449

S. C. Martiny and O. J. Jensen

An Enzyme Electrode Based on Immobilized Glucose Oxidase

(1 Figure)

Substrate-specific polarographic sensors with oxidases, applying the current from the electrochemical oxidation of the enzymatically produced hydrogen peroxide, were introduced by *L. C. Clark* in 1966. A glucose-specific electrode based on an oxygen electrode, surrounded by immobilized glucose-oxidase, and measuring by means of the decrease of the polarographic, oxygen-controlled current, has also been constructed by Hicks and Updike. Contrary to Hicks and Updike, Clark proposed an electrode with glucose-controlled current based on the reactions:

$$\text{Gluose} + O_2 \xrightarrow{\text{enzyme}} \text{Gluconic acid} + H_2O_2$$
$$H_2O_2 \xrightarrow{\text{Pt anode}} 2\,H^+ + O_2 + 2\,e^-$$

In our first attempt to bring Clark's idea into practice we used a platinum electrode surrounded with plain GOD in buffer solution soaked into filter paper and separated from the sample solution by a semipermeable membrane. However, this device showed the main problem of enzyme electrodes: poor stability. This comes partly from instability of the enzyme and partly from contamination of the metal electrode. The first problem was solved by solvent coupling of GOD to cellulose spacers and inclusion of these spacers behind a massive, dialysis membrane.

GOD is coupled to Joseph paper, also called lens paper, through a triazenyl bridge. A procedure somewhat like ours has been described by Kay and Lilly in several papers, but the direct Kay-and-Lilly-procedure gave poor activity yield in our experience. Introducing a derivative of cellulose, treating it with cyanuric chloride and reacting the triazenyl cellulose with enzyme, we produce enzyme spacers with an acitivity of 6–10 mU, i.e., each spacer is able to produce 6–10 nmole H_2O_2 per min at 25 °C, pH = 6 and 0.1 M glucose as substrate. We call the membrane-immobilized enzyme "enzyme spacer", since it fills the space between the platinum electrode and an inert dialysis membrane when it is incorporated in the electrode system. The electrode arrangement is shown in Figure 1.

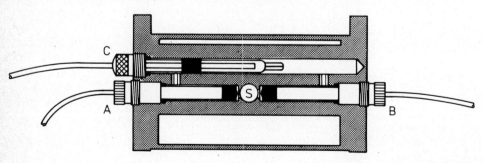

Fig. 1.

A is the enzyme-containing electrode with the "enzyme spacer" situated between the platinum tip and a Cuprophane membran. B is an electrode similar to A with an inactive piece of lens paper as spacer, and C is a common reference electrode. S is the sample cuvette, which has a magnetic stirrer in the bottom, and the whole thing is enclosed in a thermostatic housing.

When working with the system 20 µl of the sample are syringed into the sample cuvette, which holds 2 ml of diluting buffer. Twice a day you prepare a calibration curve to convert the readings of the apparatus into glucose concentrations. This is necessary for exact measurements, since the stability is specified with maximum 5% day-by-day variation. About each 2 weeks the enzyme is replaced. With a proper calibration curve a precision of typical $\pm 1.5\%$ (calculated as s.d.m.) is obtained, and each measurement takes about 50 seconds. 20 µl samples with glucose concentrations ranging from 1 mM to 30 mM = 20 — 500 mg/100 ml give a perfect linear calibration curve.

The Enzyme Electrode System, which has been described, can be used for clinical glucose determinations. For this purpose the GOD-spacers have been developed. Other substrates which can be quantified in a similar way are uric acid (uricase as the enzyme) and alcohol (with alcohol oxidase). However, in these cases oxidation of the substrate itself and similar chemical compounds may occur directly on the electrode without a previous enzyme reaction. Therefore a H_2O_2-permeable, but substrate impermeable membrane of cellulose acetate should be placed between the spacer and the tip of the platinum electrode. Further, it might be possible to use double enzyme membranes, where one membrane produces the substrate for the next, which produces H_2O_2. An example is invertase-GOD for measurements of surcrose. We have not investigated all these possibilities, but in our opinion there are many applications beside glucose measurement.

F. G. K. Baucke

Reference Electrodes for Measurements with Ion-sensitive Electrodes. The Importance of the Liquid Junction Potential

(2 Figures)

1. Introduction

In this brief paper some principles of reference electrodes which are of importance for practical work with ion-sensitive electrodes will be discussed. In the experience of the author who is involved in electrode manufacturing and dealing with users' problems, most difficulties arise at the liquid junction (which is rarely considered the source of malfunction) and the errors introduced by the liquid junction potential are always underestimated. The following will thus be limited to basic remarks on the liquid junction in potentiometric work. For an extensive treatment of reference electrodes, the reader is referred to monographs and reviews in the literature (*Bates* 1973; *Covington* 1961; *Covington* 1969; *Ives* and *Janz* 1961; *Janz* and *Ives* 1968; *Janz* and *Taniguchi* 1953).

2. The potentiometric cell

The potential difference between an ion-sensitive membrane and the adjacent solution is, in principle, not directly measurable. Since it is generated by an ion exchange between the two phases, which does not involve the fundamental step of electrochemistry, charge transfer between ionic and electronic conductors, the two-phase system must be extended by adding phases to both electrolytes so that a complete electrochemical cell with electrodes as terminating phases is formed. The potential difference of interest is necessarily only one of several Galvanic potentials the sum of which, the electromotive force of the cell, is the measurable quantity. It can, at best, be as accurate as the least accurate of the galvanic potentials involved. Although reference electrodes, in this sense, are not of primary interest, but rather are necessary complications, it is important to understand their functioning as well as that of the ion-sensitive electrode itself.

3. The liquid junction

Of the two groups into which potentiometric cells are generally divided, cells with and without transference, only the former and, more specifically, cells employing a reference electrode with fixed potential are considered here. The most general example is the cell

$$\text{Ag/AgCl (sat), KCl } (m_i) \text{ pH}_i \text{ / glass / sol'n X // KCl } (m_2), \text{ AgCl (sat)/Ag} \qquad (1)$$

containing an internal reference half-cell without transference (left) and an outer reference electrode of the second kind with fixed potential (right). Inner reference electrodes have recently been discussed (*Bates* 1969). The potential of the outer Ag/AgCl system is "fixed" by the chloride activity of its KCl solution at each temperature. Application of the electrode

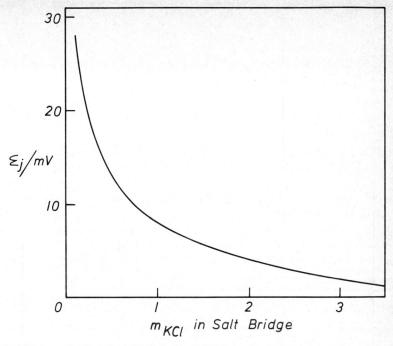

Fig. 1. Influence of KCl concentration within bridge of cell (3). Data by *Guggenheim* as plotted by *Bates* (Bates 1973).

according to Equation (1), however, introduces necessarily a junction between two electrolyte solutions, irreversible diffusion into the cell reaction, and a diffusion potential into the cell emf,

$$\varepsilon_j = - \frac{RT}{F} \int_I^{II} \sum \frac{t_i}{z_i} \, d \ln a_i, \qquad (2)$$

which cannot be calculated since single ion activities a_i, their concentration dependence, as well as that of transport numbers t_i, and the "geometry" of the boundary between solutions I and II are unknown*). Only by introducing certain assumptions can the magnitude of liquid junction potentials be estimated.

4. Practical considerations

Perhaps the most impressive experiment demonstrating the effect of concentration of the bridge solution upon the cell emf and the junction potential was conducted on the cell

$$Hg/Hg_2Cl_2, 0.1 \text{ M HCl} // X \text{ M KCl} // 0.1 \text{ M KCl}, Hg_2Cl_2/Hg \qquad (3)$$

(*Guggenheim* 1930; Fig. 1). In order to reduce the liquid junction potential, concentrated solutions of 1.1-electrolytes consisting of ions with similar mobilities are used either as bridges

---

*) Exceptions are "concentration cells" and cells according to *Lewis* and *Sargent* (see, for instance, *Ives* and *Janz* 1961).

between reference and unknown solution or as solution of the reference electrode itself, which then forms its own salt-bridge. Only few electrolytes fulfill this condition satisfactorily. KCl, KNO_3, and NH_4NO_3 are widely used. For practical reasons, standard potentials of reference electrodes with fixed potential contain liquid junction potentials and, consequently, are uncertain by this magnitude. Measurements based on standard solutions are thus in error by the residual liquid junction potential. For the most commonly used electrodes, calomel, silver-silver chloride, and Thalamid®, the data are given as functions of temperature in Figure 2.

In some cases, bridges with electrolytes other than those mentioned must be applied in order to avoid contamination of the test solution and interference with ion-sensitive electrodes. There is no great choice. 3 M RbCl was used successfully with sodium ion activity measurements by means of microelectrodes, and microdeterminations of blood pH were conducted with an equilibrium diffusate of plasma as bridge electrolyte (*Khuri* 1969). Use of 0.152 M NaCl as bridge solution resulted in pH values too low by 0.11 units (*Semple* 1961), which may be understood by considering the results demonstrated in Figure 1. Use of electrolytes with doubly charged cations is not very promising because of the greater tendency of the ions to form complexes and insoluble compounds. Application of ion-sensitive, e.g., glass electrodes (*Khuri* 1969) for reference purposes is an elegant method; it means, however, applying a cell without transference, and constant activity of the reference ion must be secured.

The liquid junction potential is constant as long as a steady state is maintained at the boundary between the two solutions; the magnitude of the potential depends on the "geometry" of the interphase (see Equation 2). The structure of the device in and at which the junction is formed is thus of practical importance. Numerous basic types of liquid junction devices have been investigated; this work has been reviewed (*Bates* 1973). For practical applications and in commercial electrodes, however, the convenient dip or immersion type junction is used almost exclusively, through which the reference or bridge solution should stream into the test solution at a low and constant rate. Turbulence of the stream lines causes fluctuating potentials and accounts for the dependence of the junction potential on stirring. We have found potential differences between stirred and unstirred 0.1 M HCl in contact with 3.5 M KCl solution up to 3 mV. Ceramic plugs, asbestos fibers, and sintered ceramic or glass disks are used. Sleeve type junctions are recommended for emulsions and protein solutions since they do not tend to clog and are easily cleaned. Threaded platinum wires melted into the glass wall ("platinum diaphragm") (*Detemple* 1966) provide well-defined leakage paths generating constant potentials; disturbing redox potentials at the smooth platinum have been observed in only a few instances. Linen fiber junctions were reported to react with tris buffer (*Ryan* 1969). Double junction type electrodes contain two diaphragms and allow a convenient and fast exchange of the bridge solution.

Particular mention must be made of reference electrodes used in micropotentiometric work and of corresponding salt bridges. Since the liquid junction should be formed between the bridge solution and the actual solution under investigation, e.g., within a single cell, extremely small capillaries of similar size as the ion-sensitive electrodes must be introduced. The resistance between measuring solution and meter is thus unusually high and may even be comparable to that of the ion-sensitive microelectrode. A 0.2 mm long and 1 μm wide thread of 2 M KCl solution, for instance, exhibits a resistance of 10^3 ohms. Source resistances

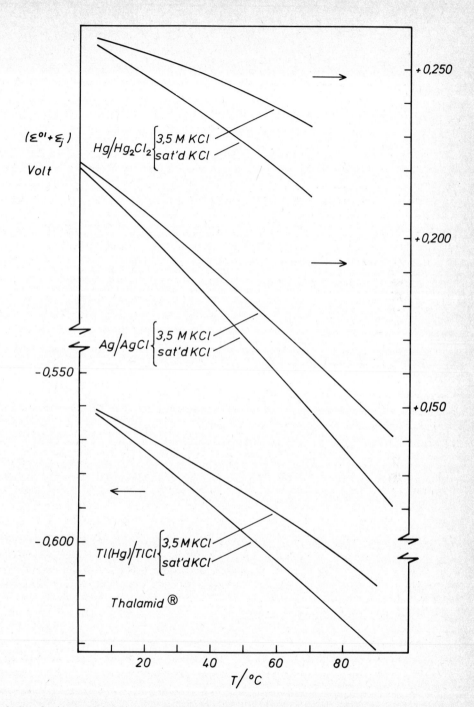

Fig. 2. Standard potentials ($\varepsilon^{o\prime} + \varepsilon_j$) of reference electrodes with fixed potential: calomel (*Ives* and *Janz* 1961), Ag/AgCl, 10–40 °C (*Bates* 1969), 0–95 °C (*Baucke*, 1975), Thalamid® (*Baucke* 1971, 1974).

of this magnitude demand the use of symmetric high-input resistance voltmeters and efficient shielding also of electrolyte bridges and reference electrodes.

Since any flow of bridge electrolyte solution into the cell as well as of cell solution into the capillary by small pressure differences should be avoided, the bridge solution should be congealed with agar-agar. Junction potentials between such KCl and adjacent HCl solutions were found to be reproducible and stable although slightly different from those measured without agar-agar (*Bates* 1973).

The temperature of a reference electrode must be well-controlled if the salt bridge separates it spatially from the test solution as in micropotentiometry. Since the temperature coefficient of electrodes with fixed potential is in the order of 0.4 to 0.8 mV K^{-1} (Fig. 2), the error introduced by temperature changes during measurements may be considerable. Constant and identical temperature of test solution and reference electrodes eliminates this source of error and also excludes thermodiffusion potentials.

References

Bates, R. G.: Inner Reference Electrodes and Their Characteristics. In: Glass Microelectrodes, pp. 1–23, ed. by *M. Lavallee, O. F. Schanne, N. C. Hebert.* Wiley, New York 1969

Bates, R. G.: Determination of pH, Theory and Practice. Wiley, New York 1973

Baucke, F. G. K.: Standard Potentials ($\varepsilon^{\circ\prime} + \varepsilon_j$) of the Thalamid® Reference Electrode, Hg, Tl (40 wt %)/TlCl (s)/KCl(s)// ..., in Aqueous Solution Between 5 and 90 °C. J Electroanal. Chem. 33 (1971) 35–144

Baucke, F. G. K.: Standardpotentiale ($\varepsilon^{\circ\prime} + \varepsilon_j$) und Polarisationsverhalten der Thalamid-Bezugselektrode (3.5 mol/l und ges. KCl) zwischen 5 und 90 °C. Chem. Ing. Techn. 46 (1974) 71

Baucke, F. G. K.: Standardpotentiale ($^{01} + e_j$)$_T$ der Silber/Silberchlorid-Elektrode in 3.5 m und in ges. KCl unter Verwendung entsprechender („Cl$^-$-ionensensitiver") Membranelektroden (0–95 °C) Chemie-Ing.-Techn. 47 (1975) 565–566

Covington, A. K.: Reference Electrodes. Academic Press, New York 1961

Covington, A. K.: Reference Electrodes. In Ion-Selective Electrodes, pp. 107–141, ed. by *R. A. Durst.* NBS Special Publication 314, Washington 1969

German Pat. 1498827, 1966 (Detemple, M)

Guggenheim, E. A.: Study of Cells with Liquid-Liquid Junctions. J. Amer. Chem. Soc. 52 (1930) 1315–1336

Ives, D. J. G., Janz, G. J. (eds.): Reference Electrodes, Theory and Practice. Academic Press, New York 1961

Janz, G. J., Ives, D. J. G.: Silver, Silver Chloride Electrodes. Ann. N. Y. Acad. Sci. 148 (1968) 210–221

Janz, G. J., Taniguchi, H.: The Silver-Silver Halide Electrodes. Chem. Rev. 53 (1953) 397–437

Khuri, R. N.: Cation and Hydrogen Microelectrodes in Single Nephrons. In: Glass Microelectrodes, pp. 272–298, ed by. *M. Lavallee, O. F. Schanne, N. C. Hebert.* Wiley, New York 1969

Khuri, R. N.: Ion-Selective Electrodes in Biomedical Research. In: Ion-Selective Electrodes, pp. 287–310, ed. by *R. A. Durst.* NBS Special Publication 314, Washington 1969

Ryan, M. F.: Unreliable Results. Science 165 (1969) 851

Semple, S. J. G.: J. appl. Physiol. 16 (1961) 576

R. Olfers-Weber, W. Stockem and K. E. Wohlfarth-Bottermann

Cytological Aspects of Membrane Regeneration After Experimental Injury of Amebae and Acellular Slime Molds

(7 Figures)

During the course of extensive experiments *Heilbrunn* (1958) came to the opinion that, after experimental injury of the plasma membrane, cells are able to close naked areas of the protoplasm within a short period of time and to survive such a catastrophe by generating new membrane. Later these observations were confirmed by experiments on egg cells, ciliates, amebae and slime molds (*Chambers* and *Chambers* 1961; *Stewart* and *Stewart* 1961; *Schneider* 1964; *Janisch* 1964, 1967; *Griffin* et al. 1969; *Wohlfarth-Bottermann* and *Stockem* 1970; *Szubinska* 1971). There were, however, differences of opinion with regard to the speed and nature of the regeneration process. The quantitative information on the time needed for the regenertion of the membrane fluctuated between "too short to be recorded" and several seconds (*Schneider* 1964; *Janisch* 1967; *Stewart* and *Stewart* 1961; *Wohlfarth-Bottermann* and *Stockem* 1970). As far as the nature of the regeneration mechanism was concerned, only two possibilities were discussed: namely an exocytotic regeneration mechanism by membrane vesiculation processes and a formation of a cell membrane by condensation or precipitation of molecular elements of the protoplasm on its surface.

The following experiments were carried out in order to contribute to the analysis of both problems. Furthermore, particular attention was paid to a problem which is relevant for electrophysiological experiments: namely, whether the type of the phase boundary (i.e. air cytoplasm or glass cytoplasm) has an effect on the speed and nature of the membrane regeneration.

Materials and Methods

a) *Experiments on slime molds*

The culture of slimemolds (*Physarum polycephalum*) was carried out in the usual way (*Camp* 1936). The regeneration of the plasmalemma was examined on drops of slime molds prepared according to the method described by *Wohlfarth-Bottermann* (1959), (Fig. 1a and b) and on isolated protoplasm. This was prepared by sucking out protoplasmic veins with the help of pipettes and finally squirting the content of the pipettes onto cellophane paper (Fig. 1c and b).

Both samples were fixed at the following times after the beginning of the experiments: 0, 1, 2, 3, 6 and 20 seconds 1, 10 and 30 minutes. Protoplasm/air, protoplasm/Ringer-solution as well as protoplasm/EDTA-solution (0.073%) were used as phase boundaries, whereby 0.5% Aerosil had been added to the latter.

The following fixatives were used:
1. 2 or 4% OsO_4 with 1 or 2% $K_2Cr_2O_7$ in KOH respectively 0.05 M cacodylate-buffer at pH 7.0–7.25; period of fixation: 1 hour.

2. 2% glutaraldehyde in 0.05 M Na-cacodylate-buffer with 0.5% CaCl$_2$ and 6.5% saccharose at pH 7.2; period of fixation: 1 h; postfixation with 1% OsO$_4$ in 0.05 M Na-cacodylate buffer with 0.5% CaCl$_2$ and 6.5% saccharose at pH 7.2; period of fixation: 1 hour.

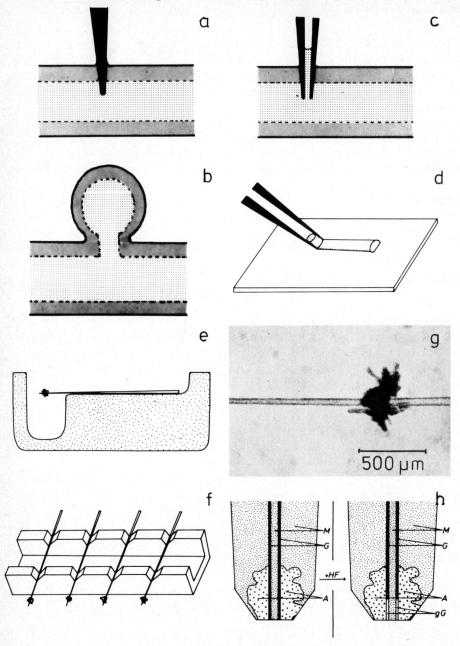

Fig. 1. Different methods for the investigation of the process of membrane regeneration after experimental injury of acellular slime molds (a–d) and of amebae (e–h). For detailed description see text.

The embedding was performed in styrol-methacrylate (*Kushida* 1961; *Stockem* and *Komnick* 1970).

b) *Experiments with amebae*

The amebae (*Chaos chaos*) were cultivated according to the method already described elsewhere (*Weber* 1972). Cells which had been starved for 1–2 days were transferred with a pipette into specially formed blocks of agar, bored through by a glass-microelectrode and fastened in the agar chamber (Fig. 1e and g). Then 4 to 6 cells were set up on a frame which guaranteed an embedment absolutely safe from vibration (Fig. 1f). This condition must be met in order to avoid any mechanical destruction of newly formed membrane in the close proximity of the microelectrode after the fixation has been accomplished.

The amebae were fixed at the following times after the beginning of the experiments: 0, 1, 2, 3, 5, 10, 20, 30 and 40 seconds. The phase-boundary always consisted of the ameba's cytoplasm on one side and the microelectrode filled with Pringsheim-solution on the other side. Os/Cr as well as glutaraldehyde-solution were used for fixation (compare Materials and Methods a. 1 and 2). It was necessary to treat the amebae embedded in styrol-methacrylate with hydrofluoric acid (HF) in order to remove the microelectrode remaining in the ameba carefully and to achieve technically good ultrathin sections (Fig. 1h).

Results

a) *Physarum polycephalum*

Earlier investigations on isolated protoplasm as well as drops of slime molds showed no differences in the mechanism of membrane regeneration (compare *Wohlfarth-Bottermann* and *Stockem* 1970). Therefore both groups of experiments will be discussed parallelly.

The surface of 0-second-old protoplasm is immediately limited by the outer medium without being protected by a membraneous layer (Figs. 2a and 3a). The bordering area contains, apart from separate mitochondria and nuclei, numerous vacuoles with slime and nutritive substances as well as a number of protoplasmic enclosures which are not limited by membranes (lipid droplets, ribosomes etc.).

After 1 second a change can already be seen in the pattern of distribution of these structures. Immediately beneath the protoplasmic surface, which still borders directly on the outer medium, a 5–10 µm deep accumulation of nutritive and slime vacuoles of different sizes has formed. This layer also contains nuclei and mitochondria (Figs. 2b and 3b) and seems to form a temporary limiting barrier against the outer medium.

After 2 seconds the formation of large vacuoles, flattened parallel to the surface of the protoplasm (Figs. 2c and 3c) occurs by progressive confluence of slime and nutritive vacuoles. Finally, these large vacuoles merge into one another until only one, continuously limiting vacuole has been formed (Fig. 3d). Beneath this vacuole a zone of approximately 10 µm that is rich in ground cytoplasm and void of vesicular material is formed (compare Figs. 2c as well as 3c and d). The limiting vacuole separates two different portions of protoplasm: the distal membrane of the vacuole represents the basic limitation of a protoplasmic portion amounting to only a few µm; it is soon destroyed (Fig. 3d, dP). The proximal membrane of the vacuole forms the final border, i.e., the new plasmalemma of the remaining mass of protoplasm (Fig. 3d, ÜP). This protoplasm is furthermore viable and survives the operation.

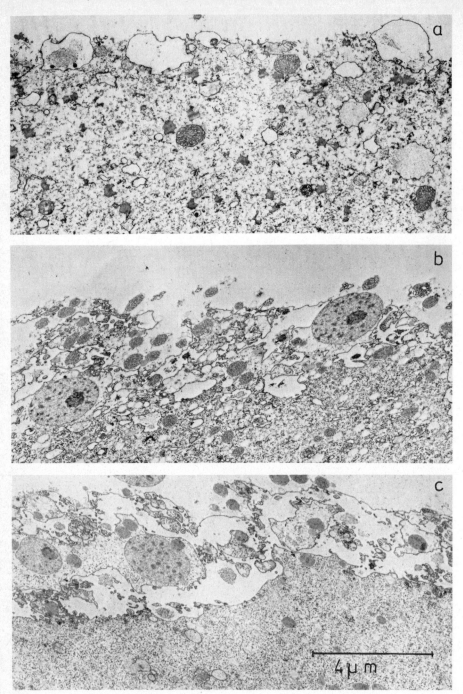

Fig. 2. Different stages of membrane regeneration in slime molds 0 s (a), 1 s (b) and 3 s (c) after preparation of naked protoplasm. For detailed description see text. (From *Wohlfarth-Bottermann* and *Stockem* 1970). 8000 ×.

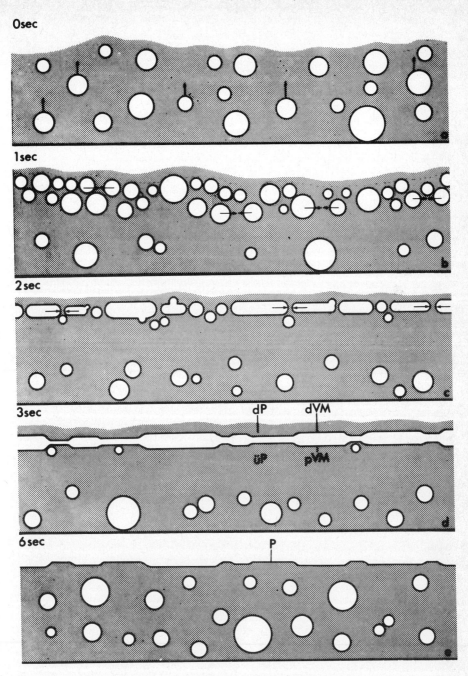

Fig. 3. Diagrammatic representation of the processes of membrane regeneration in acellular slime molds. dP degenerating protoplasm; dVM distal membrane of the limiting vacuole; pVM proximal membrane of the limiting vacuole; P new plasma membrane; üP surviving protoplasm. The arrows indicate the direction of movement of the different vacuoles (From *Wohlfarth-Bottermann* and *Stockem* 1970).

A few seconds later the distal membrane of the vacuole breaks down into single fragments. These are then washed off together with the degenerated protoplasm by the outer medium so that the surface of the protoplasm is again enclosed by one membrane, i.e., the new plasmalemma (Fig. 3e).

The composition of the outer medium used in these experiments exerted no influence on the mechanism of membrane regeneration and on the speed of this process. This can be concluded from the fact that experiments with air, Ringer solution or EDTA-solution produced the same results.

By adding Aerosil as an indicator it has been shown that the regenerated membrane functions with absolute reliability and that there was no infiltration of foreign substances even while the protoplasm remained without a surface, i.e., no particles of Aerosil penetrated the cytoplasm for the 2 to 3 seconds before the new membrane was formed.

b) *Chaos chaos*

As opposed to the investigations on slime molds, conventional microelectrodes were used in experiments with *Chaos chaos*. Here the electrodes remained within the amebae after the puncture up to the time when the ultrathin sections were made. They were carefully removed by hydrofluoric acid shortly before sectioning in order to avoid mechanical influences in the bordering area where the process of membrane regeneration takes place.

Thus the technical conditions were optimalized to investigate these processes. Figure 4 shows a cross-section of the top area of an electrode a few seconds after the puncture. The inner dark limitation (⇒) represents the former pattern of the surface of the electrode. This particular phenomenon is an instance of osmiophilic precipitation which *Szubinska* (1971) mistakenly

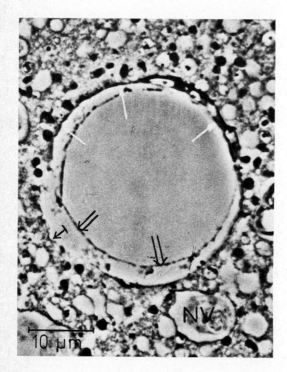

Fig. 4. Semithin cross-section through the protoplasm of *Chaos chaos* showing the channel left after removing the glass microelectrode by HF. NV nutrition vacuole; → new plasma membrane; ⇒ osmiophilic depositions indicating the original surface of the microelectrode; ↦ protoplasmic region still lacking a new membrane. 1500 ×.

Cytologial Aspects of Membrane Regeneration

interpreted as membrane regeneration. The new membrane (→) borders immediately on the protoplasm, and is at this time completely closed except for a few places (↦).
The analysis of sequential phases, each of a different age, clearly illustrates the mechanism of

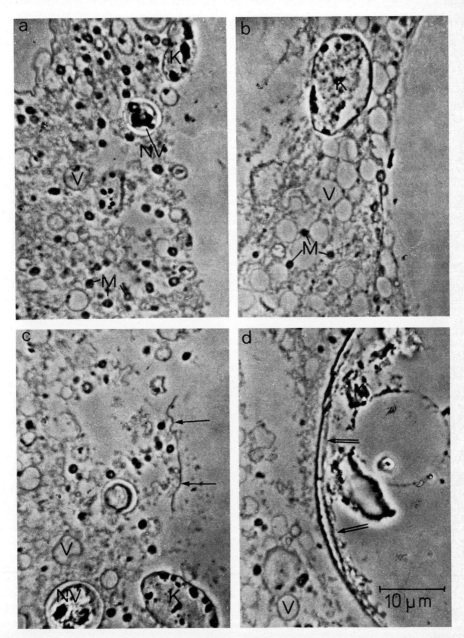

Fig. 5. Different stages of membrane regeneration in *Chaos chaos* 0 s, 1–2 s, 2–3 s and 5–6 s after experimental injury with a microelectrode, Nv nutrition vacuoles; V slime vacuoles, M mitochondria; K nuclei; the arrows and double arrows point to the new plasma membrane. 1500 ×.

membrane regeneration after the puncture of *Chaos chaos* with glass electrodes. Immediately after the introduction of the electrode, the protoplasm still borders directly on the surface of the electrode, i.e., there is no membraneous limitation (Figs. 5a and 7a). All the cell organelles are distributed in the normal way similar to the distribution in slime molds.

After 1 or 2 seconds a distinct accumulation of vacuoles as well as mitochondria and nuclei can be observed in the protoplasm near the electrode (Figs. 5b and 7b).

This layer of intact membraneous structures together with single membrane fragments (Fig. 7b, nM) forms a provisional seal against the surface of the microelectrode. After 2 to 3 seconds an increase in the number of membrane fragments can be observed. The membranes cover the surface of the protoplasm by straightening out parallel to the bordering phase (Fig. 5c, arrows). They can appear singly, folded or irregularly stacked (Fig. 7c). But, in contrast to the limiting vacuoles of slime molds, these membranes fuse in such a way that their ends always remain open. In the succeeding period a distinct enlargement of these membrane areas can be observed. The small open areas of the protoplasm, which can still be found after 3 to 4 seconds (Fig. 7d) are closed quickly so that after 5 to 6 seconds the puncture is again completely sealed by a new plasma membrane (Fig. 5d, double arrows and 7e, nM as well as Figs. 6a and b). The newly formed plasmalemma is covered by a filamentuous mucous layer typical of amebae so that it can be interpreted as a genuine cell membrane (compare Fig. 6 nM).

It is furthermore remarkable that, similar to the situation in slime molds, the space between the new cell membrane and the microelectrode, on the surface of which the osmiophilic precipitations can be seen (compare Figs. 6, arrows, and 7e, oN), is often filled up by degenerating cytoplasm which is discharged during the process of membrane regeneration (compare Fig. 5d).

Discussion

The results of the present investigation have delivered concrete morphological criteria which support the theory that the regeneration of membrane material in slime molds and amebae after experimental rupture of the plasmalemma proceeds according to vesiculation processes similar to exocytosis. Thereby, the assumption of *Stewart* and *Stewart* (1961) was confirmed which proposed that above all there are preformed membranes in the form of cytoplasmatic vacuoles or other membranous structures involved in sealing the wound. However, there were no indications found supporting the most frequent assumption on the genesis of new membranes, i.e., the de novo development of membranes by condensation processes on a molecular level, in other words, the formation of precipitation membranes (*Heilbrunn* 1958, *Chambers* and *Chambers* 1961, *Janisch* 1964, 1967, *Schneider* 1964 as well as *Szubinska* 1971). Of particular relevance here would be a comparison between the results achieved with *Chaos chaos* and the observations made by *Szubinska* (1971) on *Amoeba proteus*.

Accordingly, *Szubinska's* method of withdrawing the needle before fixation seems to be rather unsuitable as it conceals the danger of artificial changes. This was confirmed by personal observations that certain parts of the protoplasm cling to the needle after it has been withdrawn. This probably accounts for the fact that *Szubinska* was not able to find a membrane in those regions where (according to her) it would be most necessary and where it indeed has been found in the present investigations, namely, near the wound. Taking into consideration that the membrane regeneration with *Chaos chaos* as well as with *Physarum polycephalum* does not take longer than 5 seconds and comparing this period of time to the period of 35 to 40 seconds which *Szubinska* allowed between the injury and the fixation of *Amoeba proteus*, this period seems much too long to give information an the mechanism of

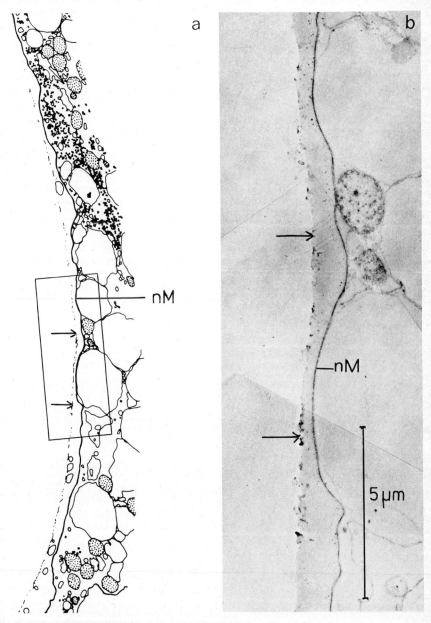

Fig. 6. a) Drawing reproduced from electron micrographs taken along the border of the microelectrode and the cell surface in *Chaos chaos* approximately 6 s after the electrode was inserted. b) Higher magnification from the region marked in 6a). The arrows point to osmiophilic depositions; nM new plasma membrane. a) 2500 ×, b) 9000 ×.

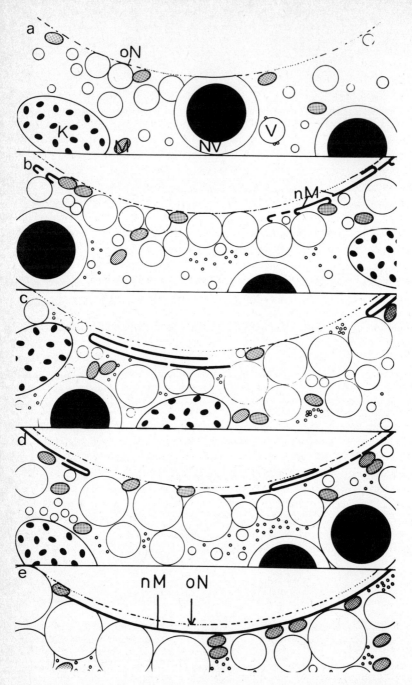

Fig. 7. Diagrammatic representation of the processes of membrane regeneration in *Chaos chaos*. K nuclei; M mitochondria; nM new plasma membrane; oN osmiophilic depositions indicating the original surface of the microelectrode; nV nutrition vacuoles. For more detailed description see text.

membrane regeneration. It may therefore be assumed that the dense droplets found by *Szubinska* were osmiophilic precipitations such as, according to personal observations, preferably occur on bordering areas. On the whole, one can conclude that there are as yet no conclusive investigations of adequate structural precision which support membrane regeneration by precipitation or condensation processes and which refute the cytotic wound-seal mechanism as postulated by *Stewart* and *Stewart* (1961).

As opposed to these contradicting theories on the nature of the regeneration mechanism, all investigations of this problem have shown that in principle the cell is capable of sealing injuries to the plasmalemma within a few seconds and even survives such major operations without permanent damage. The nature of the external phase which borders on the protoplasm seems to be of little importance. Even in mediums of low calcium content the wound-sealing functions perfectly (compare *Wohlfarth-Bottermann* and *Stockem* 1970). This is especially interesting since it seems to apply to conditions of experimental procedure in the electrophysiological field. One must therefore conclude, that when measuring membrane potentials by means of microelectrodes in cells and tissues one should not loose sight of the fact that either the whole electrode or at least large parts of it are covered by a newly formed membrane, and that therefore the measured values are not of genuine intracellular derivation. However, the immense quantity of apparently significant electrophysiological results still awaits interpretation with respect to those findings. Before the question can be answered and before evaluation of our own results, more extensive and thorough examination of this problem is necessarey. Examination of the still unclarified membrane conditions in electrophysiological measurements on single cells or tissues and the encouragement of further investigations are therefore the main aims of our present studies.

Summary

Investigations on slime molds and amebae demonstrated that experimental injuries of the cell membrane can be sealed in a few seconds by a membrane-flow mechanism similar to exocytosis. The material for the new membrane originates either from cytoplasmic vacuoles (secretion and nutrition vacuoles) or it is replaced by the disposal of mitochondria and nuclei which function only as suppliers of membrane in emergencies.

Indications for the participation of precipitation processes on a molecular level in the regeneration of new membrane could not be found. The results also point to the fact that during electrophysiological measurements with glass electrodes it must be taken into account that at least parts of the electrode are covered by a new membrane within a few seconds and therefore have no open connection to the cytoplasm.

References

Camp. W. G.: A Method of Cultivating Myxomycete Plasmodia. Bull. Torrey bot. Club 63, (1936) 205–210

Chambers, R., Chambers, E. L.: Exploration into the Nature of the Living Cell, p. 100. Harvard Univ. Press, Cambridge, Mass. 1961

Griffin, J. L., Stein, M. N., Stowell, R. E.: Laser Microscope Irradiation of *Physarum polycephalum*: Dynamic and Ultrastructural Effects. J. Cell Biol. 40 (1969) 108–119

Heilbrunn, L. v.: Grundzüge der allgemeinen Physiologie. Deutscher Verlag der Wissenschaften, Berlin 1958

Janisch, R.: Sub-Microscopic Structure of Injured Surface of *Paramecium caudatum*. Nature (London) 204 (1964) 200–201

Janisch, R.: Electron Microscopy Study of Regeneration of Surface Structures of *Paramecium caudatum*. Fol. biol. 13 (1967) 386–392

Kushida, H.: A Styrene-Methacrylate Resin Embedding Method for Ultrathin Sectioning. J. Electronmicr. 10 (1961) 16–19

Schneider, L.: Morphogenese und Dynamik cytoplasmatischer Membranen. Verh. Dtsch. Zool. Ges. Kiel 1964, pp. 243-272. Akad. Verlagsges. Geest und Portig, Leipzig 1964

Stewart, B. T., Stewart, P. A.: Electron Microscopical Studies of Plasma Membrane Formation on Slime Molds. Norelco Reporter 7 (1961) Nr. 117

Stockem, W., Komnick, H.: Erfahrungen mit der Styrol-Methacrylateinbettung als Routinemethode für die Licht- und Electronenmikroskopie. Mikroskopie 26 (1970) 199–203

Szubinska, B.: „New Membrane" Formation in *Amoeba proteus* upon Injury of Individual Cells. Electron Microscope Observations. J. Cell Biol. 49 (1971) 747–772

Weber, R.: Untersuchungen zur experimentell induzierten Membrangenese bei *Chaos chaos*. Dipl.-Arbeit, Math.-Naturwiss. Fak. der Univ. Bonn 1972

Wohlfarth-Bottermann, K. E.: Die elektronenmikroskopischen Untersuchungen cytoplasmatischer Strukturen. Verh. Dtsch. Zool. Ges. Münster i. Westf., Zool. Anz., Suppl. 23 (1959) 393–419

Wohlfarth-Bottermann, K. E., Stockem, W.: Die Regeneration des Plasmalemms von *Physarum polycephalum*. Wilh. Roux' Arch. 164 (1970) 321–340

D. Schäfer and J. Höper
with technical assistance of H. Gudat

Alterations in Rat Liver Cells and Tissue Caused by Needle Electrodes

(4 Figures)

The measurement of various parameters with needle electrodes in living tissues and cells has become an indispensible method of physiological investigation during the past few years. However, to our knowledge there are no detailed studies of the alterations caused in animal tissues and cells by the insertion and presence of glass electrodes. At the ultrastructural level, technical difficulties are probably the main reason.

These difficulties can be overcome by new preparation techniques. The glass of the electrodes can be removed from the polymerized synthetic resin embedding material with hydrofluoric acid without causing appreciable damage to the adjacent tissue (*Stockem*, 1967; *Olfers-Weber, Stockem, Wohlfarth-Bottermann* 1976). We examined the effects of unfilled glass electrodes with a tip diameter of about 2 µm on the ultrastructure of the rat liver. Electric potentials were not applied.

To prevent secondary alterations due to respiratory movements of the animal the experiments were performed on the isolated perfused rat liver (*Kessler*, 1968). The electrode was inserted into the completely motionless liver with the aid of a micromanipulator and after half an hour of continued perfusion the organ was perfused with the fixation liquid instead of the normal medium by switching a three-way valve.

The present paper is exclusively concerned with alterations of the liver in the region of the electrode shaft. A later report will be concerned with morphological conditions observed at the tip of the electrode.

Materials and Methods

Object: Isolated hemoglobin-free perfused rat liver (Wistar AF, females) (*Kessler*, 1968)
Perfusion: Krebs-Ringer-Henseleit's solution containing 3.5% bovine albumine and 35 mMol $NaHCO_3$, equilibrated with 95% O_2 + 5% Co_2 (pH 7.4), 22 °C at a perfusion pressure of 140 mm H_2O
Fixation: 2% glutaraldehyde in 0.1 M cacodylate buffer solution + 0.2 Mol saccharose (pH 7.4) 470 mosm, 20 min by perfusion (*Hündgen, Schäfer, Weissenfels*, 1971) followed by immersion of the tissue samples for one hour
Washing: 0.24 M saccharose in 0.1 M cacodylate buffer solution (pH 7.4), 4 °C, 290 mosm, 6 times 10 min
Contrasting: 1% OsO_4 + 0.24 M saccharose in 0.1 M cacodylate buffer solution (pH 7.4), 4 °C, 360 mosm for 1 hour
Dehydration: Ethylalcohol series from 15% to absolute alcohol by 7 steps, 20 °C for 2½ h in all
Postcontrasting: 1% phosphotungstic acid + 1% uranyl acetate in 70% alcohol (block); uranyl acetate and lead citrate (ultrathin section)

Embedding: Epon 812 (*Luft,* 1961)
Removal of the glass electrode from the Epon block: 40% hydrofluoric acid for 24 hours, subsequently H_2O for 24 hours (*Drum,* 1962; *Olfers-Weber, Stockem, Wohlfarth-Bottermann,* 1976)

Results

Owing to the technique of removing the glass electrode from the polymerized Epon block with hydrofluoric acid we succeeded in identifying the electrode channel clearly. This allowed us to detect the site of the puncture already in the semithin section and to follow the path of

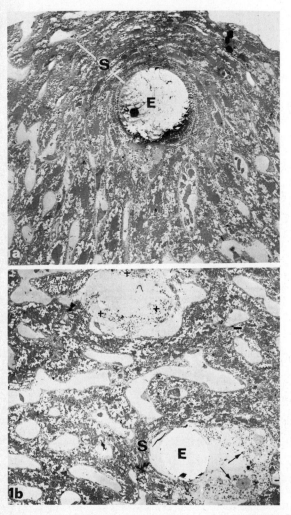

Fig. 1a and 1b. Electron micrograph surveys of an isolated perfused rat liver $1/2$ hour after insertion of a glass needle electrode, which was removed from the Epon block with hydrofluoric acid. S: compression zone, which depends on the diameter of the electrode; →: dying cells in the immediate vicinity of the electrode; +: dying cells without contact with the electrode. Magn.: 430 x.

the electrode in the tissue easily. It was found (Figs. 1a and b) that the extent of the tissue alterations obviously depends on the diameter of the shaft of the electrode. The thicker the electrode is, the more pronounced is the compression around the electrode (44 μm; S in Fig. 1a). This could be recognized already with survey electron micrographs. Microcirculation is considerably disturbed in this area, since the sinusoids are compressed more and more toward the electrode and finally obstructed completely by the compact arrangement of the liver cells surrounding the electrode channel (E in Fig. 1a).

Low magnifications also showed that the compression of the liver cells observed along the puncture channel decreases toward the tip of the electrode (S in Fig. 1b), since the diameter of the electrode was smaller (32 μm). In addition it was to be observed that some cells in the immediate vicinity of the electrode (E in Fig. 1b) died (arrows in Fig. 1b).

Higher magnification of the compression zone (Fig. 2) showed that the compact tissue around a puncture channel consists of several cell layers arranged in a characteristical manner around

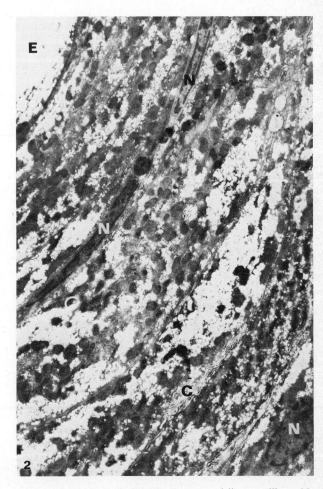

Fig. 2. Part of the compression zone around an electrode (E). C: compressed liver capillary; N: nucleus. Magn.: 6000 x.

the electrode. The diameter of the sinusoids was more or less decreased depending on the distance from the electrode (C in Fig. 2). All cells in this area loose their characteristic organization. Even the nuclei are flattened and oblong around the puncture site. The liver cells involved show cytopathological alterations: aside from the losely structured, perishing cells (*David*, 1967; *Schäfer*, 1970) in the immediate vicinity of the electrode (arrows in Fig. 1b) a distinctly stronger osmiphilia is observed. In addition, they contain a larger number of vacuoles as compared with normal liver parenchyma cells (V in Fig. 3).

At greater electron microscopic magnification of part of a cell bounding directly on the electrode channel (E in Fig. 3) one sees that a membrane (double arrow in Fig. 3) is always formed between the electrode and the contact cell. The electron micrograph also shows that almost all organelles are involved in the circular reorganization of the parenchymal cells. This is evident in the endoplasmatic reticulum (ER in Fig. 3) as well as in the mitochondria (M in

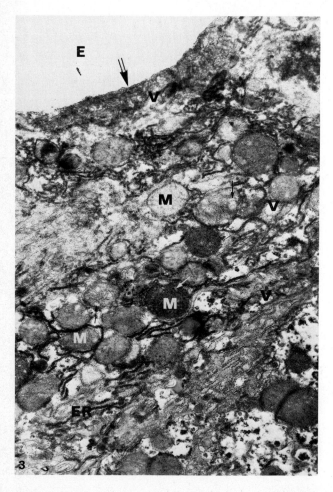

Fig. 3. Part of a liver parenchymal cell in direct contact with the electrode (E) after treatment with hydrofluoric acid. ⇒: membrane between electrode and cytosplasm; M: mitochondrium; V: vacuole; ER: endoplasmatic reticulum; →: local vacuolization of the mitochondria Magn.: 15,000 x.

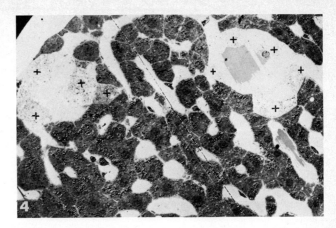

Fig. 4. Part of a rat liver with degenerated cells (+). The electrode could not be found in this ultrathin section. Magn.: 430 x.

Fig. 3), which because of their cytopathological alterations show not only various degrees of blackening but are also locally vacuolized (arrows in Fig. 3).

In addition to the results already described, one further finding seems especially interesting: we found groups of necrotic cells in areas not immediately adjacent to the electrode, sometimes at a considerable distance from it. As can be seen already in Figur 1b, groups of cells die (++ in Fig. 1b) which have had no contact with the electrode. Figur 4 furnishes a further illustration of this observation. Even after treatment with hydrofluoric acid we were not able to find an electrode channel in this liver section. We assume that the vessels supplying these cell groups pass close to the electrode at another level and are interrupted in connection with the tissue alterations around the electrode channel.

Summary

The influence of glass needle electrodes with a tip diameter of about 2 μm on rat liver cells and tissue was investigated by light and electron microscopy. Distinct alterations of morphology and ultrastructure were demonstrated.

1. Around the microelectrode the normal organization of the liver is considerably disturbed. The liver cells are compressed more and more with increasing diameter of the electrode. Consequently the lumina of the capillaries vanish or are at least restricted.
2. Half an hour after insertion of the electrode the cells in the nearest neighbourhood of the electrode are dead.
3. Furthermore, groups of cells always degenerate in areas of the liver which are not in direct contact with the microelectrode. We assume that these cells are cut off from the microcirculation by the electrode by compression of the supporting vessels.
4. In the compression area the cells and their organelles are oriented around the electrode channel. Toward the electrode the cells are vacuolized and show more and more degeneration phenomena. The same tendency is observed with respect to the disappearence of the liver sinusoids.

5. In all cases the electrode is bounded from the neighbouring cells by a membrane, with the exception of the loosely structured cells which are dead in cytopathological terms.

Possible influences of electrode filling materials and of electrical potentials were not considered, so the results we demonstrate here are caused only by puncturing of the tissue with glass capillaries.

References

David, H.: Elektronenmikroskopische Organpathologie. VEB Verlag Volk und Gesundheit, Berlin 1967

Drum, R. W.: The Non-siliceous Fine Structure of a Diatom. Fifth Intern. Congr. Electron Microscopy 2, 14, Academic Press, New York 1962

Hündgen, M., Schäfer, D., Weissenfels, N.: Der Fixierungseinfluß acht verschiedener Aldehyde auf die Ultrastruktur kultivierter Zellen. II. Der Strukturzustand des Cytoplasmas. Cytobiologie 3 (1971) 202–214

Kessler, M.: Normale und kritische Sauerstoffversorgung der Leber bei Normo- und Hypothermie. Habil.-Schrift, Marburg/Lahn 1968

Luft, J. H.: Improvements in epoxy resin embedding methods. J. Biophys. Biochem. Cytol. 9 (1961) 409–414

Olfers-Weber, R., Stockem, W., Wohlfarth-Bottermann, K. E.: Cytological aspects of membrane regeneration after experimental injury of amebae and acellular slime molds. Urban & Schwarzenberg, München this book

Stockem, W.: Die Eignung von Aerosil für die Untersuchung endocytotischer (pinocytotischer) Vorgänge. Mikroskopie 22 (1967) 143–147

Schäfer, D.: Licht- und elektronenmikroskopische Untersuchungen zum Strahlentod kultivierter Zellen. Cytobiologie 1 (1970) 383–394

Discussion of Session III

Chairman: *L. C. Clark Jr.*

Discussion of Paper by Clark and Emory:

The Measurement of Cholesterol and its Esters Using a Polarographic Anodic Enzyme Electrode

Pressman: It is not clear to me how the electrodes distinguish between oxygen and hydrogen peroxide.

Clark: You see a polarographic *anode* is totally insensitive to oxygen, the *cathode* reduces oxygen, the *anode* can only oxidize. The anode oxidizes the peroxide and a current flows.

Baum: When you did the cholesterol determinations why did you add the oxidase first and then the hydrolase?

Clark: If we are just going to do total cholesterol we add both together. By adding the oxidase first, we get the free or non-esterified cholesterol. Then the esterase is added to hydrolyze the esters and make them available to the oxidase.

Christian: Am I correct that the only enzyme you are immobilizing is the catalase and that you add the oxidase and esterase in solution form to your sample?

Clark: In the cholesterol system we use only the two enzymes, the oxidase and the hydrolase, in solution. We plan to immobilize the oxidase on the outside (e.g., in the sample) of the membrane. Immobilized catalase will be useful in the glucose electrode to destroy H_2O_2 which diffuses from the electrode, but in the cholesterol electrode we use azide to inhibit catalase which comes from very slight hemolysis. Of course, we would prefer to use enzymes between the platinum and a substrate-permeable membrane. But we have tested 30 or 40 membranes and have found none that are permeable to cholesterol. Perhaps we have one, but it gives a slow response time. I would like to ask everyone here for any help they could give in finding such a membrane. The complex with the lipoproteins can probably be "exploded" with detergents and by dilution.

Baum: Dr. Clark, would you be so kind as to make some comments about the calibration of these enzyme electrodes. Is it possible to give a slide of the calibration curve?

Clark: I don't have a calibration curve here, but there will be one in the paper. Since the cholesterol in plasma is present in a form which gives a fast and reproducible reaction in this system, we have been using me as the standard. My plasma cholesterol was measured in several independent laboratories. The average is 220 mg%.
We think the best standard will be a cholesterol (or ester) suspension in some detergent. There is a table of such detergents in the paper. It may be that a water soluble standard, such as cholesterol half succinate, can be found.

Pungor: What is the standard error of determination?

Discussion of Session III

Clark: With the glucose electrode, where the conditions are quite similar, the mean standard deviation is about 1.5% for a careful technician and about 3% on the average. It appears from the limited data we have now, that the accuracy and precision will be about the same.

Discussion of Paper by Christian: Amperometric Methods of Enzyme Assay

Buck: Can you say anything from your work about the electrochemical mechanism for anodic oxidation of NADH? Why do the current-voltage curves behave one way on platinum and another way on carbon?

Christian: Well, first of all the main problem of platinum I think is the high residual current. We can get the oxidation O.K. I suspect the oxidation mechanism may be similar to that on other electrodes. *Blaedel* and *Tenkins* reported at the recent Midwest Regional American Chemical Society Meeting on the oxidation of NADH by steady state volammetry and I haven't seen that paper yet. But a few years ago *Blaedel* and *Haas* did some studies on NADH analogues. I did not show a slide of their proposed reacting mechanism, but these are shown in the text of the manuscript. The pyridinium group is involved in the oxidation-reduction, apparently with the formation of an intermediate free radical.
The fact that we get an anodic shift in the potential with increase in concentration on carbon paste is interesting. I really don't have any evidence of what is going on, but it could be that this free radical intermediate is adsorbed by the carbon paste and builds up. This would suppress the reaction and make it more difficult.

Pungor: We have had very good results in the application of microelectrodes of carbon in silicone rubber in investigation of organic compounds generally. I would like to recommend that you try these types of electrodes, too.

Christian: Thank you, I would very much like to do it. I might mention that be carbon paste electrodes are very sensitive to oxidation problems. You cannot expose them to air when you've got a positive potential applied, otherwiese you get a very high residual current.

Jensen: When you apply the carbon paste electrode, do you renew the surface frequently or do you use the surface of the original carbon paste all the time?

Christian: We use the same surface for a period of time. It turns out the electrode response for NADH is fairly slow for a period of time until you make several runs, and it does require some pre-treatment. So once we get it pretreated we keep it until it goes bad and then we change it.

Coleman: There are a good number of clinical assays which are the result of coupling various enzymatic intermediate reactants with NADH for subsequent assay at 340 nm. Are you looking at a number of these other possibilities?

Christian: Only from the standpoint of electrochemical measurement. Our goal is to develop a general system for the monitoring of a large number of dehydrogenase and dehydrogenase-coupled enzyme reactions.

Sies: LDH is quite an active enzyme and I was just wondering about the sensitivity. With transaminases, for example, undiluted sample of serum would cause a decrease of about 8 micromolar oxygen per minute.

Christian: In the LDH analyses we have performed so far, we dilute the serum sample about 1 to 60 and under these conditions with the glucose analyzer instrument that we are using, we are working near the lower analytical measurement range. My friends at Beckmann Instruments tell me they can probably decrease this noise level maybe by an order of magnitude. It was built for measuring large signals rather than the small signals which we are measuring. In the case of tyrosine we had no difficulties, although we did have to record the signal rather than use the digital readout because again we are looking at lower signals than the instrument was built for.

Clark: Have you tried pyrolytic graphite as an electrode?

Christian: Glassy carbon I think is very similar.

Clark: Yes.

Christian: We have used this. It behaves quite nicely. It is more stable than carbon paste with respect to oxidation.

Discussion of Paper by Chan, Davies and Mosbach: Immobilized Enzymes and Ligands in Biological Research

Clark: What was your measuring electrode?

Chan: It was a *Beckman* (No. 39137) cation-sensitive electrode.

Baum: Did you immobilize native aldolase and then compare the kinetic properties of the immobilized enzyme to those of the immobilized subunit?

Chan: Yes, I have done that.

Baum: How different is it?

Chan: Apart from the urea sensitivity and the pH sensitivity that we have shown (which is very close to the situation in transaldolase) there are apparently no differences in other properties, for example, Km or V_{max}. We think now that the subunit is as active as the tetramer.

Baum: The pH response is very unusual.

Chan: The one I showed here was transaldolase but it was unusual. There seems to be a shift in the pK of a group on the enzyme.

Baum: Was it very large?

Chan: About 0.7 of a pH unit. I would like to point out that it is in fact quite easy to work with immobilized enzymes. Many people have a lot of hesitation in going from the classical enzymology into the immobilized enzyme field in the belief that special or different techniques are required. The techniques, for example, the measurement of enzyme activity are extremely simple. With sepharose, one can make a suspension of the derivative and one can simply keep it stirring and use an automatic fixed-volume pipette, to pipette any aliquot out of it extremely accurately. There are now commercially available (e.g., the Beckmann Acta series) spectrophotmeters with a stirring device in them, so that it is possible simply to add a certain amount of the sepharose-bound enzyme to the cuvette with stirring and the very small amount of sepharose which is needed to obtain enzyme activity measurements does not interfere with the optical measurements (*Mort, J. S., Chong, D. K. K.* and *Chan, W. W.-C.*, Anal. Biochem. *52*, 162–168 (1973)). Many immobilized enzymes are just as easy to work with as the soluble enzymes.

Clark: Such beads would be useful in stirred cuvettes for enzyme electrodes. One could put screen filters so as to keep them in the cuvette.
I have a question. Generally, when you immobilize an enzyme, you increase its tolerance to heat. How high could you go?

Chan: I think this must be determined also by the nature of the carrier itself (e. g., how rapid it is). Sepharose unfortunately is not very stable to heat. But glass, for example, is very good. I think it must also be determined by the length of the covalent linkage (i. e., whether one puts a spacer between the carrier and the protein or not). Unfortunately there have not been many systematic studies of what these effects are.

Clark: Could you go as high as 90 °C?

Chan: Yes, because there are some examples of stability at very high temperatures.

Eisenman: Do you have any idea for any of these enzymes what the reactive group on the polypeptide is that you are attaching?

Chan: There is some evidence that it is the amino group of lysine. This is the most reactive group under these conditions, but since we are attaching one group per enzyme molecule it is rather difficult to prove it in this particular case.

Discussion of Paper by I. A. Silver: An Ultramicro Glucose Electrode

Clark: We too have found that glucose oxidase becomes firmly attached to metal surfaces and in fact it seems that the longer it stays in contact with platinum the more difficult it is to get off. The last time I scrubbed the end of the electrode with soap and water and a brush, the oxidase was still intact. I hate to say it but I finally got it off with an electric tooth brush and toothpaste. I would like to hear some comment about what kind of bonds might be formed there. Also, another thing we found was that just contacting enzyme with a collagen membrane forms a permanent bond and the enzyme becomes stable for months.

Fleckenstein: I have only a technical question. How do you pull the two electrodes so that they fit so well? And another part of the question, how do you melt the two electrodes so that electrical connection is not broken?

Silver: The method of pulling two electrodes so that they fit is used routinely in making Thomas-type electrodes; all one does is to pull two electrodes with very closely similar tips. They must be fitted individually and you cannot get every one right. It is not particularly critical about how the diameters vary away from the tip since they do not have to be absolutely parallel. The other thing is you can cut an electrode tip off with a diamond disc. The biggest problem is not fitting the tips but not collapsing them. If it is necessary to keep them open for supplying oxygen it can be done by passing a flow of gas through the electrode during the heating process. Sometimes the electrical connection may break, but usually the

Discussion of Session III

glass will fuse to the platinum without it breaking. It does not really matter if the platinum ring is incomplete so long as there is a connection which does not have too high an impedence. The big problem is that if the platinum membrane is too thin the impedance is very high.

Jensen: You mentioned that you were putting your membrane over your platinum tip by drying a solution of glucose oxidase. Is it pure glucose oxidase solution or do you have some carrier that will form a diffusion barrier after drying?

Silver: No it is pure glucose oxidase. Of course as soon as you put it into the tissue a protein barrier is formed and I think that possibly helps to stabilize the system.

Lehmenkühler: I have a question about your recordings of ECoG activity and continuous recordings of glucose with the microelectrode. Have you any idea about the mechanism during this glucose reaction: the increase and the decrease?

Silver: The only thing I can say is that during spreading depression there is a big change in blood flow which slightly precedes the rise in the glucose as recorded. There are changes in the pyridine nucleotide redox state during spreading depression, and I assume that there must be a block in substrate utilization at this time and possibly an increased delivery of glucose due to increased blood flow which results in an increased extracellular glucose concentration. What is difficult to understand is the speed and size of the glucose change if there is a pool of extracellular glucose in the brain, with a relatively slow turnover, which one would think would act as a buffer. One can show with insulin that one is not just measuring blood flow and that the electrode is not just responding to blood flow changes.

Reneau: How fast does the glucose increase and when do you reach a point where it begins to decrease extracellularly? The time is what I was interested in.

Silver: The rise here is taking place over about half a minute or so, sometimes a little more.

Betz: Do you have an idea whether the pH changes simultaneously with the glucose? During CO_2 inhalation for example, the change of tissue pH is correlated well with the change of tissue glucose. If pH decreases glucose increases.

Silver: I have not measured it at the same time. There would be no problem if we used Hinke-type pH electrodes or one of Dr. *Simon*'s hydrogen ion exchangers.

Lehmenkühler: Have you seen changes in this glucose reaction during repeated spreading cortical depression waves?

Silver: Yes. If the spreading depression stimulus is repeated, the glucose changes usually become smaller. Normally the first two waves give about the same glucose change and then you see a but slightly more prolonged response. Sometimes after several cycles the glucose level may become permanently elevated.

Lehmenkühler: We see in our measurements that after repetitive depression waves the cortical blood flow reaches a high level, and stays constant.

Discussion of Session III

Discussion of Paper by Baum: An Automated Kinetic Analysis of Cholinesterase Activity by a Substrate-Selective Ion-Exchange Electrode

Clark: Don't nerve gases come into this picture some place?

Baum: Yes, we have also used the electrode to measure the inhibition of cholinesterase by organo-phosphates and that work is of some interest to people in environmental protection agencies for monitoring organo-phosphate pesticide residues in food crops; I published in Analytical Chemistry (43, 941 (1971)) a description of the method. The problem is that the organo-phosphates hydrolyze quite rapidly and so you get many metabolites and even though you may find some anticholinesterase products in food products you don't know what they are. Gas chromatography and mass spectroscopy are far more expensive and although they are more specific, of course, they don't tell you anything about biological activity.

Clark: Aren't these organo-phosphates widely disseminated?

Baum: I hope not, organo-phosphates hydrolyze very rapidly and after several days you really will not get any active degradation products.

Buck: What is your interpretation of the selectivity variation as a function of those solvent parameters?

Baum: There really is not enough information available to give a good answer to that. The order of selectivity is roughly related to dipole moment of the solvent. It is really related to the hydrophobic interaction between the organic ester and the solvent; however, there is a very limited amount of data available in the literature with which one can develop a reasonable model.

Discussion of Paper by Baucke: Reference Electrodes for Measurements with Ion-sensitive Electrodes. The Importance of the Liquid Junction Potential

Clark: I have one question. O would like to ask for your comments on the use of the fluoride electrode as a reference electrode, using just a lanthanum fluoride crystal electrode of the Orion type having an impedence of around 100,000 ohms. Fluoride under most circumstances is very stable in biological tissues. I have been struck with the stability of the electrode itself. I think there may be one paper in Analytical Chemistry on the use of it in vitro.

Baucke: So this would mean it is a cell without transfers.

Clark: Yes.

Khuri: I think one should not ignore the use of salts with organic anions; like with the chloride electrode we use concentrated sodium formate which forms an essentially equitransferent salt-bridge.

Simon: I fully agree and would like to add that calculations of the liquid junction potential using the *Henderson* equation show that lithium acetate is also an extremely good choice.

Coleman: A short statement with reference to Dr. *Clark*'s comment on the possibility of using other specific ion electrodes as references. This is the current approach for a number of the Orion industrial monitors.

Instead of using a conventional KCl reference we use another specific ion electrode, such as fluoride or sodium. This has worked for a number of applications.

Clark: I think we could probably use a fluoride electrode if it could be made small enough to use as a surface electrode such as placing it on the skin.

Coleman: Yes, we have a unit for cystic fibrosis screening which measures chloride directly on the skin using an $Ag_2S/AgCl$ electrode. I might add that two specific ion electrodes usually require differential amplifier inputs which are just now coming into the commercially available meters.

Baucke: That's what I mentioned in my paper.

Lübbers: By covering a part of the glass electrode with a membrane you can reduce the steepness of the glass electrodes; for example instead of 57 mV you can get 30 mV. If you then measure between these electrodes, the difference is dependent on the potential of the reference electrode.

Baucke: What you are saying is the content of a patent applied for and withdrawn.
It's by *Busch*, he was with *Ingold*, and we never believed it. He seems finally to be convinced that there is no reversible potential formed at these electrodes. This patent states that there is a linear function between the surface area covered and the slope of the response of the membrane, but I am sure this is not true.

Lübbers: Did you try to measure?

Baucke: Yes, we have tried it. We could not do anything with it. Our experience is quite contrary to what you say.

Clark: I do not blame him for withdrawing his patent application because it costs a fortune to keep a patent application viable in Germany.

Eisenman: It seems to me this is the equivalent of a partial short circuit.

Baucke: Yes, for instance, in one point in this patent it is claimed that one can use one normal pH glass electrode, and puncture a second one. One of the two is working reversibly, but the other one is not, because it has a little hole in its membrane. It may work for a short time while the shunt is still ok.

Buck: I believe there is a trade off that has to be considered in the decision whether to use an ion-selective electrode as a reference. The alternative is use of an ordinary second kind of electrode with a junction. In the use of the fluoride electrode as a reference one should add a high enough level of fluoride to the samples so that variations in the sample ionic strength do not affect the activity coefficient of fluoride. Otherwise, the fluoride electrode will not provide a constant reference potential. However, if one is willing to load the solution with electrolyte such as fluoride, ony may as well use some convenient inert electrolyte at a high and constant concentration in the sample, and then use a junction-type saturated calomel reference. The junction potential in that case is very nearly constant. One has to decide on some other basis what is the important criterion, for example, a need to exclude a particular ion.

Discussion of Session III

Discussion of Paper by Olfers-Weber, Stockem, Wohlfarth-Bottermann: Cytological Aspects of Membrane Regeneration after Experimental Injury of Amoebae and Acellular Slime Molds

Khuri: The histologists give the cell physiologists a nightmare about the hole that the microelectrode makes in the cell membrane. I would like to draw attention to the criterion that we proposed about 3 years ago, namely, that if one uses the intracellular technique with a double-barreled electrode which makes use of the simultaneous recording of the membrane potential and the intracellular ionic activity, then one can use the stability and the magnitude of the resting membrane potential as a functional index of the physiological integrity of the cell. Let me elaborate: if the cell remains electrically impermeant with the microelectrode in situ, it is not allowing molecular diffusion, e.g., a sub-analytical Na^+ influx from outside to inside to depolarize the membrane, and one can conclude that such a cell membrane is not likely to have a gap around the hole of sufficient size to allow bulk flow which could result in the mixing of intracellular and extracellular fluids, thus yielding an artefactuous intracellular reading for the ion in question. This way one can say that one is measuring the true activity of the ion in question in the cytoplasmic aqueous solution the cell contains within its membrane.

Thomas: Have you ever done a section parallel to the microelectrode? How far down towards the tip of the microelectrode does the membrane form? One is even curious as to whether with a Hinke-type, sealed-tip glass microelectrode, as used for measuring sodium or potassium, the membrane migth be forming over the entire intracellular surface of the electrode. Do you have any information on that at all?

Stockem: We sectioned parallel to the electrode in order to investigate the tip region. The electrode was filled in this case with *Ringer*'s solution, e.g., with the normal culture medium of the cell. But we found the same conditions as already described for slime molds when using different adjoining media as normal atmosphere, *Ringer* solution or *Ringer* solution with the addition of EDTA, for example.

Walker: I have somewhat the same concern about the membrane forming over the entire electrode. I wonder what the time course of that might be in vertebrate preparations, because we observe consistently in penetrations in heart muscle with potassium chloride electrodes, that if you leave the electrode in the cell for say 15 mins or longer you have a large DC offset when you come back out of the cell. There is obviously something wrong with the pipette. You have to throw the reading away, but frequently you can clear the pipette by overcompensating the capacity and ringing the circuit. It will free up and you come back to your base line again, but if you go in for just a very short time, just a couple of minutes, it doesn't happen. I wonder if this might be explainable on the basis of the sort of thing you are talking about.

Stockem: Electronmicroscopical investigations on metazoon cells are completely lacking except first experiments from Dr. *Schäfer* from the Max-Planck-Institute in Dortmund, which have been reported here. I am sure it is necessary to examine other cell types like muscle or nerve cells to get more information. We used large cells because these cells can overcome such damage much better than smaller cells. The introduction of a microelectrode into a liver cell can be compared with piercing your body with a tree trunk. I am sure that the pathological reaction in these cell types is much stronger than in amoebae and slime molds. These organisms can survive such an accident. Amoeba proteus, for example, can move through a glass needle, by which it completely pierced, without being disturbed in normal locomotion.

Discussion of Session III

Thomas: I wonder if it may be significant that both the cells you used are presumably amoeboid ones that move around and continually make new membrane. It might be more interesting to do experiments on some immobile cell that doesn't normally make new membranes, a neurone for example.

Stockem: Yes, this is right. But we measured some membrane potentials in amoebae and achieved normal values even over 30 minutes. This is difficult to explain considering the fact that the electrode is sealed by a new membrane. But you should remember that you leave rejected cytoplasm between the newly formed membrane and the surface of the microelectrode. For a short time this rejected cytoplasm may show similar characteristics compared to normal protoplasm, so that you can get normal values.

Thomas: If the newly formed membrane had a very low electrical resistance, I suppose that would also give you the same sort of results. There is no easy way of telling what the electrical resistance is from an electron micrograph.

Stockem: I agree. The only thing that I can say in this regard is that the newly formed membrane looks exactly like a normal plasma membrane, at least from the morphological point of view.

Brown: Neurones containing synaptic vesicles may be said to "make membranes" because they merge with the neuronal membrane before they empty their contents. I am curious as to whether formation occurred when you had EDTA in the surrounding medium. One comment: if this is a unit membrane, then a good number for unit membrane resistance is 1000 ohms cm^2 and for a pipette of 1 μm diam. one might expect the resistance to go up something of the order 10^{10}–10^{11} ohms. This would seriously load the usual electrometers that are used for recording and would then give you a much smaller voltage reading than you would ordinarily get.

Stockem: We used EDTA in order to remove calcium from the outer medium, because there were some indications for the involvement of calcium in the process of membrane regeneration. In this regard the paper of *Szubinska* (1972) should be mentioned. *Szubinska* argued that calcium is needed for the regeneration of new membrane material in Amoeba proteus. We therefore tested the influence of different adjoining media. But these experiments did not cause any difference in the process of membrane regeneration even when EDTA solution was used.

Brown: The fact that you don't need calcium to reform membranes is very interesting because cardiac muscle electro-physiologists are out of business if they don't have calcium in the medium after they take out one of their strips. I think this maybe an important difference.

Stockem: I think that the action of calcium on the process of membrane regeneration has yet to be clarified.

Christian: I would like to ask what the pH of your medium was to which you were adding the EDTA, because unless it is alkaline, the degree of complexation with calcium is going to be very small.

Stockem: It was in the range of 6.8 to 7.4. These pH values are physiological for amoebae and slime molds.

Discussion of Session III

Hinke: Most of my queries have already been answered; there is just one more that I would like your comments on. That is, if an electrode made a deep penetration away from the surface you would have a similar type of phenomena occurring in the depth of the cytoplasm. I am assuming that most of your observations were done relatively close to the surface of your amoeba or the slime mold.

Stockem: There is only one interesting area, namely, the bordering region between the surface of the electrode and the cytoplasm and I mentioned that we followed the electrode over its whole distance within the cell.

Hinke: Would you say that the vacuolization and the coalescence of the vacuoles would come from the surface, follow down the surface of the glass and then join up on the other side?

Stockem: No, truly not, it comes from the interior of the cytoplasm. The vacuoles move perpendicular to the surface of the electrode.

Zeuthen: Have you any experience with different surface treatments of the electrode, e. g., the degree of hydration of the glass or with the glass pipette being covered by a silicone layer?
Stockem: No, our experiments have not yet gone on so far.

Walker: On that same point, we put the treated glass pipette containing liquid exchanger into the cells, you can leave them in indefinitely and your never get an offset. You always come back to the same potential but when you use the bare glass you are in trouble if you leave it in for more than about 15 minutes. This is in heart muscle.

Stockem: I think from the cytological point of view a lot of questions could be answered if somebody would go on to investigate this problem in other cells. But this is more difficult to do than in slime molds and amoebae.

General discussion, Tuesday a. m.

Simon: I would like to make a comment in respect to the coating of glass electrodes and reducing the slope of the cell assembly. I fully agree to what *George Eisenman* said but would like to add that the shunting of the glass electrode is done in respect to the sample solution. This simply means that we cannot overcome the problem of the reference electrode and the problem of the liquid junction potential by using such shunted systems. Assuming a crack in one of the glass systems used in Prof. *Lübbers*' early experiments, such a shunt to the sample solution is obtained.

Lübbers: These measurements were done many years ago. I only can say that there was a very high resistance, so the crack must have been very small. We found this with astonishing reproducibility in one batch of 6 electrodes. With electrodes produced later we could not repeat our experiments.

Kessler: I would like to emphasize one point which I think is very important, namely, the measurement of the local DC potentials picked up by the ion-selective electrodes. Because of the limited time, I was unable to discuss this point in detail. In our measurements using multi-barreled electrodes we found, as mentioned, that there can be considerable streaming potentials as well as other DC-potentials described in the literature which are picked up at the tip of the electrode.

In order to monitor these potentials really at the tip of the ion electrode we use two channels of the four-barreled electrode filled with KCl and measure the DC-potential using differential amplifiers. Another, but less accurate method is to use a separate microelectrode inserted close to the ion electrode and to monitor the EMF between this electrode and the reference electrode of the ion electrode. These DC-potentials picked up at the tip of our electrodes have to be taken into consideration in our ion measurements.

Thomas: I would just like to say that we neurophysiologists would dearly love some way of making potassium chloride 20–30 megohm intracellular microelectrodes more reliable. I find that my pH-sensitive microelectrode is 10 times more easy to use in terms of absence of DC shifts than my ordinary microelectrodes filled with 2 molar KCl. On bad days they block soon after you put them into a cell particularly if their resistance is over 30 megohms.

Hebert: I have looked at electron mircographs of many glass micro pipettes and filled these electrodes, measured tip potentials and resistances and maybe I could make a sketch on the board of a couple of different types of electrodes that we have observed. Generally, the lower resistance electrodes had uniform conical angles all the way through the tip, those with high resistances in addition to having a uniform conical angle were cylindrical at the very tip (i.e., zero angle). These were generally 0.2 μm in inside diameter and the resistances were under 10 megohms. The micropipettes with cylindrical tip portions had resistances to ca. 100 megohms.

Hinke: Can I go back to Dr. *Stockem*'s paper, please. I am not familiar with any of the electro-physiological literature on amoebae and I am wondering whether you have done any electrophysiological measurements yourself on amoeba or slime mould and perhaps if you can tell us a little about the electrical history during a penetration.

Stockem: In electrophysiological experiments with slime moulds it was not possible to get values of the membrane potential. Measurements in amoeba, however, demonstrated the existence of a gradient in the membrane potential ranging from -70 mV to -30 mV along the cell body. This potential gradient is interpreted to be involved in the control of pseudopodium formation and amoeboid movement.

Brown: The fact that the cell is isopotential means that there is large resistance between where you got -70 and where you get -30 inside the cell. The resistance may be equivalent to the membrane resistance at the -30 mV end of the cell. The membrane resistance where you are measuring -30 is extremely unusual for at least most molluscan neurones and other preparations; the only other explanation would be that perhaps the resistance of the intracellular fluid is roughly that of the surface membrane which is about 1000 times that of the extracellular fluid.

Stockem: There may be barriers within the cytoplasm.

Hinke: Perhaps I'll reword my question because I don't think you quite answered my first question. In view of your morphological findings would you believe in your or anyones' membrane potential measurements on the amoeba?

Stockem: I think I just believe in them as well as all other people believe in the potentials they could obtain in nerve cells and other tissues. The values seem to be correct but on the other side there are our findings that a regenerated membrane exists in close proximity to the electrode.

Discussion of Session III

Zeuthen: When considering different intercellular potentials in a cell, you have to take into account the flux of ions within the cell. You could expect intracellular ion movement in this moving amoeba. This would contribute to an intracellular potential gradient.

Brown: It is possible using voltage-dependent fluorescent probes to measure membrane potential without penetrating the cell with a microelectrode. In red blood cells, where such probes have been used, the measurements of potential are very similar to those obtained with microelectrodes.

Eisenman: What strikes me about the slime moulds where the membrane heals so rapidly is that you also find that you cannot make an "effective" (as judged by potential changes) penetration with the microelectrode. Doesn't this indicate that when healing takes place this rapidly, you in fact never do get inside the cell.

Stockem: That's right, but we used the slime moulds as a model to clarify first the question whether the cell is capable of regenerating damage to the cell membrane, and how long this process takes place, and for answering this question the slime mould can be regarded as an excellent object.

Session IV A

Basic Effects of Ion Activity in Metabolic Reaction and in Membrane Functions Including Tissue Measurements

Chairman: *A. Kovách*

T. Clausen

The Role of Ions in the Control of Intermediary Metabolism

(9 Figures)

The role of ions in the control of biological processes is well established from numerous studies of excitable tissues. In nerve and muscle cells, alterations in the binding or the transport of ions constitute the signals which trigger such basic functions as the propagation of impulses and the contraction of myofibrils. This is the major example of how minute electrical signals are amplified into the very tangible events associated with muscular contractions.

On the other hand, although the role of ions in the control of the more silent events of intermediary metabolism was demonstrated already many years ago, it only just starts to be realized that ionic signals which are in some ways analogous to those seen in excitable tissues are perhaps of significance in mediating the hormonal control of organic metabolism (*Rasmussen* 1970).

A large variety of hormones produce changes in the transport of ions across the plasma membrane, but only in some instances has it been possible to ascribe these ions a specific role in the activation or inhibition of enzymatic processes in the intermediary metabolism (*Miller, Exton* and *Park* 1971; *Kurokawa, Ohno* and *Rasmussen* 1973).

Such studies are complicated because they require a multiple approach, which can not easily be coped with by a single research unit only:

First, it is necessary to establish to which extent a given enzymatic process can be influenced by changes in the concentration of an ion.

Second, it has to be demonstrated that the hormone controlling the enzymatic process considered does indeed induce a change in the transport or distribution of the ion in the target cells.

Third, and this is probably the most difficult part of the work, it is necessary to obtain rather detailed information about the time-course of the hormone-induced changes in the activity of the postulated regulatory ion at the appropriate cellular locus.

The methods generally available are still too crude to satisfy these requirements, and models operating with ions as messengers in the control of intermediary metabolism by hormones are therefore based on rather indirect evidence. The ion-selective electrodes may provide more direct evidence, and thus open shortcuts through the rather unsatisfactory experimental situation, which I shall exemplify in the following:

I have been interested in the role of calcium in the control of glucose metabolism. Already in 1939, *Lundsgaard* demonstrated that the transport of glucose across the plasma membrane constitutes the major rate-limiting step in this process. Since the cellular structures involved could be identified and in some cases even exposed to known levels of calcium ions, it seemed logical to investigate effects of this ion on the transport of glucose, the process primarily controlled by insulin, the major regulator of carbohydrate metabolism.

Almost 10 years ago, Holloszy & Narahara (1965) suggested that calcium ions might be of importance in triggering the rise in sugar transport seen during contractile activity. Following

T. Clausen

this line of evidence, I have tried to determine whether the increase in the cytoplasmic concentration of calcium ions brought about by a series of different experimental conditions, were associated with a stimulation of glucose transport.

At present, it is known that the cytoplasmic level of free Ca^{++} ions can be determined by at least the following processes:

1. Influx and efflux across the plasma membrane dependent on the Na-gradient.
2. Efflux across the plasma membrane by an ATP-requiring process.
3. Energy-dependent transport across the mitochondrial membrane.
4. Energy-dependent transport across the membranes of the endoplasmic reticulum.
5. Binding to the membranes facing the cytoplasm.
6. The formation of complexes with a variety of cytoplasmic constituents.

From this it is possible to design several different experimental situations which lead to a rise in the cytoplasmic concentration of calcium ions.

In isolated kidney cells, it was found that when the efflux of ^{45}Ca was followed during a period of incubation in the cold, the washout of the isotope proceeded at a considerably lower rate than at 37° (*Borle* 1972). When the washout was allowed to continue at 37°, the rate of ^{45}Ca efflux suddenly rose to values exceeding those measured in the control cells which had been

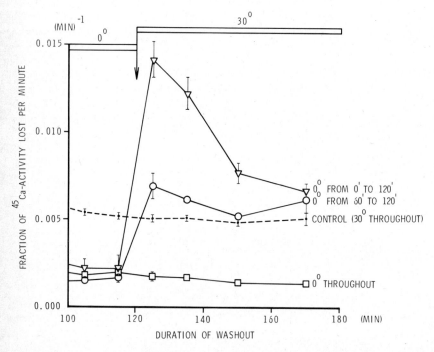

Fig. 1. Effect of cooling on the release of ^{45}Ca from rat soleus muscle. Intact soleus muscles were prepared from fed rats (60–70 g) and loaded for 60 min at 30° in Krebs-Ringer bicarbonate buffer containing ^{45}Ca. They were then transferred through a series of tubes containing unlabelled buffer, and at the end of washout, the radioactivity retained in the tissue and the amount released during each period were determined. The fraction of ^{45}Ca released during each interval was calculated (*Clausen*, 1969a). Each curve represents the mean of 3–4 observations with bars indicating S.E.M. (For details, see *Clausen, Elbrink* and *Dahl-Hansen* 1975; *Kohn* and *Clausen* 1971).

washed at 37° throughout. This was taken to indicate that during the washout in the cold, Ca^{++} ions available for transport out of the cells had accumulated in the cytoplasm, partly due to inhibition of the temperature-sensitive active transport of calcium into the extracellular phase, partly due to diminished mitochondrial accumulation. Following the return to the standard incubation temperature of 37°, part of this store of isotopic calcium retained in the cytoplasm could be transported across the plasma membrane out of the cells.

From Figure 1 it can be seen that in the isolated rat soleus muscle, the washout of ^{45}Ca responds to cooling and rewarming in a closely similar manner. As in the kidney cells, the rise in the rate constant depended on the duration of washout in the cold, again suggesting that this rebound phenomenon was related to a progressive accumulation of isotopic calcium in the cytoplasm.

When the washout of a ^{14}C-labelled non-metabolized sugar, 3-0-methylglucose, was followed under the same experimental conditions, cooling followed by rewarming was found to induce

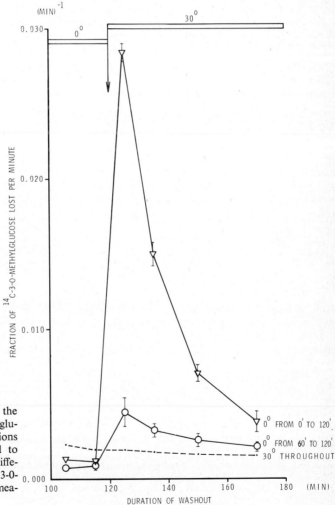

Fig. 2. Effect of cooling on the release of 3-0-(^{14}C)methylglucose. Experimental conditions as described in the legend to Figure 1, with the sole difference that the washout of 3-0-(^{14}C)methylglucose was measured.

a closely similar rebound rise in the rate constant of sugar efflux. The increase in 3-0-methylglucose efflux was correlated in time with the acceleration of ^{45}Ca release, and likewise, the response was accentuated by increasing the duration of cooling (Fig. 2). Phlorizin suppressed the rise in sugar efflux, indicating that this is not the outcome of non-specific leakage of the cells. Other experiments showed that following cooling for 120 minutes, the uptake of 3-0-(^{14}C)methylglucose was increased by 160% (*Clausen, Elbrink* and *Dahl-Hansen* 1975).

Another more obvious means of inducing an increase in the cytoplasmic level of Ca^{++} was suggested by the experience that metabolic inhibitors interfere with the mitochondrial accumulation of calcium (*Lehninger* 1970). Such compounds have been shown to produce a considerable acceleration of the release of ^{45}Ca from several preloaded intact tissues, and in the rat soleus muscle also 2,4-dinitrophenol was found to exert the same effect. This has been

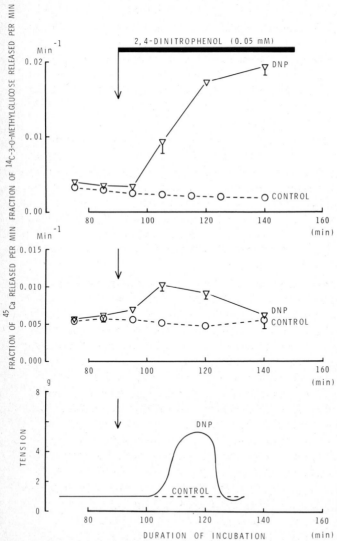

Fig. 3. Effect of 2,4-dinitrophenol on the tension and the release of ^{45}Ca and 3-0-(^{14}C)methylglucose from rat soleus muscle.

The washout of ^{45}Ca and 3-0-(^{14}C)methylglucose was determined as described in the legend to Figure 1. The lowest curve shows the isometric tension as recorded in parallel experiments with muscles suspended vertically in plastic holders (For details, see *Clausen, Elbrink* and *Dahl-Hansen* 1975). Each curve represents the mean of 3 observations with bars indicating S.E.M.

taken to indicate that the cytoplasmic level of Ca^{++} was increased, and since it seemed reasonable to expect that this could stimulate the contractile apparatus, we have measured the tension developed by the soleus muscles under experimental conditions similar to those used for the efflux experiments with ^{45}Ca and 3-0-(^{14}C)-methylglucose. Figure 3 shows the time-course of the changes induced by 2,4-dinitrophenol (0.05 mM). Whereas the rate constant of ^{45}Ca release is augmented within the first 10 minutes of exposure, a statistically significant stimulation of sugar efflux could only be detected in the second 10-minute washout period after the addition of 2,4-dinitrophenol. Around 10 minutes after the addition of 2,4-dinitrophenol, there is a marked increase in tension. This would indicate that the myofibrils have been activated by Ca^{++} ions, but it cannot be excluded that the contraction in part reflects rigor due to ATP-depletion.

Several other metabolic inhibitors as well as anoxia have been found to stimulate the transport of glucose and other sugars in muscle cells, but these effects have generally been related to ATP-depletion rather than to the often concomitant changes in cellular calcium distribution.

The phenomenon does not depend on contractility since adipocytes, the other major target for insulin action, respond to inhibition of energy-metabolism in a similar way. In the epididymal fat pad of the rat, 2,4-dinitrophenol induces an increase in the rate of ^{45}Ca release which seems to coincide with the onset of the stimulation of 3-0-(^{14}C)glucose efflux (*Clausen* 1969a; 1970).

The lipolytic hormones ACTH, adrenaline and glucagon were all found to stimulate the release of isotopic calcium from preloaded fat pads (*Clausen* 1970). This effect was found to be correlated in time with an increase in the efflux of 3-0-(^{14}C)methylglucose (*Clausen* 1969a). In muscle, the sarcoplasmic reticulum can be triggered to release free Ca^{++} ions into the cytoplasm. Following repeated electrical stimulation of the isolated rat soleus muscle, there is an immediate rise in tension and a clear-cut increase in the rate coefficient of ^{45}Ca release. This is associated with a stimulation of the 3-0-(^{14}C)methylglucose efflux, in agreement with what has been found in several previous studies of the effect of contractile activity on the transport of non-metabolized sugars (*Clausen, Elbrink* and *Dahl-Hansen* 1974).

The accumulation of calcium in sarcoplasmic vesicles can be inhibited by increasing the ionic strength (*de Meis* 1971), a condition which may also favor the release of calcium bound to the surface of membranes (*Dawson* and *Hauser* 1970). The ionic strength of the cytoplasm may be augmented by exposing the cell to a hyperosmolar environment, and measurements of the K-content of rat soleus muscles exposed to a buffer made hyperosmolar by the addition of mannitol (200 mM) showed that the intracellular concentration of this ion had increased by 52%. Hyperosmolarity has been found to induce a rise in tension and the release of ^{45}Ca (*Homsher, Briggs* and *Wise* 1974) and from Figure 4 it can be seen that in the rat soleus muscle, these effects are rather immediate and correlated to the osmolarity of the incubation medium. There is a concomitant progressive rise in the efflux of 3-0-methylglucose. Hyperosmolarity also stimulates the uptake of 3-0-methylglucose, and detailed investigations have shown that this condition produces an acceleration of sugar transport and glucose uptake which is kinetically closely similar to that seen during exposure to insulin (*Clausen* et al. 1970).

In this connection it is interesting that the reverse condition, hyposmolarity, i.e., a decrease in the cytoplasmic ionic strength, leads to a marked decrease in the stimulating effect of insulin and other agents on sugar transport in muscle (*Kohn* and *Clausen* 1972).

The collective evidence seems compatible with the idea that calcium plays a role in the regulation of glucose transport, and consequently in the overall regulation of carbohydrate metabolism.

Furthermore, it seems reasonable to assume that a rise in the cytoplasmic level of Ca^{++} ions, via an action at a specific locus, can induce stimulation of the glucose transport system. If so, it would be expected that the major regulator of glucose transport, insulin, might act via a similar mechanism. In some of the first experiments in this project, insulin was found to

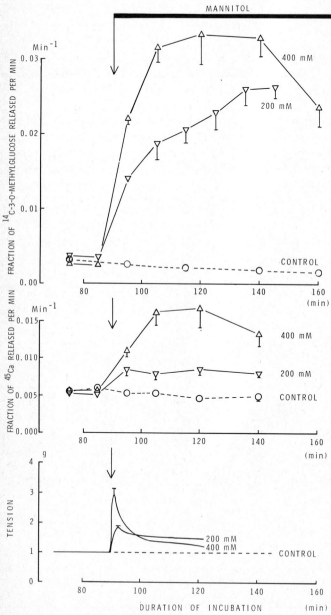

Fig. 4. Effect of hyperosmolarity on tension and the release of ^{45}Ca and 3-0-(^{14}C)-methylglucose from rat soleus muscle.
Experimental conditions as described in the legend to Figure 3. The osmolarity of the washout medium was augmented by the addition of mannitol at the indicated concentrations. Each curve represents the mean of 3–4 observations with bars indicating S.E.M.

stimulate the release of ^{45}Ca from preloaded fat pads (*Clausen* 1969b), and from other laboratories it was reported that insulin could reduce the binding of isotopic calcium to artificial lipid membranes (*Kafka* and *Pak* 1969) and plasma membranes isolated from liver cells (*Marinetti*, *Shlatz* and *Reilly* 1972). Repeated experiments with epididymal fat pads showed that insulin produces a prompt stimulation of ^{45}Ca release (Fig. 5). Later on, experiments with isolated fat cells showed that this effect was the result of an action on the adipocytes, and that concentrations of the hormone down to those seen in plasma were sufficient to elicit an acceleration of ^{45}Ca release (*Martin*, *Clausen* and *Gliemann* 1973; *Clausen*, *Elbrink* and *Martin* 1974).

Recent experiments performed with ghosts of fat cells in another laboratory have shown that insulin (10 μm/ml) diminishes the binding of ^{45}Ca (*Kissebah* et al. 1974).

Experiments with rat soleus muscles showed that insulin only induced a small (up to 20%) and somewhat variable increase in the efflux of ^{45}Ca. Since insulin does not induce contraction in muscle cells, it seemed unlikely that the hormone should bring about any major increase in the cytoplasmic concentration of calcium ions. Therefore, we tried to look at muscles in which cellular calcium had already to some extent been mobilized by incubation in a hyperosmolar buffer. From Figure 6 it can be seen that under these conditions, where the tension and the rate of ^{45}Ca-efflux are elevated (see Fig. 4), the addition of insulin produced a marked further rise in tension. This effect was completely suppressed by the addition of insulin antibody and did not depend on the availability of glucose in the incubation medium. This suggested that insulin induces a release of free calcium ions from some cellular pool into

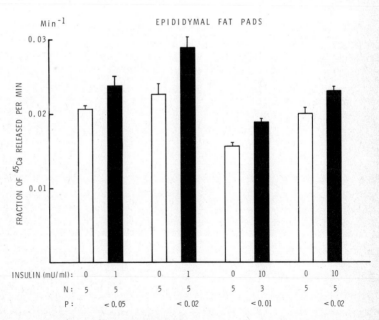

Fig. 5. Effect of insulin on the release of ^{45}Ca from fat pads.
Whole epididymal fat pads from fed rats (100–120 g) were loaded with ^{45}Ca and washed in unlabelled buffer as described elsewhere (*Clausen* 1970). 90 minutes after the onset of washing, insulin was added to the efflux medium. Each column represents the mean of rate coefficients determined in the interval from 90 to 100 minutes after the onset of washout with bars indicating S.E.M.

the cytoplasm, an effect which could also explain the above mentioned stimulation of ^{45}Ca efflux produced by the hormone.

There are additional examples of correlation between a rise in the cytoplasmic level of Ca^{++} ions and a stimulation of glucose transport (*Clausen*, 1969b, 1970; *Clausen, Elbrink* and *Dahl-Hansen* 1974), and it is by the time we start to determine whether there is any cause-effect relationship between the appearance of calcium ions and the activation of the glucose transport system.

The observation that membrane stabilizers suppress the stimulation of sugar transport induced by a variety of agents suggests that a certain mobility of the plasma membrane matrix is required either for the activation of the glucose transport system or for its continued function in the activated state (*Clausen, Harving* and *Dahl-Hansen* 1973). Conformational changes in the plasma membrane may be induced by the activation of the contractile microfilaments which have been demonstrated in various cell types. These filaments may be disrupted by cytochalasin B, and it is interesting, therefore, that this compound was recently shown to inhibit sugar transport in fat cells and other cell types (*Czech, Lynn* and *Lynn* 1973). From Figure 7 it can be seen that in rat soleus muscle, cytochalasin B (0.5–5.0 μg/ml) causes an appreciable decrease in the stimulating effect of insulin on 3-0-(^{14}C)glucose efflux. Whether this is secondary to a disruption of contractile microfilaments normally activated by Ca^{++} ions remains to be determined.

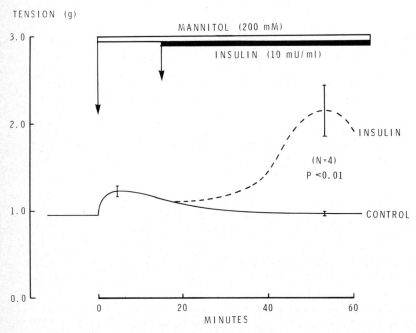

Fig. 6. Effect of hyperosmolarity and insulin on tension in rat soleus muscle.
Soleus muscles were prepared from which had fasted for 48 hours and tension was recorded as described elsewhere (*Clausen, Elbrink* and *Dahl-Hansen* 1975). Two groups of muscles were exposed to hyperosmolarity by the addition of mannitol (200 mM); one served as a control, and the other was given insulin as indicated. Each curve represents the mean of 4 observations with bars indicating S.E.M.

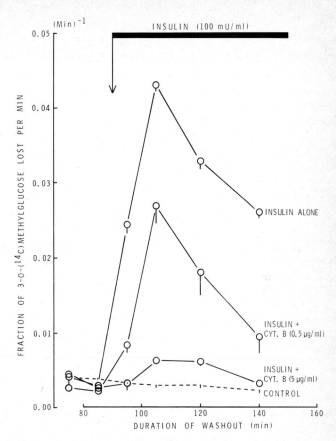

Fig. 7. Effect of insulin and cytochalasin B on the washout of 3-0-(^{14}C)methylglucose from rat soleus muscle.
Experimental conditions as described in the legend to Figure 2. Insulin (100 mU/ml) and cytochalasin B was present in the efflux medium during the time-intervals indicated by the horizontal bars.

The next question to be considered in relation to an effect of insulin on cellular calcium distribution was the possibility that this could have consequences for other cellular functions. If we accept the above mentioned evidence as indicating that the hormone induces a rise in the cytoplasmic level of calcium ions, it seems reasonable to assume that more calcium will be available for accumulation in the mitochondria. This process is closely linked to the uptake of phosphate (*Lehninger* 1970), and it has been known for more than 50 years that insulin stimulates the total uptake of inorganic phosphate in intact cells. It has generally been assumed that the hormone increases the permeability of the plasma membrane to this ion, and if this were the mechanism, one would expect that the addition of insulin should stimulate the efflux of ^{32}P from prelabelled cells. In addition, the hyperpolarizing effect of insulin would appear to favor the efflux of the negatively charged phosphate ions. However, studies with isolated perfused hearts have shown that insulin in fact inhibits the release of labelled inorganic phosphate (*Kaji* and *Park* 1961; *Sarkar* and *Ottaway* 1962). From Figure 8 it can be seen that the washout of ^{32}P from preloaded rat soleus muscles is also inhibited by insulin,

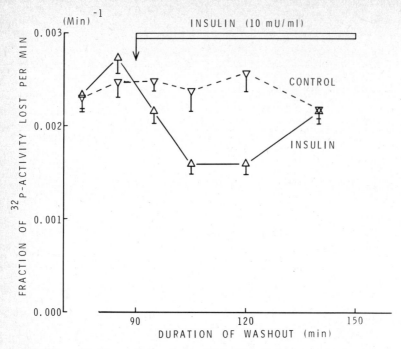

Fig. 8. Effect of insulin on the washout of ^{32}P radioactivity from rat soleus.
Rat soleus muscles were prepared with resected tendons and loaded for 60 minutes at 30° in Krebs-Ringer bicarbonate buffer containing $^{32}PO_4$ (2 µCi/ml). They were then washed in unlabelled PO_4-free buffer and the fraction of ^{32}P radioactivity released per minute determined. Each curve represents the mean of 3 observations with bars indicating S.E.M.

and experiments with whole epididymal fat pads indicated that adipocytes show the same response. These observations are compatible with the view that the hormone favors the retention of phosphate in some cellular pool rather than inducing a rise in the permeability of the plasma membrane to inorganic phosphate. If the stimulating effect of insulin on phosphate uptake were related to mitochondrial function, it should be suppressible by metabolic inhibitors. From Table I it can be seen that 2,4-dinitrophenol completely abolishes the effect of insulin on the uptake of $^{32}PO_4$ in soleus muscles and fat cells. Furthermore, it appears that hyperosmolarity, which induces a rise in the cytoplasmic calcium ion level, causes a marked increase in the uptake of $^{32}PO_4$ which is suppressed by 2,4-dinitrophenol. These observations gave some indirect support for the idea that insulin favors the mitochondrial uptake of calcium, and the next question was whether this could be detected by looking at other mitochondrial processes.

It was recently demonstrated that the mitochondrial enzyme pyruvate dehydrogenase is activated by a phosphatase which can be stimulated by calcium ions. Since the K_m for this effect is around 10^{-6} M (*Severson* et al, 1974), even a modest rise in the amount of calcium available for accumulation in the mitochonria would appear sufficient to trigger a stimulation of pyruvate dehydrogenase activity. Indeed, under conditions where the uptake of calcium in isolated fat cells was stimulated, the pyruvate dehydrogenase activity was augmented (*Clausen, Elbrink* and *Martin* 1974). Insulin might increase the cytoplasmic level of free Ca^{++}

Table 1. Effect of insulin, hyperosmolarity and 2,4-dinitrophenol on the uptake of ^{32}P.
Soleus muscles, whole epididymal fat pads or free fat cells prepared from fed rats were incubated in Krebs-Ringer bicarbonate buffer containing 0.2 µCi/ml or ^{32}PO$_4$ without or with the additions indicated. On the basis of the specific activity of the incubation medium, the total amount of ^{32}P-radioactivity taken up was calculated and expressed as µmoles $\pm$ S.E.M. The number of observations is given in parenthesis.

Rat soleus muscles:	Control	Insulin (10 mU/ml)	Mannitol (200 mM)
^{32}P uptake (µmoles/g/h)	0.69 $\pm$ 0.15 (6)	1.46 $\pm$ 0.11 (6)	1.44 $\pm$ 0.03 (6)
^{32}P uptake in the presence of 2,4-dinitrophenol (0.1 mM)	0.12 $\pm$ 0.03 (3)	0.12 $\pm$ 0.02 (3)	0.16 $\pm$ 0.02 (3)
Epididymal fat pads:			
^{32}P uptake (µmoles/ml cells/h)	0.109 $\pm$ 0.012 (4)	0.456 $\pm$ 0.064 (4)	0.538 $\pm$ 0.067 (4)
^{32}P uptake in the presence of 2,4-dinitrophenol (0.5 mM)	0.029 $\pm$ 0.010 (4)	0.015 $\pm$ 0.015 (4)	0.090 $\pm$ 0.010 (4)
Isolated fat cells:			
^{32}P uptake (µmoles/ml cells/h)	0.89 $\pm$ 0.11 (3)	1.35 $\pm$ 0.02 (3)	1.50 $\pm$ 0.04 (3)
^{32}P uptake in the presence of 2,4-dinitrophenol (0.2 mM)	0.23 $\pm$ 0.02 (3)	0.23 $\pm$ 0.01 (3)	

ions and thereby make more calcium available for mitochondrial accumulation and ensuing activation of the pyruvate dehydrogenase phosphate phosphatase. This idea is further supported by the recent observation that ruthenium red, which inhibits the mitochondrial accumulation of calcium, abolishes the stimulating effect of insulin on pyruvate dehydrogenase in fat pads (*Severson* et al. 1974).

Another enzymatic reaction which is known to be stimulated by insulin is the degradation of cyclic AMP by phosphodiesterase (*Manganiello* and *Vaughan* 1973). This enzyme is activated by calcium ions (*Kakiuchi* et al. 1973), and in view of the decisive role of cyclic AMP in the control of intermediary metabolism, it constitutes an obvious object for further investigations of the possibility that the fundamental action of insulin consists in a redistribution of cellular calcium.

In order to provide a frame for further experimental work, I have formulated a working hypothesis for the mode of action of insulin comprising the following steps (Fig. 9):

1. Insulin interacts with a receptor generating a conformational change which is propagated decrementally in the plane of the plasma membrane.
2. It is postulated that this leads to a displacement of calcium ions from the surface of membranes facing the cytoplasm.
3. An increase in the relative affintiy for sodium and calcium ions is compatible with the stimulating effect of insulin on Na-efflux.
4. A preferential stimulation of sodium efflux may account for the hyperpolarizing effect of insulin, the favoring of potassium accumulation and the stimulation of amino acid accumulation dependent on the sodium gradient.

5. A rise in the cytoplasmic calcium-ion level would favor the mitochondrial uptake of phosphate and this may account for the stimulating effect of insulin on net cellular phosphate uptake.
6. Increased mitochondrial uptake of calcium may lead to activation of pyruvate dehydrogenase.

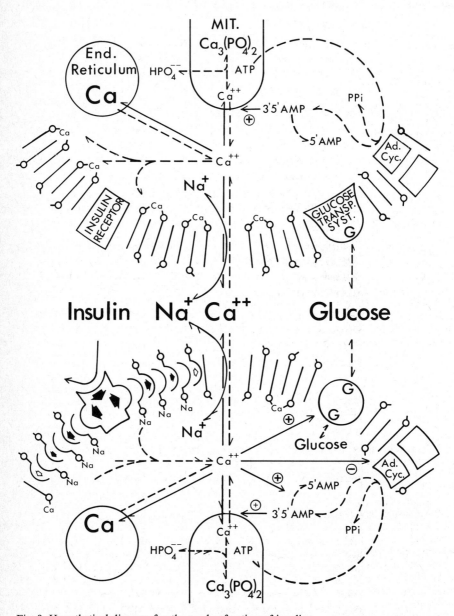

Fig. 9. Hypothetical diagram for the mode of action of insulin.
The upper part shows the basal state, the lower part the changes supposed to occur during exposure to insulin.

7. In the cytoplasm, a rise in the concentration of calcium-ion may stimulate phosphodiesterase, thereby reducing the level of cyclic AMP with ensuing changes in the metabolism of glycogen, triglycerides and proteins.
8. calcium ions made available at the inner surface of the plasma membrane may directly or indirectly induce a conversion of the glucose transport system into a conformation with higher mobility in the lipid matrix and altered affinity for glucose.

References

Borle, A. B.: Kinetic Analysis of Calcium Movements in Cell Culture. J. Membrane Biol. 10 (1972) 45–66

Clausen, T.: The Relationship between the Transport of Glucose and Cations across Cell Membranes in Isolated Tissues. V. Stimulating Effect of Ouabain, K^+—Free Medium and Insulin on Efflux of 3-0-Methylglucose from Epididymal Adipose Tissue. Biochim. Biophys. Acta 183 (1969a) 625–634

Clausen, T.: Role of Insulin in Ion Transport in Muscle and Adipose Tissue. Iugoslav. Physiol. Pharmacol. Acta 5 (1969b) 363–368

Clausen, T.: Electrolytes and the Hormonal Control of Organic Metabolism in Adipocytes. Hormone and Metabolic Research, Suppl. 2 (1970) 66–70

Clausen, T., Elbrink, J., Dahl-Hansen, A. B.: The Relationship between the Transport of Glucose and Cations across Cell Membranes in Isolated Tissues. IX. The Role of Cellular Calcium in the Activation of the Glucose Transport System in Rat Soleus Muscle Biochim. Biophys. Acta 375 (1974) 292–308

Clausen, T., Elbrink, J., Martin, B. R.: Insulin Controlling Calcium Distribution in Muscle and Fat Cells. Acta Endocrin. Suppl. 191 (1974) 137–143

Clausen, T., Gliemann, J., Vinten, J., Kohn, P. G.: Stimulating Effect of Hyperosmolarity on Glucose Transport in Adipocytes and Muscle Cells. Biochim. Biophys. Acta 211 (1970) 233–243

Clausen, T., Harving, H., Dahl-Hansen, A. B.: The Relationship between the Transport of Glucose and Cations across Cell Membranes in Isolated Tissues. VIII. The Effect of Membrane Stabilizors on the Transport of K, Na and Glucose in Muscle, Adipocytes and Erythrocytes. Biochim. Biophys. Acta 298 (1973) 393–411

Czech, M. P., Lynn, D. G., Lynn, W. S.: Cytochalasin B-sensitive 2-Deoxy-D-glucose Transport in Adipose Cell Ghosts. J. Biol. Chem. 248 (1973) 3636–3641

Dawson, R. M. C., Hauser, H.: Binding of Calcium to Phospholipids. In: Calcium and Cellular Function, p. 17, ed. by *A. W. Cuthbert.* Macmillan, London 1970

Holloszy, J. O., Narahara, H. T.: Studies on Tissues Permeability, X. Changes in Permeability to 3-Methylglucose Associated with Contraction of Isolated Frog Muscle. J. Biol. Chem. 240 (1965) 3493

Homsher, E., Briggs, F. N., Wise, R. M.: Effects of Hypertonicity on Resting and Contracting Frog Skeletal Muscles. Amer. J. Physiol. 226 (1974) 855–863

Kafka, M. S., Pak, C. Y. C.: Effects of Polypeptide and Protein Hormones on Lipid Monolayers. I. Effect of Insulin and Parathyroid Hormone on Monomolecular Films of Monooctadecyl Phosphate and Stearic Acid. J. gen. Physiol. 54 (1969) 134

Kaji, H., Park, C. R.: Stimulation of Phosphate Uptake by Insulin and its Relation to Sugar Transport in the Perfussed Rat Heart. Fed. Proc. 20 (1961) 190

Kakiuchi, S., Yamazaki, R., Teshima, Y., Uenishi, K.: Regulation of Nucleoside Cyclic 3':5'-Monophosphate Phosphodiesterase Activity from Rat Brain by a Modulator and Ca^{2+}, Proc. Nat. Acad. Sci. 70 (1973) 3526–3530

Kissebah, A. H., Hope-Gill, H. F., Clarke, P. V., Vydelingum, N., Tulloch, B. R., Fraser, T. R.: New Hypothesis on the Mode of Insulin Action. The Lancet i (1975) 144–147

Kohn, P., Clausen, T.: The Relationship between the Transport of Glucose and Cations across Cell Membranes in Isolated Tissues. VI. The Effect of Insulin, Ouabain, and Metabolic Inhibitors on the Transport of 3-0-Methylglucose and Glucose in Rat Soleus Muscles. Biochim. Biophys. Acta 225 (1971) 277–290

Kohn, P., Clausen, T.: The Relationship between the Transport of Glucose and Cations across Cell Membranes in Isolated Tissues. VII. The Effects of Extracellular Na^+ and K^+ on the Transport of 3-0-Methylglucose and Glucose in Rat Soleus Muscle. Biochim. Biophys. Acta 255 (1972) 798–814

Kurokawa, K., Ohno, T., Rasmussen, H.: Ionic Control of Renal Gluconeogenesis II. The Effects of Ca^{2+} and H^+ upon the Response to Parathyroid Hormone and Cyclic Amp. Biochim. Biophys. Acta 313 (1973) 32–41

Lehninger, A. L.: Mitochondria and Calcium Ion Transport Biochem. J. 119 (1970) 129–138

Lundsgaard, E.: On the Mode of Action of Insulin. Uppsala Läkareförenings Forhandlingar 45–46 (1939) 143–152

Manganiello, V., Vaughan, M.: An Effect of Insulin on Cyclic Adenosine 3':5'-Monophosphate Phosphodiesterase Acitivity in Fat Cells. J. Biol. Chem. 248, No. 20 (1973) 7164–7170

Marinetti, G. V., Shlatz, L., Reilly, K.: Hormone-Membrane. In: Insulin Action, p. 207, ed. by *I. B. Fritz.* Academic Press, New York 1972

Martin, B. R., Clausen, T., Gliemann, J.: Abstr. of 9th International Congr. of Biochem., Stockholm, No. 8f9 (1973) p. 376

De Meis, L.: Allosteric Inhibition by Alkali Ions of the Ca^{2+} Uptake and Adenosine Triphosphatase Activity of Skeletal Muscle Microsomes, J. Biol. Chem. 246 (1971) 4764–4773

Miller, T. B., Exton, J. H., Park, C. R.: A Block in Epinephrine-induced Glycogenolysis in Hearts from Adrenalectomized Rats. J. Biol. Chem. 246 (1971) 3672–3678

Rasmussen, H.: Cell Communication, Calcium Ion and Cyclic Adenosine Monophosphate. Science 170 (1970) 404

Sarkar, A. K., Ottaway, J. H.: Inorganic Phosphate Metabolism by the Perfused Rat Heart. Biochem. J. 84 (1962) 57

Severson, D. L., Denton, R. M., Pask, H. T., Randle, P. J.: Calcium and Magnesium Ions as Effectors of Adipose-Tissue Pyruvate Dehydrogenase Phosphate Phosphatase. Biochem. J. 140 (1974) 225–237

Berton C. Pressman

Physical and Biological Properties of Ionophores

A number of carboxylic ionophores are now known to interact with divalent ions; while *A23187* is highly selective for divalent ions, *X-537A* and several of its derivatives are virtually *universal* ionophores, readily complexing all mono and polyvalent inorganic cations, as well as primary and secondary amines. CD spectra reveal the existence of a number of concentration-dependent complexation conformers of X-537A; similar conclusions have been obtained from relative fluorescence yields which are characteristically distinct for each complexation species. Based on its ability to translocate both Ca^{++} and catecholamines, we deduced X-537A should exert striking biological effects. Accordingly we found X-537A: (1) stimulates the contraction of smooth muscle; (2) initiates parthanogenesis of sea urchin eggs; (3) alters the normal action potential of cardiac conductive (Purkinje) fibers (4) induces Ca^{++} dependent action potentials in depolarized Purkinje fibers; (5) increases the contractility of cardiac muscle strips and (6) isolated perfused heart. Many of the diverse biological effects of Ca^{++} selective carboxylic ionophores derive from their ability to stimulate Ca^{++}-dependent exocitosis in secretory cells by translocation of intracellularly bound, or extracellular, Ca^{++}. X-537A also increases cardiac contractility in anesthetized dogs accompanied by a dramatic increase in blood flow and oxygen tension in the heart muscle, an increase in renal filtration and sustained release of glucagon from the pancreas. Catecholamine blocking agents only partially inhibit these responses indicating other factors are involved. This conclusion is strengthened by finding that many carboxylic ionophores, inefficient in complexing either Ca^{++} or catecholamines, are potent cardiotonic agents. This property is not shared by *neutral* ionophores, e.g. valinomycin, which are exceedingly toxic. This discovery of the unprecedented biological and pharmacological effects of carboxylic ionophores, which have profound therapeutic implications, lend new impetus to the study of ionophores heretofore principally of interest as model membrane carriers for ions.

A. Scarpa

Metallochromic Indicator for Kinetic Measurements of Magnesium Ions and Determination of Cytosolic-free Magnesium Ions by Microspectrophotometry

(9 Figures)

The importance of magnesium in regulation of cellular properties and enzymic functions is well established (*Aikawa* 1973; *Wacker* and *Vallee* 1964, *Wacker* and *Williams* 1968). However, in spite of the importance of Mg^{++} as a major cellular regulator, the interaction of Mg^{++} with biological systems and the regulation of cellular Mg^{++} homeostasis are poorly understood. Heretofore, simple, rapid and precise methods for detecting Mg^{++} at high sensitivity and specificity were unavailable (for a discussion see *Wacker* and *Vallee* 1964; *Scarpa* 1974), which accounts for the scarcity of quantitative data on the biological distribution and and function of Mg^{++}.

The lack of suitable methods for studying Mg^{++} is in sharp contrast with the availability of several ways to measure Ca^{++}. Measurements of Ca^{++} can be performed by using easily available isotopes, specific electrodes (*Pungor* and *Toth* 1970), metallochromic (*Scarpa* 1972) and photoluminescent (*Johnson* and *Shimomura* 1972) indicators, which have provided precise and sensitive measurements of Ca^{++} and have facilitated the understanding of Ca^{++} interaction with biological systems.

This paper presents a simple technique which overcomes most of the limitations of the technique presently available for measuring Mg^{++} concentration. This method is based upon the detection of absorbance changes of a suitable metallochromic indicator by dual wavelength spectroscopy, and offers a way by which Mg^{++} binding and transport can be measured kinetically at high sensitivity and specificity. In addition the introduction of a magnesium indicator into large cells suitable for microinjection and the detection of its absorbance changes by sensitive microspectrophotometry has made possible quantitative measurements of the concentration of ionized Mg^{++} present in cytosol.

Metallochromic Indicator of Ionized Magnesium

A method was recently described (*Scarpa* 1974) for measuring ionized Mg^{++} concentrations in the presence of biological systems. This method is based on the detection of the absorption change of Eriochrome Blue or other suitable dyes by dual wavelength spectroscopy. Eriochrome Blue SE (3-[(5-chloro-2-hydroxyphenyl)azo]-4,5-dihydroxy-2,7-naphtalenedisulfonic acid) is a water soluble indicator of 519 M.W. which is easily available commercially. The differential spectra of Eriochrome Blue and Eriochrome Blue *plus* various concentrations of either Mg^{++} and Ca^{++} is shown in Figure 1. The addition of $MgCl_2$ to Eriochrome Blue at pH 7.1 produces an increase of absorbance with a ΔA_{max} at 551 nm and a decrease in absorbance with a ΔA_{max} at 580, with an isobestic point at 563 nm. In contrast, the addition of $CaCl_2$ produces a much broader absorbance decrease from 500 to 600 nm. Other divalent cations, with the exclusion of Mn^{++}, produce changes in absorbance similar to that of Ca^{++}. The different spectral response of Eriochrome Blue in the presence of Ca^{++} or Mg^{++} makes possible the selection of

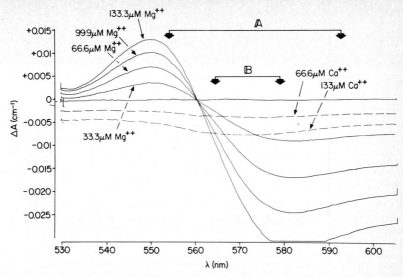

Fig. 1. The difference spectra of Eriochrome Blue SE *vs.* Eriochrome Blue SE plus Ca^{++} or Mg^{++}. The differential spectrum was obtained by adding to the measuring cuvet the concentrations of $CaCl_2$ or $MgCl_2$ reported in the figure. Both cuvets contained 30 mM Tris HCl (pH 7.1) and 200 µM Eriochrome Blue SE. A and B indicate regions of the spectra where suitable couples of wavelength pairs may be selected for (Mg^{++}) measurements without Ca^{++} interference.

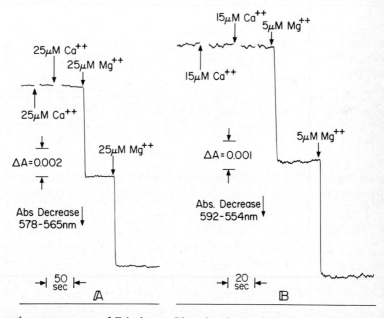

Fig. 2. Dual-wavelength measurements of Eriochrome Blue absorbance changes after addition of $CaCl_2$ or $MgCl_2$. The reaction mixture contained 30 µM Eriochrome Blue, 100 mM KCl and 30 mM Na Morpholinopropanesulfonate (MOPS) pH 7.1. The changes in absorbance were measured in 1 cm cuvet at the wavelength indicated with a double beam spectrophotometer designed and built at the Johnson Foundation, University of Pennsylvania (*Chance, B.*, 1972).

253

areas of the spectrum where Mg^{++} can be detected specifically without Ca^{++} interference. Therefore, even if the specificity of Eriochrome Blue toward Mg^{++} is not an intrinsic property of the dye, the availability of dual wavelength spectroscopy makes possible the measurements of Mg^{++} without Ca^{++} interference. Figure 2 shows that the changes in Δ absorbance produced by Ca^{++} at two couples of wavelength pairs are minimal. This is due to the fact that the small decrease in absorbance produced by Ca^{++} at the measuring wavelength is equal to the decrease of absorbance at the reference wavelength. As a result the differential readout at these wavelengths is zero. In contrast, Mg^{++} produces a significant decrease in ΔA, which is the result of a net decrease in absorbance at the measuring wavelength with respect to the reference. Several characteristics render Eriochrome Blue suitable for Mg^{++} measurements in biological systems. These are a) the high $\Delta\varepsilon$ (~ 4) (mM cm^{-1}) between the Eriochrome Blue uncomplexed and the Eriochrome Blue complexed with magnesium, b) a low affinity for Mg^{++} ($K_D = 2$ mM, higher at higher ionic strength) which results in a minimal disturbancy of free Mg^{++} by the dye and also permits measurements of high Mg^{++} concentrations, c) little or no binding to various cells or cellular organelles, d) lack of side effects on properties and functions of cells and cell fractions.

A significant limitation of Eriochrome Blue is its slow rate of complex formation with Mg^{++}, the half time of which is about 60 msec (Fig. 3) as compared with less than 3 μsec in the case of murexide-calcium (*Geier*, 1968). Therefore Eriochrome Blue allows for kinetic measurements of (Mg^{++}) transients only when rates of Mg^{++} binding and/or transport are slower than 200 msec.

Figure 4 presents same high resolution nuclear magnetic resonance spectra of Eriochrome Blue in D_2O. A pH titration (Fig. 4A) permits the assignment of the proton peaks of Eriochrome Blue molecule, the only uncertainty resting on the reciprocal position of H_1 and H_8. The similarity between pH and Mg^{++} titration (Fig. 4B) would suggest that Mg^{++} is bound to the three hydroxyl groups of Eriochrome Blue. One can speculate that the slow rate of complex formation may be due to the rotation of the chlorohydroxyphenyl moiety around the azo group to make the third hydroxyl group available for Mg^{++}. However these results do not exclude other possibilities and further studies are necessary to clarify this point.

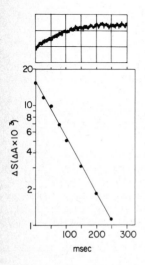

Fig. 3. Temperature jump measurements of the absorbance changes of Eriochrome Blue. The measurements were performed using an instrument similar to that described by *Eigen* and *DeMaeyer* (1963). The reaction mixture contained 100 mM KCl, 30 mM MOPS (pH 7), 100 μM Eriochrome Blue and 100 μM $MgCl_2$. 1 ml of the reaction mixture was kept at a thermal equilibrium at 20° and then perturbed in 4 μsec by a 5.2° temperature increase. The changes in absorbance were recorded at 580 nm with a 7 mm light path and the recording of the top figure shows the time course of the change in concentrations of the species (Eriochrome Blue) and (Eriochrome Blue – Mg^{++}) to a new equilibrium at higher temperature.

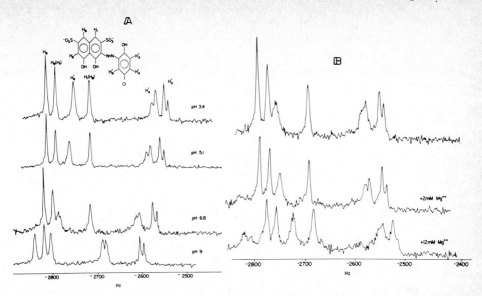

Fig. 4. High resolution nuclear magnetic resonance spectra of Eriochrome Blue at various pH (A) and in the presence of various concentrations of Mg^{++} (B). The spectra were obtained with a Varian 220 NMR spectrometer at room temperature. The tubes contained 3 mM Eriochrome Blue dissolved in 99% D_2O.

An example of kinetic measurements of (Mg^{++}) is given in Figure 5. As reported by various investigators (*Kun, Kearney, Lee* and *Wiedman* 1970; *Bogucka* and *Wojczak* 1971), the addition of uncoupling agents and ADP induces a release of endogenous Mg^{++} from

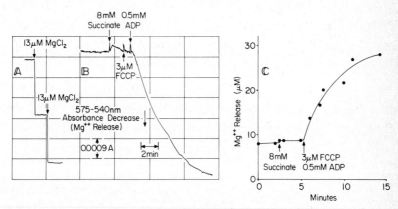

Fig. 5. Kinetics of Mg^{++} release by isolated rat liver mitochondria. The reaction mixtures contained 0.20 M sucrose, 20 mM KCl, 30 mM MOPS (pH 7.1), 30 μM Eriochrome Blue, and 3 μM rotenone. The reaction mixture was supplemented with 2.8 mg of mitochondrial protein/ml in parts B and C. Mg^{++} transients were measured in parts A and B through the detection of the absorbance changes of Eriochrome Blue at 540–575 nm. Part C represents a control experiment in which 1-ml aliquots were withdrawn at the times indicated in the figure. The samples were centrifuged 2 min at 15,000 g with an Eppendorf desk centrifuge, and the supernatants were analyzed for Mg^{++} content with absorption flame spectroscopy.

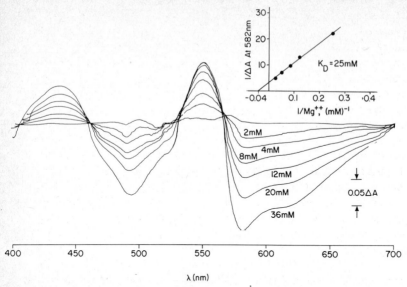

Fig. 6. Difference spectra of Eriochrome Blue *versus* Eriochrome Blue plus Mg^{++} in a reaction mixture similar to that of a squid axoplasm in pH and ionic composition. The reaction mixture contained 200 µM Eriochrome Blue at pH 7.0 in a final osmolarity of 0.74 which was made up from the following species (mM): K (350) Na (40) Cl (88), Isothianate (150) D-Aspartate (150), TES (5), Taurine (275). Temperature 21°.

mitochondria oxidizing substrates. Figure 6 shows the calibration of the reaction mixture with Mg^{++} (A) and the decrease of absorbance related to Mg^{++} efflux from mitochondria after addition of ADP and the uncoupler of respiration FCCP. As a control, Figure 6C shows that, under identical experimental conditions, similar results were obtained by measuring Mg^{++} concentration by flame absorption spectroscopy after the separation of mitochondria by centrifugation.

In summary, the measurement of ionized (Mg^{++}) with Eriochrome Blue absorption is limited by the relatively slow response, by a certain amount of binding to biological systems and by interactions with a broad range of cations. However, if suitable wavelength pairs are selected carefully with proper respect to the medium and in well-buffered conditions, Eriochrome Blue offers a unique approach to the kinetic measurements of ionized Mg^{++} in biological systems.

Microspectrophotometric Measurements of Free (Mg^{++}) in Isolated Squid Axons

A technique which is based on the use of Eriochrome Blue and microspectrophotometry was developed for measuring ionized Mg^{++} concentrations in the cytosol of large cells. The experiments reported here were the first attempt to make such measurements and were carried out in the axoplasm of giant squid axons (Leligo Pealeii) at the Marine Biological Laboratory, Woods Hole, Massachusetts.

The magnesium content of axoplasm measured by atomic absorption spectroscopy averaged 6.4 mmoles/Kg axoplasm (*Baker* and *Crawford* 1972), but little and indirect information exists in the literature on the state of this intracellular magnesium. Due to the large amount of adenine nucleotides in squid axoplasm it has been suggested that a substantial fraction of the magnesium is bound and mobility sudies of ^{28}Mg (*Baker* and *Crawford* 1972) have suggested that one third to one half of the total magnesium may be free in solution.

Figure 6 shows a differential spectrum of Eriochrome Blue versus various Mg^{++} concentrations obtained in a reaction mixture similar in pH and ionic composition to that of a squid axoplasm. With the exception of a higher K_D, Mg^{++} produces changes in absorbance similar to those shown in Figure 1: an increase in ΔA_{max} at 550 nm, a decrease in ΔA_{max} at 582 nm and an isosbestic point at 566 nm.

The absorbance of Eriochrome Blue introduced in a single axoplasm was measured *in situ* at all these wavelengths simultaneously in the way illustrated in Figure 7. The system consisted of: a lamp; four interference filters with 1 mm half bandwidth (550, 556, 575 and 592 nm) equally spaced in a rotating wheel which is driven at a frequency of 20 to 10,000 Hz by compressed air; a bridge containing two minuscule optical fibers (0.3 mm by 0.5 mm) which can be positioned in the dialysis chamber on top of the squid axon; a photomultiplier tube and sensitive electronics which made possible time sharing detection of four pulses of light at any frequency; a suitable recording system (see also *Chance* 1972). Squid giant axons were mounted in a dialysis chamber containing sea water as described by *Brinley* and *Mullins* (1967). Eriochrome Blue was injected by displacement-microinjection technique and was uniformly distributed within the axoplasm after about 15 min. A dialysis capillary of porous silica of 0.05 mm was then inserted longitudinally in the axoplasm: the center of this capillary consisted of pores of 20–40 Å diameter through which Mg^{++} but not ATP or larger molecules can freely diffuse. This technique was described in detail by *Brinley* and *Mullins* (1967).

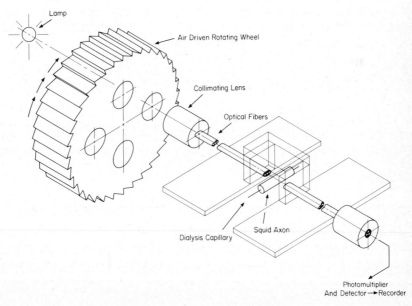

Fig. 7. Diagram of the optical assembly used for microspectrophotometric measurements of Eriochrome Blue absorbance changes within squid axons. More details in the text.

A. Scarpa

Figure 8 shows the results of an experiment of dialysis of an axon with various concentrations of Mg^{++}. The four pulses of monochromatic light were focused on a discrete area of the axoplasm which contained Eriochrome Blue and the changes in absorbance undergone by Eriochrome Blue were measured simultaneously and kinetically. The absorbance at the isosbestic point 566 nm is independent of the changes in Mg^{++} and indicates mainly the concentration of the dye present in the axoplasm at each time. Based on the spectrum of Figure 6, increase in ionized Mg^{++} concentration in the axoplasm should increase the ΔA at 550–566 nm and decrease the ΔA at 592–566 nm as 575–566 nm. Figure 8 shows that dialysis of the axoplasm with various Mg^{++} concentrations produces the Δ absorbance changes which are expected for a decrease or increase of axoplasmic Mg^{++}. Unfortunately, as is shown in trace D, there was a slow but continuous decrease of absorbancy at 566 nm which indicates a loss of dye from the axoplasm during dialysis. This does not permit reliable recording for the total time of dialysis (about 40 min) necessary for the ionized Mg^{++} to reach diffusional equilibration. On the other hand loss of dye was undetectable during the first 5–10 minutes of dialysis so that the Δ absorbance measurement provides an accurate determination of whether the ionized magnesium of the axoplasm is greater or smaller than the concentration of ionized Mg^{++} in the dialysate. The initial absorbance changes undergone by Eriochrome Blue were recorded in 24 different squid axons dialyzed with Mg^{++} concentrations ranging from 0 to 10 mM. Figure 9 shows some of these recordings obtained at 592–565 nm. There was no appreciable change in absorbance by Eriochrome Blue when the ionized Mg^{++} in the dialysate ranged from 3 to 3.2 mM. Increases and decreases in ΔA are evident when axoplasm

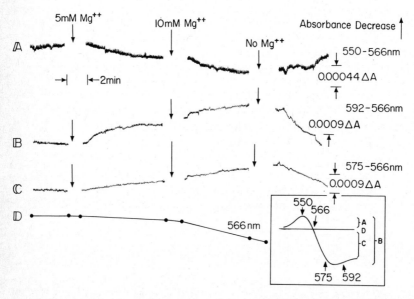

Fig. 8. Simultaneous measurements of dual-wavelength absorbance changes undergone by Eriochrome Blue at three wavelength pairs and at an isosbestic point during the dialysis of a squid axon. Squid axon was prepared and mounted in the dialysis chamber according to the procedure of *Brinley* and *Mullins* (1974). Eriochrome Blue was introduced into the axon by microinjection (*Hodgkin* and *Keynes* 1956) 20 min after dialysis to give an axoplasm concentration of about 4 mM. The axon was then dialyzed with the concentration of Mg^{++} indicated in this figure dissolved in a reaction mixture similar to that of Figure 6.

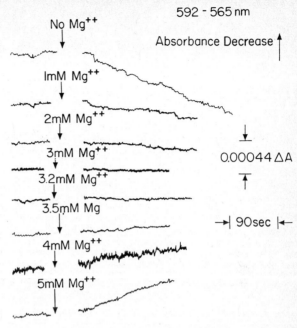

Fig. 9. Recording of the changes in absorbance undergone by Eriochrome Blue within the axoplasm of various squid axons dialyzed with increasing concentrations of Mg^{++}. Experimental conditions similar to those of Figure 8.

was dialyzed with concentrations of Mg^{++} outside this range. These results therefore indicate that the concentration of ionized Mg^{++} in squid axons, isolated and impaled, but electrically excitable and not metabolically poisoned, is close to 3 mM. Although preliminary, these results also indicate the potential of this method, which can be applied not only to absolute measurement of ionized Mg^{++} in squid axons but to kinetic measurements of Mg^{++} movement in response to electric or metabolic events. In the same preparation, with suitable filters and indicators, H^+ or other cations can also be measured kinetically or at equilibrium. In addition this technique offers a realistic approach to the quantitative measurements of cation concentration and cation movement in a variety of large cells suitable for microinjection.

Acknowledgements

The experiments of Figures 7–9 were done in collaboration with Dr. *F. J. Brinley*, Johns Hopkins University in May, 1974 at the Marine Biological Laboratory, Woods Hole, Massachusetts. The author would like to express his gratitude to Dr. *B. Chance* for his continuous encouragement and assistance and to Mr. *A. Bonner* and *N. Graham* for the assembly, testing and modification of the optical and electronic system used for the measurements. Many thanks are also due to Miss *N. Hogan* and Mr. *K. Ray* for the preparation of the manuscript and of the figures, respectively. This work was supported by grants 73-743 from the American Heart Association, GM 12202 from U.S. Public Health Service and HL 15835 from the National Institute of Health. A. Scarpa is an Established Investigator of the American Heart Association.

References

Aikawa, J. K.: The Relationship of Magnesium to Disseases in Domestic Animals and Humans, pp. 3–49. Thomas, Springfield/Ill. 1973

Baker, P. F., Crawford, A. C.: Mobility and Transport of Magnesium in Squid Giant Axons. J. Physiol. (London) 227 (1972) 855–874

Bogucka, K., Wojtczak, L.: Intramitochondrial Distribution of Magnesium. Biochem. Biophys. Res. Commun. 44 (1971) 1330–1337

Brinley, F. J., Mullins, L. J.: Sodium Extrusion by Internally Dialyzed Squid Axons. J. Gen. Physiol. 50 (1967) 2303–2331

Chance, B.: Principle of Differential Spectrophotometry with Special Reference to the Dual Wavelength Method. Meth. Enzymol. 24 (1972) 322–346

Chance, B., Oshino, N., Sugano, T., Mayevsky, A.: Basic Principles of Tissue Oxygen Determination from Mitochondrial Signal. In: Oxygen Transport in Tissue, pp. 277–292, ed. by *H. I. Bicher* and *D. F. Bruley.* Plenum Press, New York 1973

Eigen, M., DeMaeyer, L.: Relaxation Methods. In: Technique of Organic Chemistry, Vol. VIII pp. 894–1054, ed. by *S. L. Friess, E. S. Lewis, A. Weinberger.* Wiley, New York 1963

Geier, G.: Die Kinetik der Murexid-Komplexbildung mit Kationen verschiedenen Koordinations-Charakters. Helv. chim. Acta 51 (1968) 94–105

Hodgin, A. L., Keynes, R. D.; Experiments of the Injection of Substances into Squid Giant Axons by Means of a Microsyringe. J. Physiol. (London) 131 (1965) 592–616

Johnson, F. H., Shimomura, O.: Preparation and Use of Aequorin for Rapid Microdetermination of Ca^{++} in Biological Systems. Nature (London) 237 (1972) 287–288

Kun, E., Kearney, E. B., Lee, N. M., Wiedman, I.: Interaction of a Cytoplasmic Factor with Electron and Ion Transfer Coupled Function of Mitochondria. Biochem. Biophys. Res. Commun. 38 (1970) 1002–1008

Punger, E., Toth, K.: Ion Selective Membrane, Electrodes: a reviews. Analyst 95 (1970) 625–648

Scarpa, A.: Spectrophotometric Measurements of Calcium by Murexide. Meth. Enzymol. 24 (1972) 343–351

Scarpa, A.: Indicators of Free Magnesium in Biological Systems. Biochemistry 13 (1974) 2789–2794

Wacker, W. E. C., Vallee, B. L.: Magnesium, In: Mineral Metabolism, Vol. II, pp. 483–521 ed. by *C. L. Comar* and *F. Bronner.* Academic Press, New York 1964

Wacker, W. E. C., Williams, R. J. P.: Magnesium/Calcium Balances and Steady States of Biological Systems. J. Theor. Biol. 20 (1968) 65–78

Helmut Sies

Ion-selective Electrodes in the Study of Metabolic Steady States of Potassium and Ammonium Ions in Isolated Perfused Rat Liver

(5 Figures)

Considerable advances have been made in recent years in the understanding of dynamic steady states of metabolite distribution between the cellular and extracellular spaces, a problem involving mainly uncharged or anionic molecules. Due to methodological reasons, accompanying movements of cations, equally important from the cell-physiological standpoint, have received somewhat less attention. With the availability of electrodes selective for cations, information about net changes in rates of uptake or release of important ions like potassium or ammonium can be obtained at a sensitive scale. Obviously, the continuous monitoring of cation activity offers a number of experimental advantages compared to discontinuous sampling methods, in particular with respect to small and transient effects.
In this report, a few examples will be given on the application of ion-selective electrodes in an open metabolic system. The system under study is the isolated intact rat liver and an artificial extracellular fluid, the perfusate. With an open, non-recirculating perfusion, the composition of the extracellular fluid entering the liver can be experimentally controlled at all times, so that concentration changes in the fluid leaving the liver can be directly related to net rates of uptake or of release. In the absence of added erythrocytes, stepwise concentration changes in the influent perfusate can be used to induce transitions between cellular metabolic steady states, and the dependent concentration changes in the effluent perfusate reveal effects occurring across the liver cell plasma membrane.

Methods and Materials

Hemoglobin-free perfusion of rat liver (Open System)

Livers from male Wistar rats of 150–180 g body weight, fed on stock diet (Altromin), were perfused as described previously (*Bücher, Brauser, Conze, Klein, Langguth* and *Sies* 1972). Perfusate flow was 3.5–4.0 ml per min per gram of liver, and the temperature was 36.5–37°. Stepwise additions of substances were performed by infusion of neutralized solutions into the perfusion fluid entering the liver using precision micropumps.

Perfusion fluid

Standard perfusion fluid was the following: 115 mM NaCl, 1.2 mM $MgCl_2$, 5.9 mM KCl, 1.2 mM NaH_2PO_4, 1.2 mM Na_2SO_4, 2.5 mM $CaCl_2$, 25 mM $NaHCO_3$, equilibrated with a CO_2/O_2 mixture 5/95, v/v. In addition, L-lactate and pyruvate were added as sodium salts to give a concentration of 0.3 mM each. For anoxia, CO_2/O_2 was replaced by CO_2/N_2.
In the perfusion fluid termed 'HEPES medium', the bicarbonate was replaced by N-2-hydroxy-ethylpiperazine-N'-2-ethanesulfonate [HEPES], 4.6 mM, and made isoosmotic by appropriate addition of NaCl. It was equilibrated with 100% O_2.

Electrode Measurements

Effluent perfusate was collected by a cannula inserted in the vena cava and directed to a lucite electrode block where it flowed past the sensing surfaces of electrodes to monitor O_2, pH, K^+, NH_4^+. The dead volume was kept low [approx. 0.5 ml altogether], so that the responses refer to practically the same times.

O_2 concentration was monitored with a Clark-type platinum cathode of 20 μm wire diameter. pH was measured with an Ingold Ag 7.0 glass electrode. K^+ activity was followed with a Philips IS 560-K electrode, using valinomycin in diphenyl ether on a Millipore filter [*Simon, Pioda* and *Wipf* 1969 and *Pioda, Simon, Bosshard* and *Curtis*, 1970]. The reference electrode was Philips R 44/2-SD 1 with 0.1 M $NaNO_3$ as electrolyte between sample and half-cell. NH_4^+ activity was followed with a Philips IS 560-NH_4 electrode, using nonactin and monactin in tris[2-ethylhexyl]phosphate on a Millipore filter; the reference electrode was the same as that used for K^+.

For K^+ and NH_4^+, the antilog of the potentiometer outputs was first obtained before recording on a strip chart recorder. Thus, the signals afforded responses linear with K^+ and NH_4^+ activitites. A special 3-channel potentiometer with the antilog circuits was designed and constructed by Dr. *A. Schwab* and *M. Strobel* of the Electronics Department of Sonderforschungsbereich Medizinische Molekularbiologie und Biochemie, Munich. For a given condition, the signals were linear for at least 2 orders of magnitude including the physiological ranges of 5–7 mM for K^+ and 0.02–2 mM for NH_4^+, allowing changes of the order of 10^{-5}–10^{-4} M to be detected.

Results and Discussion

Potassium Ions

A representative example of the application of K^+ activity measurement in a metabolic transition is given in Figure 1. K^+ in influent perfusate was 5.9 mM throughout, and a stable baseline was established in effluent K^+ [top panel], pH [center panel] and O_2 [bottom panel] after 15 to 20 min of perfusion. As indicated on the top of the figure, t-butyl hydroperoxide was infused at 40 min, and its concentration in influent perfusate was kept constant at 0.42 mM until 50 minutes.

In all three of the traces there is a characteristic response which has been described in more detail elsewhere [*Sies, Gerstenecker, Summer, Menzel* and *Flohé* 1974]. It may suffice here to say that the primary metabolic response is the reduction of the hydroperoxide to the corresponding alcohol in a reaction catalyzed by glutathione peroxidase, resulting in increased rates of formation of glutathione disulfide and its release into the effluent perfusate [*Sies, Gerstenecker, Menzel* and *Flohé* 1972]. Of particular interest in the present context is the observation of a substantial transitory release of potassium into the effluent perfusate, peaking at about 3 minutes after the onset of t-butyl hydroperoxide infusion. At that time, there is an extra K^+ output of about 0.5 mM, close to 2 μval $K^+ \times min^{-1} \times g$ liver^{-1}. After about 8 minutes, effluent K^+ is back to the original level. Following the stop of t-butyl hydroperoxide infusion, there is a return of the parameters to the original baseline. Interestingly, there is a transitory uptake of potassium, indicating that a substantial part of the K^+ is taken back up by the liver cells, as may be judged by the areas under the two curve deflections. A detailed biochemical interpretation of the effect is beyond the scope of this presentation. It may be mentioned, however, that glutathione disulfide (although at high concentration) was found to inhibit adenosine triphosphatase in sheep red cell membranes (*Dick*, *Dick* and *Tosteson* 1969). On the other hand, measurement of liver tissue levels of adenine nucleotides revealed a decrease of the ATP/ADP ratio from 4.1 to 1.9 3 minutes after onset of t-butyl hydroperoxide infusion and a return to above 4 at 5 minutes (*Summer* 1973). The effect observed with the hydroperoxide is not found in the absence of a liver in the system. Furthermore, flame photometric analyses of samples taken at 30 second intervals also showed the transitory increase of K^+ in the perfusates. There was no change in effluent Na^+.

In the center part of Figure 1, two additions of KCl are shown, 0.36 and 0.72 mM. The half times are 10–15 seconds, and the responses are linear with concentration under the specific experimental conditions.

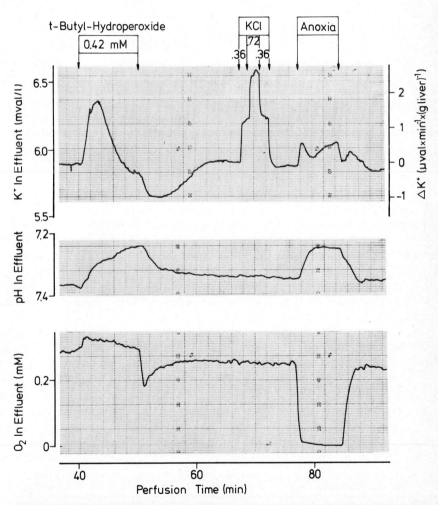

Fig. 1. Effluent potassium (top), pH (center) and O_2 (bottom) in a system of isolated, hemoglobin-free perfused rat liver. Three experimental changes were performed: an infusion of t-butyl hydroperoxide is followed by a KCl standard and by an anoxia interval.

Another metabolic effect is shown on the right part of the figure. The transition from normoxia to anoxia is followed by a biphasic response in potassium concentration. There is an initial fast phase of K+ release, peaking within the first min, and a second slow phase [Fig. 1]. After restoring the supply of oxygen, both processes occur with an inverse sign. This anoxia response is somewhat variable, and particularly the second slow phase is variable in its extent. In preliminary experiments, we were unable to demonstrate effects of ouabain (30 µM) on these responses.

Ammonium Ions

An example for a successful application of the ammonium-selective electrode is given in Figure 2. As indicated at the top of the figure, NH_4Cl was infused at three different concentrations during the given time intervals. After establishment of a steady state at each concentration, oxygen was replaced by nitrogen in the gas mixture, and the anoxia response was observed as an increase of effluent ammonium concentration approaching influent ammonium concentration.

This experiment demonstrates that the liver is capable of removing metabolically part of the infused ammonia, and that net ammonia uptake is abolished under anoxic conditions. When NH_4^+ uptake, calculated from the data of Figure 2 and an additional step of 2.68 mM from the same experiment [not shown], is plotted against NH_4Cl addition, a maximal uptake of about 2.2 µmol × min^{-1} × g liver^{-1} is observed [Fig. 3]. This very closely matches extra urea formation measured by enzymatic optical tests. Maximal extra urea formation is about 1.1 µmol × min^{-1} × g liver^{-1} [Fig. 4]. It has been found that effluent urea and ammonia add up to about 95% of influent ammonia in the normoxic steady state [*Häussinger, Weiss* and *Sies* 1974]. The major compound accounting for the residual 5% is glutamine, which is also released into the effluent perfusate. Thus, it may be concluded that in a first approximation the monitoring of effluent ammonium is useful in observing changes in ammonia disposition into urea. One point of interest, for example, is the oxygen sensitivity of the process. In the

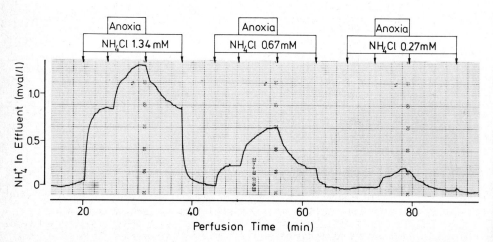

Fig. 2. Ammonium concentration in effluent perfusate. As indicated at top, ammonium chloride was infused over the given time periods, and an anoxia interval was included at the respective steady states.

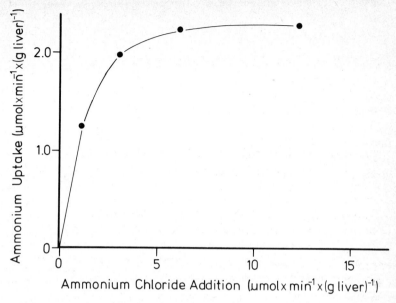

Fig. 3. Ammonium uptake by the liver as dependent upon NH_4Cl supply in influent. Data from experiment in Figure 2.

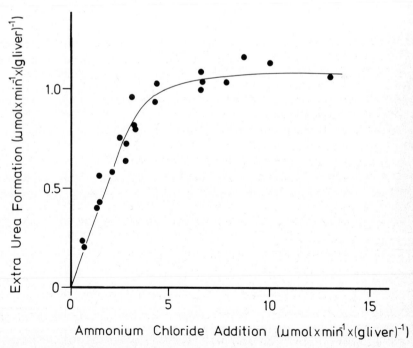

Fig. 4. Extra urea formation by the liver as dependent upon NH_4Cl supply in influent. Data from enzymatic analyses from six livers. Compare Figure 3. (*After Häussinger, Weiss* and *Sies* 1974).

experiment of Figure 2, ammonia began to rise at a time when O_2 concentration was approx. 20 µM in effluent perfusate. However, this was not a steady state, and for a determination of oxygen dependence of urea formation the experiment should be carried out with O_2/N_2 gas mixtures and steady states allowed to establish.

Ornithine stimulation of urea formation [Hems, Ross, Berry and Krebs 1966] can be readily demonstrated by a decrease of effluent ammonium concentration upon addition of ornithine [not shown].

Cation-Cation Exchange

Passive cation distribution between the cellular and extracellular spaces is governed by the Donnan relationship. When gradient equilibration has occurred with ammonia, addition of methylammonium chloride will lead [a] to a proton release in the extracellular space because uncharged methylammonia is the permeant species, [b] to an intracellular alkalinization due to methylammonium formation on the other side of the plasma membrane, and [c] to a net cation-cation exchange. This type of consideration has been previously performed for monocarboxylates where the effects are inverse [Sies, Noack and Halder 1973]. The process [c] is demonstrated in Figure 5 where 1.3 mM NH_4Cl is present throughout, and a 4 minute pulse of methylammonium chloride [7.1 mM] is given. During gradient equilibration of the methylammonium, there is a transient release of ammonium into the extracellular space, amounting to about 3.5 µmol per gram. The effect is reversible. This experiment was carried out in HEPES medium, a condition where extra urea formation from ammonia is suppressed [Sies and Häussinger 1974]. Therefore, the observed changes reflect a redistribution in the absence of metabolic transformation of ammonia.

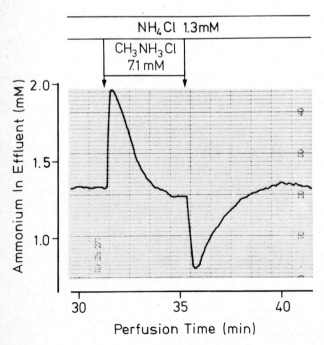

Fig. 5. Ammonium movement between intra- and extracellular spaces upon addition of methylammonium chloride. Influent ammonium was 1.3 mM throughout.

Commentary

The use of ion selective electrodes in an open metabolic system allows to detect small and transient changes, and such electrodes, therefore, are valuable analytical tools. Effects of the electrode system on the cells, as observed in closed systems, e.g., valinomycin contamination of the object, are excluded. However, caution must be exerted in the design of the experiment, and it is mandatory that for each experimental change a proper control is performed. In fact, small changes in ionic composition may have pronounced effects on the electrode response. Just one example is that a variation in bicarbonate concentration alters the response of the ammonium electrode. On the other hand, the long-term stability of the electrodes permits the performance of one- or two-hour experiments without any problems.

The few examples of metabolic effects disclosed with ion selective electrodes in this report clearly demonstrate the suitability of these new tools in the study of complex metabolic systems like a perfused organ.

Acknowledgements

Skillful technical assistance was provided by Miss *I. Linke* and Mrs. *A. Marklstorfer*.
The ion-selective electrodes were kindly provided by *A. Kühnau*, Philips Industrie Elektronik, Hamburg.
This investigation was supported by Deutsche Forschungsgemeinschaft, Sonderforschungsbereich 51, Medizinische Molekularbiologie und Biochemie, Grant D/8.

References

Bücher, T., Brauser, B., Conze, A., Klein, F., Langguth, O., Sies, H.: State of Oxidation-Reduction and State of Binding in the Cytosolic NADH-System as Disclosed by Equilibration with Extracellular Lactate/Pyruvate in Hemoglobin-Free Perfused Rat Liver. European J. Biochem. 27 (1972) 301–317

Dick, D. A. T., Dick, E. G., Tosteson, D. C.: Inhibition of Adenosine Triphosphatase in Sheep Red Cell Membranes by Oxidized Glutathione. J. gen. Physiol. 54 (1969) 123–133

Häussinger, D., Weiss, L., Sies, H.: Activation of Pyruvate Dehydrogenase During Metabolism of Ammonium Ions in Hemoglobin-Free Perfused Rat Liver. European J. Biochem 52 (1975) 421–431

Hems, R., Ross, B. D., Berry, M. N., Krebs, H. A.: Gluconeogenesis in the Perfused Rat Liver. Biochem. J. 101 (1966) 284–292

Pioda, L. A. R., Simon, W., Bosshard, H. R., Curtius, H. Ch.: Determination of Potassium Ion Concentration in Serum Using a Highly Selective Liquid Membrane Electrode. Clin. Chim. Acta 29 (1970) 289–293

Sies, H., Gerstenecker, C., Menzel, H., Flohé, L.: Oxidation in the NADP System and Release of GSSG from Hemoglobin-Free Perfused Rat Liver during Peroxidatic of Glutathione by Hydroperoxides. Fed. Eur. Biochem. Soc. Lett. 27 (1972) 171–175

Sies, H., Gerstenecker, C., Summer, K. H., Menzel, H., Flohé, L.: Glutathione-Dependent Hydroperoxide Metabolism and Associated Metabolic Transitions in Hemoglobin-Free Perfused Rat Liver. In: Glutathione, pp. 261–275, ed. by L. Flohé, H. C. Benöhr, H. Sies, H. D. Waller and A. Wendel. Thieme, Stuttgart 1974

Sies, H., Häussinger, D.: Lack of Urea Formation in Hydrogen Carbonate Free Media. Hoppe-Seyler's Z. physiol. Chem. 355 (1974) 1256

Sies, H., Noack, G., Halder, K. H.: Carbon Dioxide Concentration and the Distribution of Monocarboxylate and H^+ Ions between Intracellular and Extracellular Spaces of Hemoglobin-Free Perfused Rat Liver. European J. Biochem. 38 (1973) 247–258

Simon, W., Pioda, L. A. R., Wipf, H. K.: Cation Specificity of Inhibitors. In: Inhibitors-Tools in Cell Research, pp. 356–364, ed. by Th. Bücher and H. Sies. Springer, Berlin 1969

Summer, K. H.: Pyridinnucleotide und Glykolyseintermediate in der hämoglobinfrei perfundierten Rattenleber bei Umsatz von exogenem Hydroperoxyd. Diplomarbeit, Fachbereich Chemie, Univ. Tübingen 1973

Daniel D. Reneau

A Mathematical Analysis of Simultaneous Transport Phenomena in the Microcirculation (Ions, Substrates, Gases, Metabolites)

The general purpose of this work is an attempt to obtain a better understanding of the simultaneous transport phenomena in the microcirculation by means of both theoretical analysis and experimental measurement. The specific purpose of this paper is to present the results of a continuing theoretical analysis of the simultaneous diffusion and interaction of respiratory gases, substrates, metabolites and ions at the cellular level. Adaptation of the model is given to brain under conditions of partial and total anoxia.

The intent is to seek basic understanding and consequent identification of parameters that are the chief factors associated with the regulation of the microenvironment and changes which may lead to nerve cell network damage, destruction and total irreversible decay. The approach to the problem is as follows:

1. Use previously developed equations based on the Krogh geometry to determine the spatial distribution of oxygen and glucose in all parts of the capillary and tissue;
2. Theoretically calculate and subsequently geometrically map anoxic tissue regions surrounding the capillary which occur under conditions of specified quantitative changes in biological functions (i.e., flow rate, metabolic rate, anoxic-anoxia);
3. Solve interconnected equations that describe:
   a) the production and diffusion of acids in the mathematically defined anoxic region,
   b) the diffusion through the non-anoxic region,
   c) the diffusion into and removal by blood flowing through adjacent capillaries.
4. Based on the above results solve ion transport equations to predict changes in ion concentrations (K^+, Na^+ and Cl^-) and resting membrane potential that may occur during periods of developing anoxia.

Mathematical simulation of the above conditions can give insight into helping answer some very important questions. For instance, reductions in rates of blood flow through capillaries can lead to the development of localized areas of cerebral tissue anoxia which become sites of anaerobic glycolysis. What reduced magnitude of capillary flow rate yields minute regions of tissue anoxia but is still sufficient to remove acidic metabolic waste material? And, for each progress reduction of determined degree below this critical flow rate, what is the magnitude and rate of increase of tissue acidosis? What is the progressive shift of ion concentration? What is the initiator of anoxic depolarization?

Mathematical Description

Oxygen Diffusion. Based on the Krogh geometry, known phenomena, and certain assumptions which are outlined in *Renau, Bruley* and *Knisely* (1967), a mathematical model has been developed which describes the change in oxygen partial pressure in capillary blood and tissue as a function of time, position, flow rate, pH, oxygen capacity, metabolic rate, and various constants such as diffusion coefficients and solubility. Certain details concerning the model are given in previous publications (*Reneau, Bruley* and *Knisely* 1967, and *Reneau, Bruley* and *Knisely* 1969).

Capillary:

$$\left(1 + \frac{NknP^{n-1}}{c_1(1+kP^n)^2}\right)\frac{\partial P}{\partial t} = D_1\left(\frac{\partial^2 P}{\partial r^2} + \frac{1}{r}\frac{\partial P}{\partial r}\right) + D_1\frac{\partial^2 P}{\partial x^2} - V_x\frac{\partial P}{\partial x}$$
$$- \frac{V_x NknP^{n-1}}{c_1(1+kP^n)^2}\frac{\partial P}{\partial x} \quad (1)$$

Interface

$$P_i\bigg|_{Blood} = P_i\bigg|_{Tissue} \quad (2)$$

$$D_1 c_1 \frac{\partial P}{\partial r}\bigg|_{\substack{r=R_1 \\ Blood}} = D_2 c_2 \frac{\partial P}{\partial r}\bigg|_{\substack{r=R_2 \\ Tissue}} \quad (3)$$

Tissue

$$\frac{\partial P}{\partial t} = D_2\left(\frac{\partial^2 P}{\partial r^2} + \frac{1}{r}\frac{\partial P}{\partial r}\right) + D_2\frac{\partial^2 P}{\partial x^2} - \frac{A}{c_2} \quad (4)$$

Glucose Diffusion. Based on the Krogh geometry, known phenomena, and certain assumptions which are outlined in *Reneau* and *Knisely* (1968) and *Reneau* et al. (1971), a companion mathematical model has been developed which describes the changes in glucose concentration in capillary blood and cerebral tissue. The model is given below:

Capillary:
Plasma

$$\frac{\partial c}{\partial t} = D_1\left(\frac{\partial^2 c}{\partial r^2} + \frac{1}{r}\frac{\partial c}{\partial r^2}\right) + D_1\frac{\partial^2 c}{\partial x^2} - V_x\frac{\partial c}{\partial x} + K\left(\frac{L_2 c'}{L_2 c' + \emptyset} \frac{L_2 c}{L_1 c + \emptyset}\right) \quad (5)$$

Erythrocyte

$$\frac{\partial c'}{\partial t} = -V_x\frac{\partial c'}{\partial x} - k\left(\frac{L_2 c'}{L_2 c' + \emptyset} \frac{L_1 c}{L_1 c + \emptyset}\right) \quad (6)$$

Blood-Tissue Interface:

$$\frac{\partial c}{\partial t} = B\left(\frac{\lambda_1 c}{\lambda_1 c + \emptyset'}\bigg|_{\substack{r=R_1 \\ Capillary \\ side\ of \\ Interface}} - \frac{\lambda_2 c}{\lambda_2 c + \emptyset'}\bigg|_{\substack{r=R_1 \\ Tissue \\ side\ of \\ Interface}}\right) \quad (7)$$

Tissue:

$$\frac{\partial c}{\partial t} = D_2\left(\frac{\partial^2 c}{\partial r^2} + \frac{1}{r}\frac{\partial c}{\partial r}\right) + D_2\frac{\partial^2 c}{\partial x^2} - A \quad (8)$$

Acid Diffusion

Based on the Krogh geometry, known phenomena, and very idealizing assumptions including non-ionic diffusion, the following equations are presented to describe the steady-state, geometric selective, acid production, diffusion and removal by flowing capillary blood during partial anaerobic glycolysis in tissue:

Tissue Equation

$$D_t \left(\frac{\partial^2 c}{\partial r^2} + \frac{1}{r} \frac{\partial c}{\partial r} \right) = -r_A \tag{9}$$

Blood Equation

$$V_x \frac{\partial c}{\partial x} = D_b \left(\frac{\partial^2 c}{\partial r^1} + \frac{1}{r} \frac{\partial c}{\partial r} \right) \tag{10}$$

Interface Equations

$$D_b \frac{\partial c}{\partial r} \bigg|_{\substack{\text{Blood side} \\ r = \text{interface}}} = D_t \frac{\partial c}{\partial r} \bigg|_{\substack{\text{tissue side} \\ r = \text{interface}}} \tag{11}$$

$$C_t \bigg|_{\substack{\text{tissue} \\ r = \text{interface}}} = \lambda C_b \bigg|_{\substack{\text{blood} \\ r = \text{interface}}} \tag{12}$$

The interface equations describe conditions in tissue at the anaerobic-aerobic interface and at the capillary wall for the blood-tissue interface.

Standard boundary conditions have been used which describe the tissue as a closed system interacting only with the capillary system.

Ion Transport

Mathematical description of ion transport between intracellular and extracellular environments was developed by viewing the intracellular and extracellular regions as two compartments separated by a selectively permeable membrane. The description is based on the method of *Brace* and *Anderson* (1973) and was applied in this work to conditions of total anoxia after cessation of capillary blood flow. The development and equations are summarized in the following paragraphs.

Exchange of ions between compartments are assumed to be governed by concentration gradients, voltage gradients and active transport. From the Nernst-Planck equations, passive diffusion across a membrane due to electro-chemical gradients can be described by the equation;

$$J_i = \frac{E_m F}{RT} P_i Z_i \frac{C_{i,I} e^{Z_i E_m F/RT} - C_{i,E}}{1.0 - e^{Z_i E_m F/RT}}$$

Where i refers to a specific ion species, J; is in flux, P_i is permeability, Z_i is ion valance, E_m is resting membrane potential, F is the *Faraday* constant, R is the universal gas constant, T is absolute temperature $C_{i,I}$ is intracellular ion concentration and $C_{i,E}$ is extracellular ion concentration.

The flux of water across the membrane is given as

$$J_{H_2O} = P_{H_2O} (osm_i - osm_e)$$

Where, P_{H_2O} is the hydraulic permeability, osmi and osme is the intra and extracellular osmolarity, respectively.

Active fluxes of potassium and sodium ions are calculated according to the scheme of *Brace* and *Anderson* (1973) as

$$J_i, \text{Active} = J_{i,0} \frac{A}{A_0}$$

where

$$i = Na, K$$

$$A = A_K A_{Na}$$

$$A_K = 1.5 \left[1.0 - \left(\frac{C_{K,E} + 2.5}{2.5} \right) e^{(-0.45 C_{K,E})} \right]$$

$$A_{Na} = 5.7 \left(1.0 - e^{(-0.35 C_{Na,I})} \right)$$

The flux $J_{i,0}$ is equal to the passive flux of i at steady-state. Transient intracellular changes are calculated as,

$$\frac{dn_{i,I}}{dt} = (J_i \text{ passive} + J_i \text{ active}) A$$

$$V_i = \frac{N_{H_2O,I} \cdot MW_{H_2O}}{pH_2O}$$

$$C_{i,I} = \frac{N_{i,I}}{V_i}$$

where, N_i is moles of i, MW is molecular weight, V_I is intracellular volume, and A is membrane area and
$i = K^+, Na^+, Cl^-, H_2O$ but only K^+ and Na^+ are to be considered transported actively.
Transient extracellular changes are calculated as,

$$\frac{dn_{i,E}}{dt} = \frac{dn_{i,I}}{dt}$$

$$V_E = \frac{N_{H_2O} \cdot MW_{H_2O}}{pH_2O}$$

$$C_{i,E} = \frac{N_{i,E}}{V_E}$$

Brace and *Anderson*, 1973, suggested that changes in resting membrane potential can be determined by transient ionic fluxes as follows,

$$\frac{dE}{dt} = \frac{dQ}{c}$$

where Q is change and C is membrane capacitance

$$dQ = F \sum_i Z_i (J_i \text{ passive} + J_i \text{ active})$$

Consequently, the above equations can be solved simultaneous and values predicted as a function of time in both the intracellular and extracellular compartments.

Solution Techniques

The distributed parameter equations describing oxygen diffusion, glucose diffusion and acid diffusion were solved by numerical analysis techniques and computer computation. These methods have been outlined in the literature and are available. (*Reneau* et. al, 1967, 1968, 1970, 1973).

The ion transport equations outlined here were solved on an IBM 360 Computer using numerical techniques available with the CSMP simulation language. Ion concentrations and resting membrane potentials were predicted as a function of time following a specified departure from steady-state.

Application

Tissue pH during Anoxia. Recent reports in the literature indicate that tissue deterioration during anoxia is more rapid during ischemic anoxia than during norm flow anoxia. Perhaps the rate of removal of lactic and pyruvic acids by means of blood flowing through capillaries is sufficient to prevent appreciable decrease in cellular pH. A simulation in steady-state of this condition was conducted for both partial anoxia and total anoxia. Acid accumulation was predicted in tissue as a function of increased, normal, and decreased capillary blood flow rates.

The results of the simulation indicate that during total anoxia, *normal* rates of capillary blood flow can remove acidic metabolic by-products and maintain a low tissue concentration of acid. This effect produces a stabilizing factor for tissue pH. *Increases* in flow rates above the standardized normal provide an additional benefit that is of small advantage. (Normal flow rates are much more than adequate.) As capillary blood flow rates are *reduced* below normal, tissue acid concentrations begin to increase. The increase is not dramatic until flow rates are reduced to below 25–30 per cent of the normal rate. Further reductions theoretically lead to an exponential increase in tissue acid concentrations, and significant pH changes can be expected.

These caculations are based on several assumptions among with (1) it is assumed that lactic and pyruvic are free to diffuse through brain and are not blocked at the capillary wall and (2) hypothesized metabolic reaction rates in acid production are reasonable.

Ion Transport During Anoxia. During anoxia the literature indicates that experimental measurements in brain with microelectrodes in the extracellular environment indicate that significant and relatively rapid changes occur in the extracellular concentrations of ions and that the cells go through a period of rapid anoxic depolarization. The ion transport model presented here was solved to simulate these conditions.

Assuming that anoxia would lead to a rather rapid depletion of ATP, it was postulated that the energy mechanisms necessary for maintenance of the active pump for sodium and potassium would fail. In such a condition, what shift in ion concentrations would occur?

The ion transport model was standardized with normal, accepted values for cellular and extracellular concentrations and checked for accuracy by simulating known experimental response to known changes. The techniques suggested by *Brace* and *Anderson*, 1973, were used. Following a theoretical and total failure of the active pump, the concentration of both the sodium and chloride ion decreased extracellularly, the concentration of the potassium ion increased, and the cell began to depolarize. However, the rate of ionic concentration change and the rate of depolarization was found to be a function of transport area and the ratio of extracellular volume to intracellular volume.

Rapid responses similar to experimentally measured values for anoxic depolarization and ion movements, can be shown to occur if the magnitude of the cellular volume is one to five percent of the intracellular volume. These volumes are consistent to literature values for extracellular brain volume; however, no firm conclusion can be drawn.

Additional theoretical studies with the other mathematical models demonstrate that anoxia develops differentially and then spreads throughout a typical region. It would seem that some areas would become depleted of energy stores before others and a progressively increasing deterioration would occur. Also diffusion of species produced intracellularly during ionic anoxia must be incorporated into the model, when data become available, before firm conclusions can be drawn. However, the responses at present indicate that failure of the active transport mechanism in brain during anoxia can play a significant role in ionic exchanges but is not the exclusive factor.

References

Brace, R. A., Anderson, D. K.: Predicting Transient and Steady-State Changes in Resting Membrane Potential. J. appl. Physiol. 35 (1973) 90–94

Reneau, D. D., Bruley, D. F., Knisely, M. H.: A Mathematical Simulation of Oxygen Release, Diffusion, and Consumption in the Capillaries and Tissue of the Human Brain. In: Chemical Engineering in Medicine and Biology, pp. 135–241, ed. by *D. Hershey.* Plenum Press, New York 1967

Reneau, D. D., Bruley, D. F., Knisely, M. H.: A Digital Simulation of Transient Oxygen Transport in Capillary-Tissue Systems (Cerebral Gray Matter). A. I. Ch. E. J., 15 (1969) 916–925

Reneau, D. D., Bruley, D. F., Knisely, M. H.: A Computer Simulation for Prediction of Oxygen Limitations in Cerebral Gray Matter, JAAMI 4 (1970) 211–223

Reneau, D. D., Knisely, M. H.: A Mathematical Simulation of Glucose Diffusion and Consumption in Brain. Proceedings of the 21st Annual Conference on Engineering in Medicine and Biology. 10 (1968) 29. 2

Reneau, D. D., Knisely, M. H., Bicher, H. I.: Glucose Diffusion and Consumption in the Human Brain. Transport Process in Biology and Medicine, Preprint 33a, 70th National Meeting of A. I. Ch. E., Atlantic City/N. J., August 1971

Reneau, D. D., Lafitte, L. L.: Calculation of Concentration Profiles of Excess Acid in Human Brain Tissue During Conditions of Partial Anoxia. In: Oxygen Transport to Tissue, Vol. 2, pp. 849–858, ed. by *D. F. Bruley, H. I. Bicher.* Plenum Press, New York 1973

Discussion of Session IV A

Chairman: *A. Kovách*

Discussion of Paper by Clausen: The Role of Ions in the Control of Intermediary Metabolism

Christian: I think this work is very interesting. I would just like to comment that *Mertz* and coworkers have found evidence for depressed insulin effectiveness in the presence of chromium. For example *Mertz* and *Feldman* a few years ago found a rise in plasma chromium along with glucose rise, and there is other more direct evidence. I wonder if you feel there might be a relationship with calcium.

Clausen: I have no experience with chromium but the work you refer to was related to intact organisms.
The problem is really that if you start to work with multivalent ions you have inhibitory effects of these ions on the insulin response. For instance lanthanum ions and nickel ions and other heavy metal ions would prevent the effect of insulin on sugar transport. They are membrane stabilizers and act in a very similar way to local anaesthetics which would also prevent the effect of insulin. I think that these data are difficult to assess because there are some changes in the surface properties of the plasma membrane with heavy metal ions.

Kessler: I would like to ask one question: You mentioned that when you add mannitol the effects you observed may be unspecific and caused just by hyperosmolarity. Could it be possible that you induce an influx of sodium into the cell by this big increase in extracellular osmotic pressure?

Clausen: We have measured the sodium influx in the hyperosmolar condition (200 mM mannitol). A more than two-fold increase in the influx of sodium is seen, and the membrane potential doesn't seem to be too much changed because the efflux of sodium is increased by about the same magnitude. There are other studies showing that the membrane potential is not changed very much by increasing osmolarity in frog muscles. So I think that we have changes in monovalent ion permeability with osmolarity and this could be related to the effect of ion strength on the calcium bound to the surface of the membrane; this would be one possible interpretation. The other question you raised was whether the state of anoxia occurring when you stop the flow is associated with a rise in calcium in the extracellular phase. I think this is consistent with the stimulating effect of metabolic inhibitors on calcium release from the intact tissue.

Brown: I know that the sodium gradient hypothesis may not be the only explanation. Nonetheless many of the results you have shown might possibly be explained on the basis of an increased sodium gradient. For example when you rewarm the cell the sodium pump is activated which may increase the sodium-calcium and sodium-sugar exchange mechanisms. Are they of sufficient magnitude to explain the calcium exchange?

Clausen: During the period of cooling the efflux of calcium is diminished. I don't know which proportion of calcium is getting out by an ATP-dependent mechanism as suggested by *Schatzmann*, and which proportion is dependent on the sodium gradient, but at least those processes would depend on energy supply. Probably much more important is the net loss of calcium from the mitochondria which would be favored by cooling. Cooling really would inhibit the mitochondrial uptake of calcium so much that you easily build up a high concentration of free calcium in the cytoplasm. This is not seen in the cold, but when you return to the normal temperature you would have more free calcium here available for efflux and that could start immediately the moment you rewarm.

Brown: What is the evidence that cooling results in a release of calcium from the mitochondria?

Clausen: I think that you block the active uptake of calcium in the mitochrondria and that, even though you may not accelerate the efflux, the net loss of calcium from the mitochondria will be increased and since you cannot in this situation prevent calcium from getting in from the extracellular phase, I think that many of the mechanisms normally maintaining a low cytoplasma calcium will no longer function in the cooling. The mitochondrial accumulation may be decreased, as well as the uptake in the endoplasmic reticulum and the extrusion of calcium across the plasma membrane. But I think we can dissociate changes in sugar transport from those seen in the active pumping of sodium and potassium because blocking the pump with ouabain does not change the rate of sugar transport immediately, at least not in this system.

Scarpa: There is no evidence of release of Ca^{++} by mitochondria induced by cooling, at least I don't know of any with isolated mitochondria.

Eisenman: Is there a direct effect of insulin on mitochondrial uptake of calcium?

Clausen: I have never seen any convincing data for it. We measured the uptake of calcium in whole fat cells and the major part of that uptake seems to be mitochondrial uptake. I think it is very hard to look at, because, when you isolate the mitochondria you may lose a lot of their calcium so you don't really see it. There is no direct effect as far as I know.

Eisenman: Do you see any variations due to external changes of the hydrogen ion concentration or do you keep the hydrogen ion concentration constant?

Clausen: We haven't got into that yet.

Hinke: I would like to go back to your hypertonic experiments again, when you expose any kind of cell to hyperosmotic shock you are obviously going to get lots of movement of water out and your free ions or free glucose or free anything is just going to automatically increase in concentration. Are you able to calculate such concentration increases from your knowledge of the cellular water loss?

Clausen: The shrinkage amounts to about 66%. The rise in sugar efflux is about tenfold and the uptake of sugar is also increased.

Hinke: The second point that I would like to make is that with hypertonic exposure you have to be very careful whether you are doing it with the sugar or with an ion, i.e., whether you are changing the ionic strength outside because the response of the cell may be different. Not only can you have water movement during hypertonic shock, but you may also have considerable solute movement.

Clausen: The increase in sugar transport is about the same whether you increase osmolarity by 200 mM mannitol or 100 mM sodium chloride which would be the equivalent osmolarity increase.

Discussion of Session IV A

Scarpa: I have a very naive question: Can you explain part at least of the inorganic phosphate uptake in the presence of insulin as due to an increase in efficiency of glycolytic ATP production?

Clausen: Insulin does not stimulate glycolysis very much. You can find in some studies reports of an increase of glycolysis but would you think of an increased efficiency of the mitochondrial oxidative phosphorilization? Was that what you thought?

Scarpa: No, I was thinking in terms of increasing inorganic phosphate incorporation into ATP via glycolysis. This condition seems to be true in heart muscle where you can drive anaerobic heart in the presence of glucose and insulin.

Clausen: I think there is an increased incorporation of inorganic phosphate into ATP and that probably accounts for the major part of the increased uptake in the presence of insulin we have here. These experiments were done in the absence of glucose, so it is not related to glucose transport changes, but it is true that it has been suggested also by others that insulin affects the efficiency of oxidative phosphorilization.

Discussion of Paper by B. C. Pressman: Physical and Biological Properties of Ionophores

Thomas: Where on earth does this magic compound come from?

Pressman: It is the natural product of streptomycis fermentation of a particular strain.

Lübbers: Have you seen a concentration-dependent effect of this drug in your animal experiments?

Pressman: Yes, it shows the typical log dose response curve. We have used the criterion of DP/DT versus dose.

Hinke: Do you know whether this drug passes the blood brain barrier and if so what does it do to the brain cells?

Pressman: We put it into a couple of monkeys whose behavior patterns had been studied by long observation and which also had implanted cardiac flow probes. We saw some cardiac response but there was no overt behavioral change in these monkeys. We don't know whether it passes the blood brain barrier but if it did it produced no obvious effect.

Clausen: Are the vascular effects secondary to an increased adrenaline secretion?

Pressman: The vascular effects are partially blocked by beta adrenergic blockers but not completely. The other point is that we don't know to what extent catecholamines are involved there or how they are involved. The effects we get cannot be mimicked by any known combination of catecholamines so even if it is due to the release of endogenous catecholamines, these must be carried to places that they normally would not have access to if they were administered i.v. exogenously.

Discussion of Paper by Scarpa: Metallochromic Indicator for Kinetic Measurements of Magnesium Ions and Determination of Cytosolic-Free Magnesium Ions by Microspectrophotometry

Brown: Can you expand upon the calcium ion-sensitive indicator you mentioned?

Scarpa: One indicator that can be used is (2, 2'-[1,8 Dihydroxy-3-6-Bisulfo-2,7-Naphtalene)-Diazo] Dibenzlarsonic Acid. This indicator is water soluble, has kinetics in the range of microseconds and a sensitivity which is a couple of orders of magnitude greater than murexide. It can be used quite specifically for Ca^{++} by dual wavelength spectroscopy at 665–685 nm and, at the sensitivity shown for Mg^{++} measurements, measurements of 0.1 μM ionized Ca^{++} may be possible.

General Discussion (24 b)

Clark: Dr. *Pressman*, I was wondering what the total oxygen consumption was in the dog and whether you measured mixed venous Po_2, and what anaesthetic was used, because that was a sizeable increase in cardiac output.

Pressman: Yes, we did measure mixed venous oxygen tension, but nothing spectacular happened there; it was mostly in the cardiac venous blood that we got this tremendous rise – we saw hardly anything at all in the total mixed venous blood. This is barbiturate anaesthesia; there is some depressant effect from that undoubtedly. We did not measure total O_2 consumption.

Clark: But if mixed venous blood stayed about the same and cardiac output went way up, then total oxygen consumption must have gone up.

Pressman: I would presume yes, but we didn't measure it directly. Sometimes we got a very small rise in mixed venous O_2 but not uniformly. There must have been a uniform compensatory increase in oxygen consumption.

Clark: But then this would indicate that this compound stimulated oxygen consumption and that could account for some of the effects you got at any rate.

Pressman: We did measure in the heart, and we certainly know quantitatively that cardiac consumption of oxygen went up 50%.

Clark: Yes, right, but this, you know, in a way is like giving a large dose of thyroxine or something like that. It seems that your ionophore may increase oxygen consumption, at least of some tissues.

Lübbers: If I understood correctly you said that the blood had not enough time to give off all its oxygen, but this is certainly not true, since the time should be long enough to release chemically bound oxygen even if the blood runs rather quickly through the heart. If the flow increases, you get a higher venous oxygen tension. It is a matter of balance, and I do not think kinetics are involved.

Pressman: There is no discrepancy, maybe I didn't express myself well, I should have expressed myself exactly the way you did.

Grote: Would you be so kind as to make some further comments on the results of your heart muscle experiments. If I remember correctly, at the end of the experiment the blood flow values were back to the initial value, but at the same time, although there was an increased oxygen consumption rate, the oxygen tension in the coronary sinus blood was also increased. The reason for this is not clear to me. Which additional factor changed: the oxygen tension in the arterial blood, the oxygen capacity of the blood or the oxygen affinity of the blood?

Discussion of Session IV A

Pressman: You can't assume anything from that because the oxygen tension alone doesn't tell you the true oxygen content; the pH of the blood is changing, carbon dioxide is changing and the hemoglobin content of the blood is changing. There is a huge amount of hemoglobin discharged from the spleen, so there is no direct correlation between the oxygen tension and the total oxygen contents. You have to calculate it.

Clausen: I think that these are all very dramatic effects, but there is one new point here, the suggestion about the relation of catecholamines to the ionophore. Catecholamine concentration of the blood is probably much lower than that of the ions which will also be bound by ionophore. Would you think the binding of ionophores and catecholamines could mean something for the actions we see here?

Pressman: Undoubtedly it does, because of the fact that we can suppress part but not all of the effects with beta adrenergic blockers. The results seem to be a complicated mixture of effects some of which are ascribable to catecholamines, some of which are ascribable to mobilisation of calcium and there are additional factors that we still don't understand. Perhaps I was overenthusiastic when I stressed the rationality of what we were doing. There were rational reasons for doing the experiment and we got the results we were hoping for but we don't know quite why. It will take years to unravel all of the various effects that happened. We didn't realize that the response would be so complicated when we attempted a very naive experiment. I might point out in this connection something I didn't have time to go over, that the cardiac action potential as determined in Purkinje fibers can shift too which is indicative of calcium inward movement into the heart. This is prolonged so that is evidence for a direct implication of calcium in the contractility but in a surprising fashion; it comes only from the cardiac action potential.

Brown: If the complex is neutral how can it be carrying inward current?

Pressman: I said that the application of X537A to a Purkinje fiber increases the amount of calcium that goes inward during phase 2 but I didn't say why and I'm not drawing the logical conclusion that you do; I don't know what the mechanism is.

Eisenman: How does the ionophore produce its effect if its aqueous complexation constants are so small? It is not necessary to have a high stability constant in the aqueous phase in order to produce a substantial transport of the material across the membrane. In particular, valinomycin may effectively transport ions across membranes but its stability constants for complexation in the aqueous phase are very small.

When one compares stability constants from one solvent to another, or across one organic phase versus another, you have always to remember that your reference state is important. For example if you refer to stability constants measured in methanol then your reference states are methanolated ion versus the methanolated complex. Your complex may still have substantial interaction with the solvent of the membrane; and this probably is the reason for the differences in selectivity sequence.

Pressman: We originally did this to see what would happen under dynamic conditions where kinetics would actually have an effect, but we were struck later on by the correlation that in all of the water systems the constants for calcium were higher, just as we see it in transport. It was only the ethanol system that showed calcium less active and we think that has something to do with the lowered energy of solvation in the ethanol. It is true that we are measuring in the three-phase system, a kinetic system, while all the others were measuring an equilibrium and the two systems need not correlate precisely.

Kovách: Is there any possibility of measuring what is bound in the dog?

Pressman: No, because everything bound to the ionophores is moving on and off very dynamically. What would end up being bound would depend on the method that we used to extract the ionophore in dynamic equilibrium with everything presented to it.

Discussion of Paper by Sies: Ion-selective Electrodes in the Study of Metabolic Steady States of Potassium and Ammonium in Isolated Perfused Rat Liver

Simon: I was a little surprised that acetate and especially bicarbonate should give that much interference. I believe that it is not really an anion interference as described by *Boles* and *Buck*.

Sies: I agree, but it is not a large effect at all, amounting to only about 1 % interference with Krebs-Ringer-solution. But since we are looking at small metabolic effects, such controls are certainly required.

Baucke: You may have mentioned it but what was the scale in terms of mV in your case?

Sies: The scales were the antilog of the potentiometer outputs, so we can compare the responses on linear scale. In this case it is not so important whether we have *Nernstian* behavior or not, as long as the complex system, consisting of a mixture of different ions, gives linear responses.

Kessler: I think it is very interesting that you observe only a small and transient increase in potassium during anoxia. This indicates that anoxia can be tolerated well by the isolated perfused liver.

Sies: It is surprising to us, too, that there is such a small K^+ effect of anoxia. The second slow phase is variable in its extent and it cannot be suppressed by quick reoxygenation. There is a chain of events which appears to take place, and it is altogether unclear what really causes the different phases of potassium efflux.

Kovách: How long was the anoxic cycle? Did you get larger K^+ elevation in the perfusate after longer lasting hypoxia?

Sies: No, the second phase is completed after about 7 or 8 minutes of anoxia. Our experiments were done only as short-term transitions. The longest period of anoxia lasted about 15 minutes.

Klingenberg: Can you calibrate your anoxia effect. I mean the efflux of potassium which you would expect, by perfusing valinomycin?

Sies: We have not done this so far. This would be an interesting experiment.

Discussion of Paper by Reneau: Mathematical Analysis of Transport Phenomena in the Microcirculation

Zeuthen: This is just to say during anoxia the extracellular space is decreased due to cell swelling, so it is probably valid to reckon on a very small extracellular space.

Discussion of Session IV A

Reneau: A small extracellular space in brain will yield rapidly changing concentrations in extracellular substances; however, other factors also contribute.

Zeuthen: How can you incorporate in your model the fact that the cells have a definite size so that actually you may have only one cell between two capillaries?
That means that most of your gradients really are intracellular so that different parts of the cells suffer from different degrees of anoxia or different glucose activities.

Reneau: At present the simulation calculates lumped intracellular and extracellular concentrations and does not predict spatial distribution for ions. Our previous work on oxygen allows mapping of partial anoxic areas that can be related to the production of the anoxic byproducts.

Buck: In my group we are working on digital simulation of the very same sort as you described in the latter part of your talk. I compliment you on having gone so far in this field and would like to ask some questions and make a comment. We find that the calculation of the internal diffusion potential is best done using the Nernst-Planck equations with the electric vector term to account for Poisson's Equation. We calculate the potential by integrating the electric field with the membrane potential, then we add the interfacial potentials determined the same way in the bathing solutions. We can see the space charge build up and the time and spatial dependence of concentration, field and potential. Would you comment on the reasons for your particular boundary conditions? Why do you choose the boundary conditions and then secondly – do you solve these equations by approximations, by exact analytical methods or by essentially finite difference digital simulation methods?

Reneau: Boundary conditions are not necessary – only initial conditions. I picked this particular situation because it is a situation with which we are very familiar and because of the computational time and the techniques involved. I feel it will simulate the problem. In addition we want to add the effect of capillary flow which will further complicate computation. Also this work is a first attempt and can be modified later, if necessary. We have developed our own techniques for solving the partial differential equations. We are using simulation languages that we have accumulated for solving the differential equations. With respect to ion calculations, one can double check with pencil calculations to make sure that the principle of electroneutrality has not been violated.

Betz: In one of your graphs you show the relation between the decrease of O_2 and the decrease of the glucose. You mentioned that this occurred during a complete stop of the circulation. At the same time an acidosis occurs and I wondered about the linearity of the decrease of glucose for the acidosis inhibits the breakdown of glucose, therefore the decrease of glucose concentration should not be linear.

Reneau: The rate may not be linear. The point I wanted to make is that glucose is available much longer than oxygen.

Betz: Do you think that it takes longer with brain until there is no glucose?

Reneau: Yes. Some glycogen is also present but was not accounted for in the simulation.

Session IV B

Basic Effects of Ion Activity in Metabolic Reactions and in Membrane Function Including Tissue Measurments

Chairman: *C. W. Lübbers*

A. M. Brown, J. M. Russell and D. C. Eaton

Resting Ionic Permeabilities in a Neuron

(1 Figure)

Introduction

Ionic permeation mechanisms have been extensively studied in axonal preparations such as squid giant axon and frog myelinated nerve. The geometry is favorable for using voltage-clamp techniques which are very useful in the investigation of voltage-dependent ionic currents. Moreover the number of ionic currents is only about three, namely the resting or leak current and the voltage-dependent Na^+ and K^+ currents. Equivalent studies on nerve cell bodies have not been forthcoming partly because of technical limitations of the voltage clamp resulting from the use of microelectrodes and partly because of the greater numbers of more complex ionic permeation mechanisms. Thus, in addition to the voltage-dependent Na^+ and K^+ currents found in axons, the following currents have been identified in neurons: inward Ca^{2+} current (*Geduldig* and *Junge* 1968), fast outward K^+ current (*Connor* and *Stevens* 1971; *Neher* 1971), inwardly rectifying current (probably K^+) (*Kandel* and *Tauc* 1966; *Marmor* 1971), slow outward K^+ current (*Brodwick* and *Junge* 1972), Ca^{2+}- activated K^+ current (*Meech* 1972) and light-activated K^+ current (*Brown* and *Brown* 1973). Moreover electrogenic Na^+ pumping contributes significantly to neuronal resting potentials (*Thomas* 1969) and active transport of Cl^- as well as Na^+ and K^+ occurs (*Russell* and *Brown* 1972 a and b).

As a prelude to quantitative studies of neuronal ionic selectivities we have examined the ionic permeabilities of an *Aplysia* giant neuron (R_2) at or near its resting potential. Two methods were used: chemical, in which net fluxes of Na^+, K^+, and Cl^- across the neuronal membrane were measured using intracellular ion selective microelectrodes and electrical in which measurements of changes in membrane potential or current were made under circumstances where a single ionic species was the major contributor.

Methods

Measurements of intracellular K^+ and Cl^- activities were made using ion-selective liquid ion exchanger microelectrodes (*Russell* and *Brown 1972 a and* b). Intracellular Na^+ activity was measured using Na^+-selective glass microelectrodes (*Brown* and *Brown* 1973). Membrane potential was measured using glass micropipettes with a variety of filling solutions (*Russell* and *Brown* 1972a). Voltage clamp was applied by two intracellular micropipettes according to the method described by *Brown* and *Brown* (1973). Neurons whose axons were ligated were also used. A complete account of the application of these techniques to this particular investigation is in preparation (*Eaton, Russell* and *Brown* 1974).

Results

1. Chemical Measurements

Net fluxes of Na^+, K^+ and Cl^- were measured by inhibiting metabolic processes either by cooling to 3°C or by applying ouabain. The initial movement of ions towards equilibrium followed first order kinetics so that net fluxes were determined and adjusted for cell volume and surface area as previously described (*Russell* and *Brown* 1972a and b). The relevant equations are:

$$M_i F = I_i = g_i (E_M - E_i) \tag{1}$$

where $M =$ is net of ion i flux, F the *Faraday*, g_i the chord conductance, E_M the resting potential and E_i the electrochemical equilibrium potential.
From the constant field equation:

$$P_i = g_i (E_M - E_i) \frac{ZRT}{E_M F^2} \left(\frac{\exp E_M F^2/RT - 1}{a_i^i \exp E_M F^2/RT - a_i^o} \right) \tag{2}$$

where Z is valence, R is gas constant, T is °K, a_i^i and a_i^o refer to the intra- and extracellular activities respectively. The values for "chemical" selectivities obtained in the present experiments were:

K: $P_K = (4.8 \pm 1.5) \times 10^{-8}$ cm sec^{-1}
$g_K = (1.1 \pm 0.4) \times 10^{-5}$ ohm^{-1} cm^{-2}
Cl: $P_{Cl} = (11.4 \pm 3.2) \times 10^{-8}$ cm sec^{-1}
$g_{Cl} = (4.6 \pm 1.7) \times 10^{-5}$ ohm^{-1} cm^{-2}
Na: $P_{Na} = (6.1 \pm 3.3) \times 10^{-9}$ cm sec^{-1}
$g_{Na} = (4.7 \pm 3.1) \times 10^{-6}$ ohm^{-1} cm^{-2}

The permeability ratios were:

$P_{Na}/P_K = 0.13$
$P_{Cl}/P_K = 2.2$

2. Electrical Measurements

a) Dilution experiments

In these experiments the neuron was exposed to a Na Me SO_3, $Na_2 SO_4$ or Tris Me SO_3 solution and equimolar substitutions of K^+ or in the case of Tris solutions K^+ or Na^+ were made. According to the constant field equation the relationship between the change in membrane potential elicited by an equimolar substitution and substituted cation is:

$$\Delta V_M = \frac{RT}{F} \ln 1 + \frac{(P_K/P_{Na} - 1) [K^+]_o}{500 + P_a/P_{Na} [Cl_i]_i} \tag{3}$$

Hence a plot of $\exp(F\Delta V/RT)$ versus $[K]_o$ should be a straight line with an intercept of one on the y-axis. This proved to be the case as shown in Figure 1. The Cl_i term could be ignored since a_{Cl}^i fell to zero in these solutions. The calculated permeability ratios were:

$P_{Na}/P_K = 0.15$ and
$P_{Tris}/P_K = 0.14$

With this value of P_{Na}/P_K an estimate that $P_{Cl}/P_K = 0.50$ (versus 2.2 from the "chemical" measurements) was obtained from curve C in Figure 1. However, this assumes that $M_e SO_3^-$ is impermeant whereas subsequent experiments demonstrated that this was not the case.

b) Step changes

In these experiments the neuron was voltage clamped at -57 mV (near E_M) in a solution adjusted so that $E_{Cl} = E_K = E_{Na} = E_M = -57$ mV. Then it was exposed to a solution in

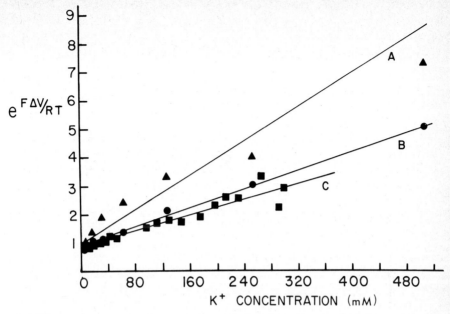

Fig. 1. $e^{F\Delta V/RT}$ vs. [K$^+$] (mM) for (A) Cl$^-$-free and Na$^+$-free solution (Me SO$_3^-$ and Tris$^+$ replacement) (each point represents the mean of 5 cells); (B) Cl$^-$-free with Na$^+$ (each point represents the mean of 12 cells); and (C) external solutions that contain both Cl$^-$ and Na$^+$ (each point represents a single value obtained in one of 37 cells).

which a single ionic species was changed and the resultant current changes were measured. The "electrical" selectivities were:

K: $P_K = (8.2 \pm 5.1) \times 10^{-8}$ cm sec^{-1}
$g_K = (0.7 \pm 0.4) \times 10^{-5}$ ohm^{-1} cm^{-2}
Cl: $P_{Cl} = (5.7 \pm 2.8) \times 10^{-8}$ cm sec^{-1}
  in $M_e SO_3^-$
$g_{Cl} = (1.0 \pm 0.4) \times 10^{-5}$ ohm^{-1} cm^{-2}
$P_{Cl} = (11.8 \pm 5.0) \times 10^{-8}$ cm sec^{-1}
  in $So_4^=$
$g_{Cl} = (2.0 \pm 0.6) \times 10^{-5}$ ohm^{-1} cm^{-2}
Na: $P_{Na} = (9.2 \pm 3.8) \times 10^{-10}$ cm sec^{-1}
$g_{Na} = (6.2 \pm 2.5) \times 10^{-7}$ ohm^{-1} cm^{-2}

The P_{Cl}/P_K ratio was 0.70 where $M_e SO_3^-$ was the substituted anion and 1.5 where $SO_4^=$ was used. This discrepancy and the lower Na$^+$ permeability and conductance values will be discussed.

Discussion

The "chemical" and "electrical" measurements of P_K and g_K are in good agreement. However discrepancies exist for both Cl$^-$ and Na$^+$. For Cl$^-$ the differences arose when Me SO$_3^-$ was the substituted anion in "electrical" experiments but not when $SO_4^=$ was employed. The most reasonable explanation is that the resting neuronal membrane is not impermeable to $Me\ SO_3^-$ so that an inward $Me\ SO_3^-$ electrical current would effectively mask an outward Cl^- current. Our calculations of $P_{Me\ SO_3}/P_{Cl}$ give a value of 0.63. This raises serious questions about the permeability of other widely used Cl^- substitutes such as propionate and acetate in this preparation. These ions are not significantly larger than $M_e\ SO_3$, are singly charged, and produce little or no change in membrane potential when substituted for Cl^-.

Alteratives might be larger organic anions such as p-toluenesulfonate or salicylate. Although they may be quite impermeable, these anions may have a pharmacological effect on the membrane (*Barker* and *Levitan*, 1971). Consequently, it would appear that despite its divalent nature and Ca^{++} complexing ability, $SO_4^=$ may still be the most acceptable replacement anion for Cl^- if an impermeant ion is required.

While the results from electrical measurements using the dilution method may be misleading, if the assumptions made are inaccurate, these types of experiments can nevertheless yield interesting information that may not be available by other methods.

For example, an interesting finding disclosed by the dilution method that was not apparent from the other methods was the relatively high permeability to $Tris^+$ ion. This ion has generally been considered a rather impermeable ion in other preparations, but like $Me\ SO_3^-$ appears to be a poor choice for an impermeant ion in *Aplysia* neurons. The permeability to $Tris^+$ helps to explain the other major disparity between the "chemical" and "electrical" experiments. That is, the difference between the Na^+ permeability calculated from flux measurements and the dilution experiments and that obtained by measuring the current associated with a step change of Na^+ concentration. The difficulty is similar to the problem of the apparent reduction of P_{Cl} in $Me\ SO_3$ solutions. Here $Tris^+$ moving into the cell masks the Na^+ outward current; and since $Tris^+$ is of comparable permeability to Na^+, the P_{Na} value appears quite small. On the basis of this apparent reduction in permeability, an estimate of P_{Tris}/P_{Na} is 0.84; this compares favorably with the value of 0.89 obtained from the dilution experiments.

Table 1. Data for comparison of other preparations with *aplysia*.

Preparation	cm sec^{-1} × 10^{-7}					Source
	P_K	P_{Na}	P_{Cl}	P_{Na}/P_K	P_{Cl}/P_K	
Squid axon	18	0.7	7.9	0.04	0.45	Hodgkin and Katz (1949)
	6.6	0.13	1.7	0.02	0.26	Tasaki (1963)
	–	–	7.0	–	–	Caldwell et al. (1960)*
	10	–	11	–	1.1	Shanes and Berman (1955)
					0.29	Hurlbut (1970)
Lobster axon	4.0	0.05	0.93	0.01	0.19	Brinley (1965)
Frog muscle	10–20	0.1–0.2	40	0.01–0.02	2–4	Hodgkin and Horowicz (1959)
	13	–	9–24	–	0.7–1.8	Adrian and Freygang (1962)
	–	–	50	–	–	Moore (1969)
Barnacle muscle	–	–	1.9	–	–	Di Polo (1972)
Crab muscle	–	–	8	–	–	Richards (1969)
Aplysia giant cell	0.82	–	1.2	0.10	1.5	this paper, electrical measurements
	0.48	0.06	1.1	0.13	2.2	this paper, chemical measurements

Taking the permeabilities of R_2 as a whole, one notes that the absolute magnitudes are unusual compared with other preparations. Table 1 compares the permeability and conductance data from this preparation with several others. It is clear that the membrane permeability of the *Aplysia* giant nerve cell body is extraordinarily low. As previously noted the presence of the axon did not greatly affect the results (*Russell* and *Brown*, 1972 a and b). The low permeability agrees with high values of specific membrane resistance, R_m, reported for neurons of *Aplysia* (*Carpenter* 1970), *Anisodoris* (*Gorman* and *Marmor* 1970), and squid (*Carpenter* 1973) somal membranes. R_m in the present experiments was about 10^5 ohm/cm^2. Moreover the P_{Na}/P_K ratio is almost ten times greater than that found in preparations other than nerve cell bodies. It is not clear whether these differences represent an actual difference in the permeation mechanisms or only a scarcity of K$^+$-selective sites in the somal membrane.

References

Barker, J. L., Levitan, H.: Salicylate: Effect on Membrane Permeability of Molluscan Neurons. Science 172 (1971) 1245

Brodwick, M. S., Junge, D.: Post-Stimulus Hyperpolarization and Slow Potassium Conductance Increase in *Aplysia* Giant Neurone. J. Physiol. 223 (1972) 549

Brown, A. M., Brown, H. M.: Light Response of a Giant *Aplysia* Neuron. J. gen. Physiol. 62 (1973) 239–254

Carpenter, D. O.: Membrane Potential Produced Directly by the Na$^+$ Pump in *Aplysia* Neurons. Comp. Biochem. Physiol. 35 (1970) 371

Carpenter, D. O.: Electrogenic Sodium Pump and High Specific Resistance in Nerve Cell Bodies of the Squid. Science 179 (1973) 1336

Connor, J. A., Stevens, C. F.: Inward and Delayed Outward Membrane Currents in Isolated Neural Somata under Voltage Clamp. J. Physiol. 213 (1971) 1–19

Eaton, D. C., Russell, J. M., Brown, A. M.: Ionic Permeabilities of an *Aplysia* Giant Neuron. J. Membrane Biology 21 (1975) 353–374

Geduldig, D., Junge, D.: Sodium and Calcium Components of Action Potentials in *Aplysia* Giant Neurone. J. Physiol. 199 (1968) 347–365

Gorman, A. L. F., Marmor, M. F.: Contributions of the Sodium Pump and Ionic Gradients to the Membrane Potential of a Molluscan Neurone. J. Physiol. 210 (1970) 897

Kandel, E. R., Tauc, L.: Anomalous Rectification in the Metacerebral Giant Cells and its Consequences for Synaptic transmission. J. Physiol. 183 (1966) 287

Marmor, M. F.: The Effects of Temperature and Ions on the Current-Voltage Relation and Electrical Characteristics of a Molluscan Neurone. J. Physiol. 218 (1971) 573–598

Meech, R. W.: Intracellular Calcium Injection Causes Increased Potassium Conductance in *Aplysia* Nerve Cells. Comp. Biochem. Physiol. 42 A (1972) 493–499

Neher, E.: Two Fast Transient Current Components during Voltage Clamp on Snail Neurons. J. Gen. Physiol. 58 (1971) 36–53

Russell, J. M., Brown, A. M.: Active Transport of Potassium by the Giant Neuron of the *Aplysia* Abdominal Ganglion. J. gen. Physiol. 60 (1972a) 519–533

Russell, J. M., Brown, A. M.: Active Transport of Chloride by the Giant Neuron of the *Aplysia* Abdominal Ganglion. J. gen. Physiol. 60 (1972b) 499–518

Thomas, R. C.: Membrane Current and Intracellular Sodium Changes in a Snail Neurone during Extrusion of Injected Sodium. J. Physiol. 201 (1969) 495

R. C. Thomas

The Effects of CO₂ and Bicarbonate on the Intracellular pH of Snail Neurones

(5 Figures)

In this paper I intend to show that the application of a CO_2 solution at pH 7.5 to a snail neurone previously in a CO_2-free Ringer of the same pH causes only a transient fall in intracellular pH. Within 15 minutes the internal pH returns to its previous value of about 7.4. During the exposure to CO_2 the cell accumulates bicarbonate. The expected bicarbonate concentration can be calculated from the intracellular pH and the CO_2 level (which is presumed to be equal inside and outside the cell, since CO_2 readily crosses cell membranes). Independent evidence that the cell does accumulate the expected amount of bicarbonate comes from the internal pH change seen on return to CO_2-free Ringer, and from the increased buffering power in CO_2-containing solutions.

Methods: The experimental arrangement is illustrated in Figure 1. The "brain" was removed from a dormant snail (Helix aspersa), mounted in a small bath, and the neurones in the visceral and right pallial ganglia exposed as previously described (*Thomas*, 1974). The bath was continually perfused with a snail Ringer containing: KCl, 4 mM; NaCl, 80 mM; $CaCl_2$, 7 mM; $MgCl_2$, 5 mM; and tris maleate buffer, pH 7.5, 20 mM. A selected large neurone (preferably the large "Dinhi" cell in the right pallial ganglion) was penetrated with a recessed-tip pH-sensitive micro-electrode (see earlier paper in this book) and up to three conventional

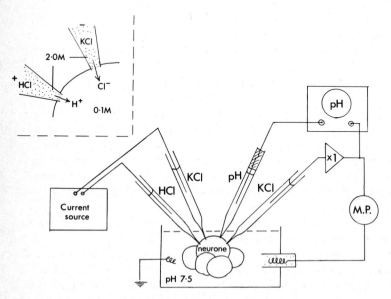

Fig. 1. Diagram of the experimental arrangement. The inset illustrates the inter-barrel iontophoretic technique used to inject ions.

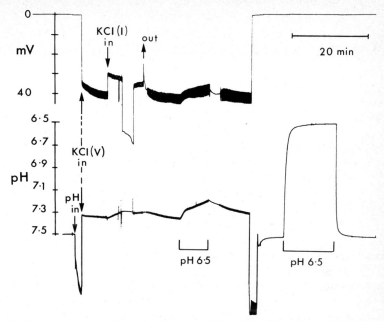

Fig. 2. Pen-recordings of an experiment to determine the intracellular pH and the effects of changing the membrane potential and external pH. The voltage recorded by the KCl-filled membrane potential microelectrode is shown at the top. The voltage recorded by the pH-sensitive micro-electrode is shown at the bottom. The spontaneously-occurring action potentials were greatly reduced in size by the long time constant of the pen recorder.

micro-electrodes for recording the membrane potential and for passing current. A multiway tap enabled the solution flowing through the bath to be changed without disturbing the electrodes.

Intracellular pH. Figure 2 illustrates a short experiment which shows the procedure for measuring internal pH, and the effects on the internal pH of changing the membrane potential and the external pH. At the beginning the micro-electrodes were outside the cell. Then, as indicated in the lower part of the figure, the pH micro-electrode was pushed into the selected nerve cell. By itself this electrode recorded the sum of the pH gradient and the membrane potential: the 20–30 mV change in potential after penetration shows that H^+ ions were not passively distributed across the cell membrane. Then the KCl voltage-recording microelectrode was pushed into the cell; since its output was fed into the pH meter low impedance input the membrane potential was then subtracted from the pH trace. The intracellular pH was revealed to be about 7.35, and the membrane potential about 40 mV. (If H^+ ions were passively distributed, the internal pH would have been about 6.8).

A third micro-electrode was then inserted, so that the membrane potential could be changed. A brief pulse of 5 nA was tried first, then a current of 10 nA was applied for 3 minutes. It can be seen that the resulting hyperpolarisation of over 25 mV had very little effect on the internal pH. Some time later, after the current-passing micro-electrode had been withdrawn, and the cell had recovered from the injury, the Ringer was changed to one of pH 6.5. This had only a small, slow effect on the internal pH. Both the pH and KCl-filled micro-electrodes were then simultaneously withdrawn from the cell, and their pH responses tested.

This experiment clearly shows that the H+ ions were not passively distributed, and that the internal pH is only slowly affected by changing the external pH or the membrane potential.

Effect of CO_2. The normal tris maleate buffered Ringer contained no CO_2. When the solution bathing a neurone was changed to one equilibrated with 2.2% CO_2, 97.8% O_2 and containing 20 mM bicarbonate (giving a pH close to 7.5) there was a rapid internal acidification. This can be seen in Figures 3 and 4. These levels of CO_2 and bicarbonate were chosen because they are close to those of snail blood (*Burton, 1969*). The pH fall of about 0.25 units was presumably caused by a large and rapid influx of CO_2 into the cell. Once inside, it reacted with water to produce carbonic acid, which then dissociated to produce H+ and HCO_3^- ions:

$$CO_2 + H_2O \rightleftharpoons H_2CO_3 \rightleftharpoons H^+ + HCO_3^-$$

The H+ ions produced caused the observed fall in pH. But this pH change was only transient; some active process brought the internal pH back to its pre-CO_2 level. (This only occurred when the external pH was not changed. When CO_2 was applied in a Ringer of pH 6.9, the internal pH did not return to normal until the CO_2 was removed, as shown on the right of Figure 3.)

As well as producing H+ ions, the dissociation of carbonic acid produces bicarbonate. Once the internal pH has stabilised at its pre-CO_2 value the internal bicarbonate concentration can be calculated. For the experiment of Figure 3, the calculated internal bicarbonate when the cell had equilibrated in 2.2% CO_2, pH 7.5, was about 17 mM.

On returning to the mormal CO_2-free Ringer, the internal pH rose, transiently, by about 0.5 units. This was presumably caused by the intracellular bicarbonate taking up protons and

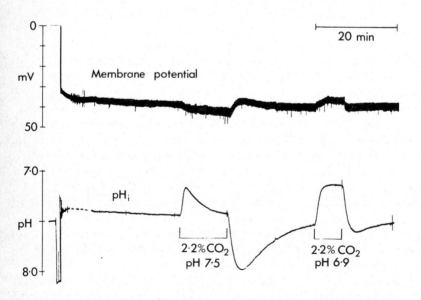

Fig. 3. Pen-recording of an experiment to show the effects of CO_2 on the intracellular pH. The bicarbonate concentrations of the two test solutions were 20 mM and 5 mM respectively.

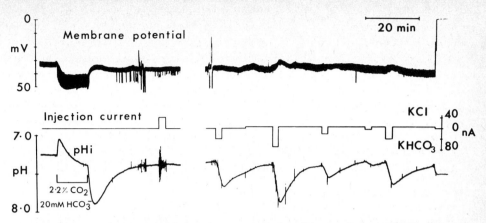

Fig. 4. Pen-recordings from an experiment to compare the effects on the internal pH of removing external CO_2 with those of injecting bicarbonate ions. Part of the record (where the current-passing electrodes were inserted) has been removed. The middle trace is a record of the injection current.

then leaving the cell as CO_2. The same sort of pH change occurs when bicarbonate ions are injected iontophoretically, as shown in Figure 4.

Bicarbonate injection. In this experiment, the cell was first exposed for a short period to 2.2% CO_2, pH 7.5 Ringer, then returned to normal. The calculated internal bicarbonate just before the return to normal was 15.4 mM. Then a series of iontophoretic injections of potassium bicarbonate were made by interbarrel iontophoresis. The injected bicarbonate would be expected to react in the same way as the accumulated bicarbonate; with no CO_2 outside the cell it would take up protons and leave the cell as CO_2. The quantity of bicarbonate injected was calculated from the injection charge, by assuming that the cell was spherical, and that all the current flowing through the $KHCO_3$ micro-electrode was carried by HCO_3^- ions moving into the cell. It is clear from Figure 4 that the largest of the bicarbonate injections, calculated to give an internal concentration of 14 mM, produced roughly the same pH change as that seen on removal of external CO_2.

This experiment, then, confirms that the pH change seen on return to CO_2-free Ringer was caused by intracellular bicarbonate, and shows that the bicarbonate concentration was close to that calculated from the internal pH.

Intracellular buffering power. The buffer power of a solution is a measure of its capacity to resist pH changes resulting from the addition of acid or alkali. In physiology the usual units are ml of Normal acid per litre per pH unit, or "slykes" (*Woodbury*, 1965). The buffer value of a bicarbonate solution when the dissolved CO_2 is kept constant is much higher than that of a solution of the same concentration of a normal buffer. The molar buffering power is 2.3 per pH unit for the bicarbonate solution, whereas for ordinary buffers it is only 0.575. This is because the PCO_2 and thus H_2CO_3 is held constant: in ordinary buffers the undissociated acid concentration varies as acid is added.

If the cell membrane is sufficiently permeable to CO_2, any intracellular bicarbonate will be an important intracellular buffer when the cell is maintained in a constant CO_2 environment. By measuring the intracellular buffering power before and after applying CO_2 it should thus be possible to estimate the accumulated bicarbonate.

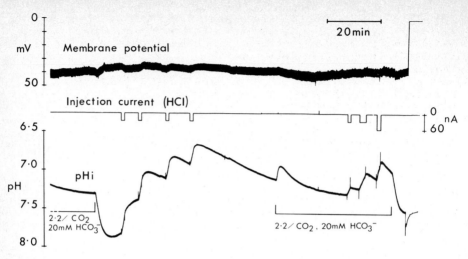

Fig. 5. Pen-recordings from an experiment to measure the intracellular buffering power. The middle line is a record of the current used to inject HCl.

The most direct way of measuring the intracellular buffer power is by injecting known amounts of acid or alkali and measuring the pH change. Figure 5 illustrates an experiment in which HCl was injected into a snail neurone before and after exposure to CO_2. The cell was first exposed to 2.2% CO_2 for 30 min. Then the CO_2 was removed, so that the internal pH was increased to well above its normal value. Then four injections of HCl were made so that the internal buffer power could be measured over a range of values of internal pH. The four values for buffer power obtained were 10.2, 9.2, 9.8 and 10.4 slykes.

About 32 minutes after the last injection, the Ringer was changed to one equilibrated with 2.2% CO_2 and buffered to pH 7.5 with 20 mM bicarbonate. The internal pH fell briefly, and then stabilised at a value of 7.32. HCl was injected as before, causing a much smaller pH change. From the first injection, a buffer value of 33 slykes was calculated.

Thus CO_2 caused the buffer value to increase by 23 slykes, equivalent to an accumulation of intracellular bicarbonate to a level of 10 mM. At that internal pH the bicarbonate calculated from the pH and CO_2 was 13.2 mM. Two further HCl injections were made, giving pH changes corresponding to buffer values of 30 and 25 slykes. These lower values are as expected from the lower pH levels.

Conclusion: Two indirect methods of estimating intracellular bicarbonate thus give values quite close to those calculated from the measured intracellular pH and the external CO_2. This confirms that intracellular bicarbonate must be primarily determined by the pH and CO_2 levels, and is good evidence that the pH measurements are accurate. Hopefully it will be possible to measure internal bicarbonate directly, using a bicarbonate sensitive microelectrode as described by *Khuri, Bogharian & Agulian* (1974).

If H^+ and HCO_3^- ions were passively distributed, the internal pH and bicarbonate concentration would both be much lower. Clearly, H^+ or HCO_3^- ions or both must be actively transported. The internal pH is well maintained in CO_2-free Ringer. The speed of the internal pH response to CO_2 shows that it is rapidly converted to bicarbonate. It is thus more probable that H^+ and not bicarbonate ions are actively transported.

References

Burton, R. F.: Buffers in the Blood of the Snail Helix Pomatia. Comp. Biochem. Physiol. 29 (1969) 919–930

Khuri, R. N., Bogharian, K. K., Agulian, S. K.: Bicarbonate in Single Skeletal Muscle Fibres. Pflügers Arch. 349 (1974) 285–294

Thomas, R. C.: Intracellular pH of Snail Neurones Measured with a New pH-Sensitive Glass Micro-Electrode. J. Physiol. (London) 238 (1974) 159–180

Woodbury, J. W.: Regulation of pH. In: Physiology and Biophysics, pp. 899–934, ed. by T. C. Ruch and H. D. Patton. Saunders, Philadelphia 1965

E. Dora and T. Zeuthen

Brain Metabolism and Ion Movements in the Brain Cortex of the Rat During Anoxia

(1 Figure)

The brain cortex responds to stresses such as anoxia by an early shutdown of the spontaneous electrical activity and by changes in extracellular potassium activity (*Vyskočil* et al. 1972). But this occurs much earlier than appreciable changes in some of the chemical compounds related to metabolism (*Albaum* et al. 1953). Most earlier work on correlating functional with chemical changes in the brain have been based on chemical analyses of brain biopsies but due to the rapid nature of the initial changes these may have been missed. The purpose of this study was therefore to monitor continuously *in vivo* the metabolic status of the brain by means of pyridine nucleotide fluorometry, extreacellular activities of potassium, sodium and chloride and electrical potential by microelectrodes, and brain activity through the ECoG.

Methods

In White Wistar rats anaesthetized with Nembutal®, tracheotomy was performed and the animal was immobilized with Flaxedil® and artificially respired with either atmospheric air or $30\% \ O_2 + 70\% \ N_2$, or in anoxia with $100\% \ N_2$. The mean arterial pressure (BP) was recorded from the femoral artery. A hole was drilled on each side of, and with centres 2.5 mm from the midline and 3 mm behind the bregma suture. The diameter of the right hole was 5 mm and that of the left, 2 mm. A quartz light pipe (diameter 5 mm) was inserted into the large hole so that it just touched the dura. The underlying tissues were excited with ultraviolet light (360 nm) and fluorescent light (F) emitted from the NADH (456 nm) and reflected light (R) were recorded (*Chance* et al. 1962, 1973). Changes in emitted light caused by changes in blood flow were compensated by subtracting the changes on reflected light from the amount of fluorescent light which gave the corrected fluorescence (CF) expressed as a percentage of the preanoxic reflected light (*Jöbsis* et al. 1971, *Harbig* et al. 1973). A rise in CF was associated with increased NADH fluorescence. The maximum level of NADH fluorescence corresponded to anaerobic respiration since this level was maintained in dead animals. Via the smaller hole we measured electrical potential (DC) and either extracellular potassium activity $[K^+]$, chloride activity $[Cl^-]$, or sodium activity $[Na^+]$. $[K^+]$, $[Cl^-]$ and DC were measured by means of a double-barrelled microelectrode (*Zeuthen* et al. 1974) (total tip diameter $<0.6 \ \mu m$) 1 mm below the pia-arachnoid membrane. One barrel was filled with K^+-sensitive ion exchanger (Corning 477317) or with Cl^--sensitive ion exchanger (Corning 477315), and the other barrel was filled with 2M KCl to record the potential. $[Na^+]$ was recorded by a Hinke-type electrode (*Hinke* 1959) in which case the electrical potential was recorded by a single micropipette, the tip of which was less than 1 mm from the Na^+-electrode. The electrical potential recorded with this pipette closely approximates that at the site of the Na^+-electrode (*Jennings* 1974). Electrocorticogram (ECoG) was recorded between two stainless steel screws one 5 mm in front of the microelectrode and one 5 mm behind it.

Results

The results are compiled in Table 1 and an example of an experiment in which [K+] and DC were measured is shown in Figure 1. During approximately the first 2.5 minutes the brain is in a predepolarization phase with only small changes in ion activities and DC; after about 2.5 minutes the brain exhibits an anoxic depression (AD) (*Leão* 1947).

In the predepolarization phase both the BP, CF and the ECoG show maximum changes within the first minute. The CF increased linearly and reached a plateau indicating anaerobic respiration. That the CF is affected only after 17 seconds is partly a result of the dead space in the mechanical respirator system. The chemical potentials of the ions increased linearly through the predepolarization phase although the chemical potential of K+ in about half of the experiments increased initially at a speed of 78.7 µV/sec $\pm$ 12.5 until about t = 73.7 sec $\pm$ 4.2 and then at a slower rate 26.7 µV/sec $\pm$ 6.6. Due to the low signal to noise ratio for the [Cl-] and [Na+] recordings it was difficult to assign any time for the onset of the changes of these ions, although they seemed to start simultaneously with the beginning of the change in K+. If the animal was given atmospheric air at t = 135 sec normal values were typically re-established within 10 minutes, during which the [Na+] had an undershoot of about 2% and [Cl-] of about 4%.

If the anoxia was prolonged beyond 2.5 minutes, an anoxic depolarization (i.e., a tenfold increase in activity) started (Table 1 column 4) and the chemical potential of K+ increased

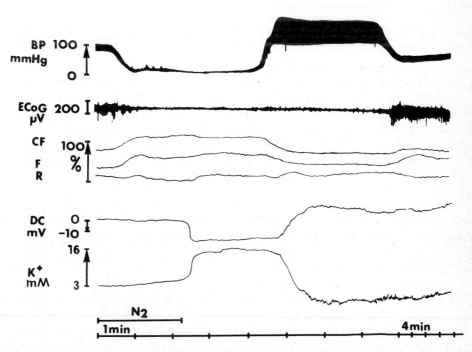

Fig. 1. Changes in blood pressure (BP), electrocorticogram (ECoG). corrected fluorescence (CF), fluorescence (F), reflectance (R), extracellular electrical potential (DC) and extracellular potassium activity (K+) during anoxia. At 2.5 minutes an anoxic depolarization started. The preceding period is called the predepolarization phase. After 5 minutes the parameters began to recover. The bar (N$_2$) indicate the period in which the animal was ventilated with N$_2$.

Table. 1. Changes in the recorded parameters during the predepolarization phase and the anoxic depolarization phase. The second number in columns 3 and 4 is the change on the logarithm of the ion-activity as compared to the normoxic level. The numbers in parenthesis are the number of experiments/numbers of animals. Numbers are ± standard error of mean. The values in mM are the recorded concentrations, assuming equal activity coefficients of electrolytes in brain and in water.

	Normoxic level	Time of onset of change	Extreme value during predepolarization phase reached after t sec.	Extreme value during anoxic depolarization
CF	Aerobic	17.3 sec. ± 1.3 (13/6)	Anaerobic 77.4 sec ± 3.6 (13/6)	Anaerobic — (7/7)
$[K^+]$ $\Delta \log (K^+)$	3.0 mM ± 0.01 (5/5)	37.8 sec ± 2.7 (13/6)	3.7 mM 0.091 ± 0.005 135 sec. (13/6)	26.1 mM 0.94 ± 0.23 (7/7)
$[Na^+]$ $\Delta \log (Na^+)$	142.5 mM ± 7.5 (2/2)	—	148 mM 0.017 ± 0.003 135 sec. (6/4)	55.4 mM −0.41 ± 0.07 (5/5)
$[Cl^-]$ $\Delta \log (Cl^-)$	127.5 mM ± 5.9 (21/4)	—	134 mM 0.021 ± 0.003 135 sec. (16/7)	49.6 mM −0.41 ± 0.04 (3/3)
DC	0	35.7 sec ± 3.5 (13/6)	−2.1 mV ± 0.4 73.7 sec ± 4.2 (13/6)	−46.1 mV ± 7.8 (7/7)
ECoG	200 µV	52.2 sec ± 4.2 (13/6)	zero 95.4 sec ± 5.0 (13/6)	zero
BP	100 mmHg	26.1 sec ± 1.8 (13/6)	64.8 mmHg ± 4.7 69.1 sec ± 2.6 (13/6)	45.8 mmHg ± 13.0

with a maximum speed of about 2.5 mV/sec as much as 50 mV; similarly the $[Cl^-]$ was reduced to about 40% and the $[Na^+]$ to about 30% of the normoxic level. Since the time constant of the sodium electrode plus electrometer system was slower than the fastest rate of change of the DC (about 2.5 mV/sec) and since $[Na^+]$ is obtained from the difference between the DC and the potential recorded with the sodium electrode, $[Na^+]$ is not recorded correctly at the onset of the anoxic depolarisation. We therefore only assessed the values of $[Na^+]$ during periods when the voltages recorded with the Na^+-electrode and the DC-electrode changed less than about 100 µV/sec. Sometimes the animals could recover from the AD even if the extracellular $[K^+]$ had been as high as 30 mM. In cases of recovery all parameters usually recovered within 15 to 30 minutes.

Discussion

[K$^+$], [Na$^+$] and [Cl$^-$] all increased during the predepolarization phase. As this is due to one or more of the following mechanisms (a) movements of ions, (b) movements of water or (c) release of bound ions, we cannot, on the basis of our measurements alone, deduce which of these are involved. However we can hypothesize on the mechanism of the changes in ion activities if we consider these in relation to changes in cortical impedance which has been supposed to be a measure of the conductivity and therefore of the extent of the extracellular space (for references see *Harreveld* 1971). In asphyxia this impedance increases about 20% during the first few minutes and then 2 to 3 fold during what is probably an anoxic depolarization. As the extracellular tonicity and therefore the conductivity of the extracellular fluid only changes about 5% (Table 1 column 3) in the predepolarization phase, the increase of 20% in the impedance can be explained if we postulate that 20% of the water is removed from the extracellular space. A possible explanation of observed [K$^+$] changes could be that the number of K$^+$ ions in the extracellular space remains constant to a first approximation since this activity increases about 23%. Na$^+$ and Cl$^-$ must be removed from the extracellular space together with water during predepolarization since the activities of these ions increase only about 5%. Water and NaCl probably move into the cells as a result of an imbalance in osmolarity across the cell walls due to an accumulation of the products of anaerobic metabolism. During anoxia the brain quickly resorts to anaerobic respiration (Table 1 column 2) which leads to an increased intracellular production of lactate (*Kaasik* et al. 1970).

During anoxic depolarization the tonicity in the extracellular space decreases one third to about 50 mM. This is in agreement with the increase in cortical impedance by a factor of 2 to 3 if we assume that the extracellular volume remains the same as that in the predepolarization phase. Cell swelling and intracellular accumulation of NaCl has been demonstrated by chemical and histological methods to occur during anoxia (*Harreveld* 1971). This study suggests that these changes occur either before or at the onset of anoxic depolarization and a mathematical model of brain anoxia (*D. Reneau* in this meeting) indicates that the extracellular space must be as small as 1 to 2% at the beginning of depolarization if the fast rise times of [K$^+$] and [Cl$^-$] are to be explained. The changes in [Na$^+$] during anoxic depolarization are not reflected in the subarachnoid fluid, whereas those of [K$^+$] are (*Bito* and *Myers* 1972). The diffusion of Na$^+$ through the depolarized brain into the subarachnoid fluid is probably much slower than the diffusion of K$^+$.

Acknowledgements

The work was supported by NIH Grant 1 POL NS 10,939-02. Our thanks are due to Drs. *B. Chance, A. Kovach* and *I. A. Silver* for facilities and encouragement. *T. Zeuthen* holds a Junior Research Fellowship from the Faculty of Medicine of the University of Copenhagen; present address Department of Pathology, University of Bristol, Great Britain.

References

Albaum, H. G., Noell, W. K., Chinn, H. I.: Chemical Changes in Rabbit Brain During Anoxia. Amer. J. Physiol. 174 (1953) 408–412

Bito, L. Z., Myers, R. E.: On the Physiological Response of the Cerebral Cortex to Acute Stress (Reversible Asphyxia). J. Physiol. 221 (1972) 349–370

Chance, B., Cohen, P., Jöbsis, F., Schoener, B.:

Intracellular Oxidation-Reduction States *In Vivo*. Science 137 (1962) 499–508

Chance, B., Oshino, N., Sugano, T., Mayevsky, A.: Basic Principles of Tissue Oxygen Determination from Mitochondrial Signals. In: Oxygen Transport to Tissue, pp. 277–292, ed. by *H. I. Bicher* and *D. F. Bruley*. Plenum Press, New York 1973

Harbig, K., Reivich, M.: Corrected PN Fluorescence from the Cat's Brain Surface *in Vivo*. (Abstract, 6th International Conference on Cerebral Blood Flow). Stroke 4 (1973) 341.

Hinke, J. A. M.: Glass Micro-Electrodes for Measuring Intracellular Activities of Sodium and Potassium. Nature 184 (1959) 1257–1258

Jennings, P. G.: An Investigation into Potassium Movement in Rat Cerebral Cortex and Lateral Ventricles in Reversible Hypoxia. B. Sc. Thesis, Univ. of Bristol (Great Britain) 1974

Jöbsis, F. F., O'Connor, M., Vitale, A., Vreman, H.: Intracellular Redox Changes in Functioning Cerebral Cortex. I: Metabolic effects of epileptiform activity. J. Neurophysiol. 34 (1971) 735–749

Kaasik, A. E., Nilsson, L., Siesjo, B. K.: The Effect of Asphyxia upon the Lactate, Pyruvate and Bicarbonate Concentration of Brain Tissue and Cisternal CSF, and upon the Tissue Concentration of Phospho-Creatine and Adenine Nucleotides in Anaesthetized Rats. Acta Physiol. Scand. 78 (1970) 333–447

Leão, A. A. P.: Further Observations on the Spreading Depression of Activity in the Cerebral Cortex. J. Neurophysiol. 10 (1947) 409–414

Van Harreveld, A.: The Extracellular Space in the Vertebrate Central Nervous System. In: The Structure and Function of Nervous Tissue, Vol. IV, pp. 447–511. Academic Press, New York 1971

Vyskočil, F., Křiž, N., Bureš, J.: Potassium-Selective Microelectrodes Used for Measuring the Extracellular Brain Potassium During Spreading Depression and Anoxic Depolarization in Rats. Brain Res. 39 (1972) 255–259

Zeuthen, T., Hiam, R., Silver, I. A.: Recording of Ion Activities in Brain with Ion-Selective Microelectrodes. In: Ion Selective Microelectrodes. Plenum Press, New York 1974

W. Crowe, A. Mayevsky, L. Mela and I. A. Silver*)

Measurements of Extracellular Potassium, D. C. Potential and ECoG in the Cortex of the Conscious Rat During Cortical Spreading Depression

Introduction

Potassium microelectrodes using a K^+-selective liquid ion exchanger have been applied to measurements of extracellular K^+ activity in the brain by *Vyskočil, Kříž,* and *Bureš* (1972), *Prince, Lux,* and *Neher* (1973), and *Zeuthen, Haim,* and *Silver* (1974). *Mayevsky, Zeuthen,* and *Chance* (1974) have used double-barrelled K^+ microelectrodes in conjunction with NADH fluorometry to measure changes in energy metabolism in the cortex of the anesthetized rat during cortical spreading depression. In addition to the characteristic cycles in the ECoG and NADH, corresponding cycles in extracellular K^+ activity and d. c. potential were recorded. It is desirable to study the nature of the spreading depression disturbance and especially the recovery from the leakage of K^+ into the interstitial space, in the unanesthetized brain. In the present work, a procedure is presented for extending the previous measurements to the freely moving conscious rat.

Methods

Male Wistar rats (180–200 g) were operated on while (lightly) anesthetized with Equi-Thesin (Jensen-Salsbery Laboratories), 0.5 ml. The skull was exposed; a 2 mm diameter hole was drilled in the parietal bone 3 mm lateral and 3 mm posterior to the Bregma sutures; the dura was later removed to allow insertion of the microelectrode. A second hole 1 mm in diameter was drilled 3 mm posterior to the first to allow implantation of a double-lumen cannula for epidural application of a KCl solution. Steel screws for a bipolar ECoG recording were implanted in the skull.

The method of *Silver* and *Zeuthen* (personal communication) has been followed to produce double-barrelled microelectrodes for simultaneous measurements of K^+ activity (Corning 477317 ion exchanger was used) and d. c. potential with a tip size of less than 1 μm. The d.c. potential measurement is used to correct the signal from the K^+-sensitive barrel in addition to providing another channel of information.

The electrode tip was inserted into the parietal cortex to a depth of 1 mm (± 0.2 mm). Waves of spreading depression (SD) were initiated by drawing a small quantity of a KCl solution over the dural surface (0.5 M KCl, or sometimes stronger, was used). After a check that waves of SD could be initiated and recorded, the electrode was fixed in place using dental acrylic cement; the animal was placed in a cage that permitted free movement and was allowed to recover from the anesthetic while recording continued. In about 3 hours from the time of the dose of anesthetic, the animal was fully conscious, and SD could be recorded although occasionally obscured by electrostatic artifacts due to movements of the animal.

---

*) Harrison Department of Surgical Research and Johnson Research Foundation, University of Pennsylvania Medical School, Philadelphia, Pa. 19174, U.S.A.

W. Crowe, A. Mayevsky, L. Mela and I. A. Silver

Results

In preliminary experiments in the conscious rat (7 animals) we observed the characteristic phenomena of SD to be similar to those observed in the anesthetized animal: the d.c. potential showed a negative wave with amplitudes of up to 20 or 30 mV; the extracellular potassium activity (K^+o) increased by up to 1 pK or a fraction more, i.e., up to 30 to 50 mM. The rise in extracellular potassium started synchronously with the change in d.c. (± 1 sec), or, at times, followed it by a few seconds. In every case observed, the d.c. recovered its baseline before Ko^+. The half-width of the potassium wave was about 30 seconds with values commonly seen ranging 50% to either, side. Interestingly, the potassium release during the first SD wave was invariably the most prolonged. In addition, the first wave of subsequent cycles was longer than those which followed, but not as long as the very first wave. The rising phase of the d.c. wave and of the K^+o wave showed slopes of up to 10 mV/sec and 18 mV/sec (0.4 pK/sec) respectively, and each reached 90% of its amplitude within 6 sec. The recovery was slower, and was affected by the state of the animal, oxygen supply, metabolism, and body temperature (See review by Bureš, Burešová, and Křivánek (1974)). Possibly, anesthesia is also a factor in K^+o recovery. The recovery of the resting extracellular

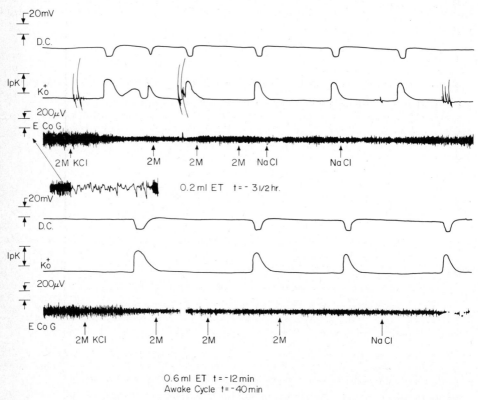

Fig. 1. Cortical spreading depression in the conscious (upper portion) and anesthetized (lower portion) rat. Both portions were recorded from the same animal: the first, 3-½ hours after the anesthetic (0.2 ml Equi-Thesin); the second, 40 minutes later and 12 minutes after an 0.6 ml anesthetic dose. In each portion, the traces are d.c. potential, extracellular potassium activity, and ECoG.

potassium level may be faster in the conscious animal (Fig. 1), however, this difference does not appear to be great, and more measurements are needed to separate this difference if it exists from the variations in K$^+$o recovery time encountered during the course of measurements.

Acknowledgements

This paper was supported by the portion of the Program Project Grant NINDS-10939-02 under the direction of Dr. *Leena Mela*.

References

Bureš, J. Burešová, O., Křivánek, J.: The Mechanism and Applications of Leao's Spreading Depression of Electroencephalographic Activity. Academic Press, New York 1974

Mayevsky, A., Zeuthen, T., Chance, B.: Measurements of Extracellular Potassium. ECoG and Pyridine Nucleotide Levels Durin Cortical Spreading Depression in Rats. Brain Res. 76 (1974) 347–349

Prince, D. A., Lux, H. D., Neher, E.: Measurement of Extracellular Potassium Activity in Cat Cortex. Brain Res. 50 (1973) 489–495

Vyskočil, F., Kříž, N., Bureš, J.: Potassium-Selective Microelectrodes Used for Measuring the Extracellular Brain Potassium During Spreading Depression and Anoxic Depolarization in Rats. Brain Res. 39 (1972) 255–259

Zeuthen, T., Haim, R., Silver, I. A.: Recording of Ion Activities in Brain with Ion Selective Microelectrodes. Plenum Press, New York 1975

H. D. Lux

Simultaneous Measurements of Extracellular Potassium-ion Activity and Membrane Currents in Snail Neurones

(4 Figures)

Introduction

Studies using radioactive tracer (*Hodgkin* and *Keynes* 1955) gave rise to the accepted view that potassium efflux is responsible for the delayed outward current produced in response to depolarization of the nerve cell membrane (*Hodgkin* and *Huxley* 1952). This has been corroborated with recent techniques utilizing potassium-selective microelectrodes (*Neher* and *Lux* 1973) which permitted direct demonstration that the delayed current coincides with an increase in potassium ion concentration [K^+] outside the nerve cells. An inward transport of calcium ions during the phase of the delayed current has been demonstrated in the squid axon by the reaction of the protein aecquorin with Ca (*Baker*, *Hodgkin* and *Ridgway* 1971; *Baker*, *Meves* and *Ridgway* 1971). This entry of Ca occurs in addition to that during the early inward current (see also *Meves* 1968; *Geduldig* and *Gruener* 1970; *Krishtal* and *Magura* 1970; *Meves* and *Vogel* 1973). A slow inward current component during the late outward current in snail neurones was recently inferred from the finding that the net charge efflux during the second of two depolarizing pulses is proportionally lower than the measured increment in extracellular potassium activity which accompanies the second pulse (*Lux* and *Eckert* 1974). This led to the conclusion that the outward charge transfer during a subsequent pulse is partially short-circuited by a slow inward membrane current, producing a net current which is smaller than the partial current carried by potassium ions. Thus, the use of indicators for partial ions appears useful for separating membrane current components. The inferred conditioned slow inward current component could serve as a model for facilitation at the membrane level and may have particular implications for an understanding of neuronal pacemaker activity.

Methods

Electrodes with tips of 2 to 5 μm outer diameter were manufactured using the modification of *Lux* and *Neher* (1973) of the *Walker*'s (1971) technique (see also *Khuri*, *Agulian* and *Kalloghlian* 1972; *Vyskočil* and *Kříž* 1972). Double-barrelled electrodes were drawn from theta capillary glass, i.e., pipettes with a straight middle partition (Duran glass, Schott, Germ.), in which one side served as the reference for the ion-sensitive side (Fig. 1). The reference side was filled with 0.1 M NaCl, the indicator side with 0.1 M KCl solution. A filling procedure was used which results in almost 'instant' combined electrodes. Holding the reference barrel under pressure, a 5% solution of trimethyl-chlorosilane in carbon tetrachloride which was drawn by suction into the indicator barrel from the tip up to a height of 200 to 500 μm. After the siliconizing medium is ejected, the ion-exchanger fluid (Corning Potassium Exchange Resin 477317) flows spontaneously into this tip up to the height of the inner hydrophobic coating. After introducing Ag-AgCl leads, the upper parts of the barrels were carefully isolated with wax.

Simultaneous Measurements of Extracellular Potassium-ion Activity and Membrane Currents

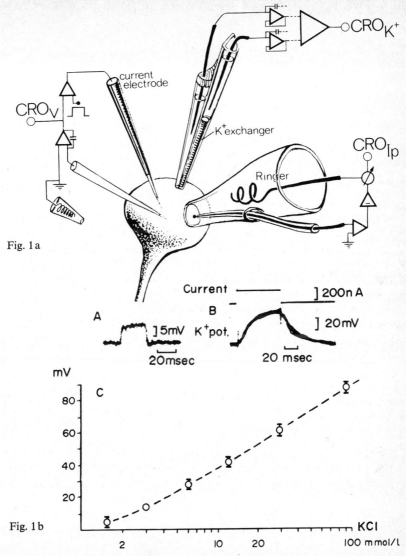

Fig. 1a

Fig. 1b

Fig. 1. Diagram of the relation between nerve cell and instruments. The patch clamp pipette is shown at the right. Local potential is measured with the Ag-AgCl wire (introduced through a lateral opening) and kept at ground level by potentiostatic control through the coiled current injecting electrode. Liquid K-exchanger-reference microelectrode assembly is shown in the middle. The potentials of both barrels are preamplified at unity gain by two identical FET amplifiers of high input impedance and differentially amplified in a second stage to cancel the local electric field. Left: Electrodes and circuitry of the intracellular voltage clamp. A: voltage response of a K^+-specific electrode barrel to a ramp pulse of 30 msec duration applied to the electrode via a capacitance of 1.7 pF. Response is shown at optimal capacitance neutralization. B: Rise and decay of the potential of the K^+-specific-reference pair (lower beam) during an outward going electrophoretic current pulse (upper beam) from a micropipette which contained 1 M KCl solution. The K^+ source was in the immediate vicinity of the K^+ electrode. The measurement was made in a test solution of 3 mmol/l KCl and 150 mmol/l NaCl. C: Calibration of 14 K^+-specific double-barrelled electrodes. Electrode potentials (ordinate) are plotted against log KCl concentration of the test solutions.

The combined electrodes were calibrated with solutions containing known concentrations of potassium chloride ranging from 1 to 60 mmol/l. Sodium chloride was added to approximate the ionic activity of the bathing medium of the preparation. The calibration curves obtained in this way were used to convert the measured K^+ activity to a K^+ concentration value. Calibration curves were quite reproducible with a slope of about 56 mV for a ten-fold change in $[K^+]$; Na^+ interference reduced the slope to about 40 mV at potassium concentrations of 3 to 10 mmol/l. In this range, the potential change at the reference site were neglegible (<0.2 mV). The data were compatible with selectivity constants for Na^+ of about 0.7 to 1.1×10^{-2} at $[K^+]$ below 10 mequ/l. The other constituents of the media caused no significant misreading. The combined electrodes showed no obvious differences in sensitivity from pairs of single electrodes (*Lux* and *Neher* 1973). The ion exchanger was sensitive to acetylcholine (ACh) and to other quarternary ammonium compounds (see *Neher* and *Lux* 1973). Competitive interaction of ACh and K^+ was observed if ACh concentrations of 10^{-7} to 10^{-4} mol/l were added to the $[K^+]$ standard solutions. About 0.5×10^{-6} mol/l ACh in solutions containing 2 to 10 mmol/l KCl were equivalent to an increase of $[K^+]$ of 1 mmol/l. While it is thought unlikely that comparable concentrations of ACh can occur extracellularly, this possibility cannot be excluded.

We found it necessary to check the K^+ transients also for stronger hyperpolarizing voltages and to ascertain that the inward (leakage) current is accompanied only by an appropriate small negative K^+ signal (compare Fig. 2). The ratios between currents and K^+ transients should always be of the same order of magnitude. A voltage-dependent signal of the K^+ electrode can occur as an artefact. Furthermore, the electrode response can be symmetrical for positive and negative membrane voltage, regardless of current size and direction. This artefact was observed only when the electrode tip was in close contact with or even visibly indenting the cell surface; it can possibly be explained by the creation of a K^+-specific, localized leak conductance (see *Neher* and *Lux* 1973).

Potassium ion signals were recorded with a 'neutralized'-capacitance d.c. preamplifier with short leads to the electrodes; the main component of this device was an operational amplifier (Teledyne Philbrick 142902) which provided a 10^{13} ohm input resistance and a low capacitance input. Figure 1A gives an example of the speed of the electrical response with optimal compensation for capacitive losses by feeding back an appropriate part of the amplified signal through a 1.7 pF capacitance to the K^+-sensitive electrode. Response time constants of less than 3 msec could be obtained with electrode resistances of 10^8 to 10^9 ohm. An identical preamplifier was used for the reference electrode. The output signals of both amplifiers were differently amplified with a common mode rejection better than 1:100. In addition to the differential recording, single ended recordings against the remote common silver-silver chloride ground electrode were made to control the potential at the reference electrode.

For step changes of electrophoretic current delivered when the K^+ source and the sensitive electrode tips were separated by less than 10 μm (Fig. 1B), half values for the rising phase of the potential generated at the K^+ electrode were between 5 and 20 msec. When steady-state $[K^+]$ levels were used to determine source-electrode distances from diffusion equations, it was also found that half times of the rise in $[K^+]$ were predicted within 5 msec (*Neher* and *Lux* 1973). It was concluded from these tests that the equilibration process at the boundary of the ion-exchange test solution is fast and that $[K^+]$ variations with a time course of 10 msec or longer were faithfully reproduced.

Simultaneous Measurements of Extracellular Potassium-ion Activity and Membrane Currents

The Helix preparation was bathed in Ringer according to *Gainer* (1972) with 5 mmol Hepes buffer in place of Tris at a temperature of 13 °C to 15 °C. The voltage clamp design has been described in detail (*Neher* and *Lux* 1969). Clamp currents from a limited patch of soma membrane were measured by applying a conventional two-needle clamp (two intracellular microelectrodes) to the cell interior, and additionally placing a semi-micropipette (50 to 100 μm diameter, Ringer-filled) on the exposed soma (see diagram in Fig. 1). The pipette interior was held close to ground potential by feedback control. Through adjustment of the feedback amplification of the pipette clamp, usually by a factor of 30 to 50, the effective internal resistance of the pipette was reduced to an appropriate value at which the current field in the vicinity of the pipette showed minimal disturbance. Thus, during a voltage pulse to the cell interior, currents which cross the patch of membrane covered by the pipette can be measured and their density can be estimated. Experiments were made on bursting pacemaker cells of the right parietal ganglion (cell A, *Kerkut* and *Meech* 1966).

Depression of delayed outward current by a conditioning pulse

The depression of delayed outward current can be characterized as follows: Long depolarizing pulses produce a delayed outward current which gradually decays to about 60% or less of its maximum intensity with time constants of several tenths of a second (*Alving* 1969; *Leicht, Meves* and *Wellhöner* 1971; *Neher* and *Lux* 1971). The presentation of paired pulses of equal amplitude demonstrates that longterm depression of the slow outward current persists (*Neher* and *Lux* 1971; *Lux* and *Neher* 1972; *Gola* 1974) even though the voltage returns to the holding level between a conditioning pulse (pulse 1) and subsequent test pulses (pulses n). For a time interval of up to one second, the delayed current produced by the second pulse may even fail to reach the level of the current produced at the same point in time by the prolonged first pulse (*Neher* and *Lux* 1971; *Lux* and *Eckert* 1974). Recovery from depression of subsequent pulses of outward current takes about 20 to 30 seconds. The length of pulse 1 necessary to depress the test pulse current corresponds to the rise of pulse 1 outward current. Prolongation of pulse 1 beyond its peak does not produce much further depression of the pulse 2 current (*Lux* and *Neher* 1972). This, plus the failure of subthreshold conditioning pulses to produce the phenomenon, suggests that the activation of the current gates during pulse 1 may be causally related to the time-dependent depression of pulse 2 outward current. Increasing pulse 1 further reduces pulse 2 currents, but only slightly. The maximum effect appears to be achieved at positive membrane voltages (up to $+100$ mV) of pulse 1. With repeated pulses outward currents are further reduced with a maximum of reduction reached at the third to fourth repetition (Fig. 2). The kinetics of closing of activated gates at the termination of pulse 1 can be determined from the instantaneous current steps which appear at the onset of subsequent pulses at short intervals (see *Lux* and *Eckert* 1974). The depression of subsequent pulse currents lasts much longer and must therefore be caused by a different mechanism. In bursting pacemaker cells the rising phase of the outward currents of subsequent pulses is always considerably slower than that of pulse 1 (see Fig. 2 and 3).

The long-lasting depression of outward current provides a convenient situation for comparing the depression of net outward current during pulse n with the concomitant reduction in K^+ accumulation at the outer neuronal surface using the K^+-specific microelectrode (Figs. 2 to 4). In the present experiments the pause between pulses was standardized at one second, and pulse sequences were delivered at a repetition rate of one every 30 seconds to allow time for complete recovery from depression. The ratio $Q_{n:1}$ is obtained by dividing the time integral

H. D. Lux

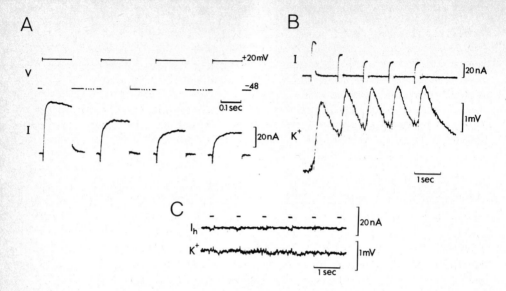

Fig. 2. Repeated voltage clamp pulses (1/sec) of equal amplitude and duration. A: Currents (I) during the first four voltage pulses with omitted parts of the interpauses. Note the decrease and slowed rise of net outward currents during the second and the following pulses. B: Simultaneous records of currents and K^+ signals. Amplitudes of net outward currents of third and following pulses obviously reveal stronger reductions than peaks of K^+ signals. C: Hyperpolarizing voltage pulses indicated by bars of same amplitude as in A and B with proportionally small currents and K^+ signals.

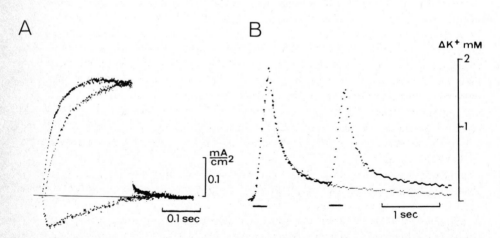

Fig. 3. A: Currents of pairs of voltage pulses (+ 55 mV each from holding potential of — 42 mV) delivered at an interval of one second are normalized to same peak amplitude to show retardation in rising phase of pulse 2 current (digitalized display). Pulse 1 current subtracted from (enlarged) pulse 2 current; result below baseline. B: Digitalized display of the sample of extracellular K^+ signals (n = 8) during the clamp pulses. Bars denote duration of pulses. The slope of a similar sample of single pulses at the same electrode position is superimposed.

of net outward current during pulse n by the equivalent integral during pulse 1 and was found to be in the range of 0.25 to 0.66.

The dissipation time of accumulated potassium was tested by approaching the cell as closely as possible. After one second, at the onset of the second pulse, the K^+ signal of the first pulse had declined in the minimum to about 15% of its peak value (Fig. 3). Nevertheless, the outward current of subsequent pulses was depressed for even longer periods when the K^+ signal had come closer to prestimulus level. This finding speaks against a major effect of extracellular accumulation of potassium in depressing the outward current. In general, the K^+ signals were recorded at some distance from the cell (at about 5 to 30 μm) and showed a slower rise and decay as anticipated from the longer diffusional pathway.

A slow inward current shortcircuits part of the delayed outward potassium current

If it is assumed that the net outward current recorded under voltage clamp conditions is entirely a potassium current, it follows that potassium activity transients recorded near the surface of the cell should parallel quantitatively the net charge transfer produced by the outward current. That is to say, the ratio $Q_{n:1}$ should equal the ratio $FK_{n:1}$ (K flux during pulse n/K flux during pulse 1) for any pair of pulses. From casual inspection of Figures 2, 3 and 4 it is evident that $FK_{n:1} < 1$; that is, K^+ efflux is reduced, as would be expected if the K-channels are partially inactivated during subsequent pulses. Quantitative comparison, however, indicates that there is a 'deficit' in the net outward current of pulse 2 relative to the K^+ signal. In Figure 2B this is obvious from the peaks of the K^+ signal of pulses 3 to 5 which are more than half of the K^+ peak of pulse 1, whereas the ratios of the amplitudes of the concomitant currents are less than one half. This deficit ($FK_{n:1} - Q_{n:1}$) ranged from 0.10 to 0.30 in various units studied. The deficit in net current is most readily explained as the result of an inward current which makes its appearance during subsequent pulses and shortcircuits a portion of the K^+ efflux. Such an inward current would result in a net charge transfer during pulses n lower than that transferred by K^+ efflux.

The presence of a previously unrecognized slow inward current receives further support from a comparison of the kinetics of the outward currents of pulses 1 and n (Figs. 2A and 3). The trajectory of the outward current characteristically has a slower rising phase during pulses n than during pulse 1 when the currents were normalized. This was done by amplifying the pulse n currents either to the peak amplitude of pulse 1 (Fig. 3) or by the ratio $FK_{n:1}$. The first approximation implies complete inactivation of the inward current at times of the peak currents. The second assumes equal distribution over time and neglects for example also a slow rise to steady state of the inward current. The true slope of the inferred current is probably within both estimates.

The difference in the trajectory of the outward current of pulses 1 and n was displayed by electronically subtracting, after normalization, the averaged trajectories of pulse 1 currents from averaged trajectories of the enlarged pulse n currents from the same neuron (Fig. 3). The result of the subtraction is a trajectory with half times up to maxima of 8 to 30 msec after onset of the current pulse. The time integral of the deficit in pulse 2 current in the different cells was comparable in size to that of $FK_{n:1} - Q_{n:1}$. This deficit in outward charge transfer usually increased from the second pulse to a maximum during the third or fourth subsequent pulse (in the experiment of Fig. 2 from 11.5 to 27%). It was also present, however smaller, when the pulse durations were reduced to approximate the duration of action potentials (Fig. 4A and B). While a difference in FK-Q could not be detected during pulses of smaller

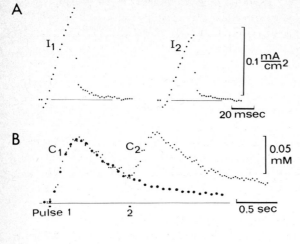
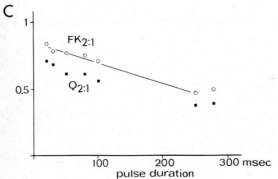
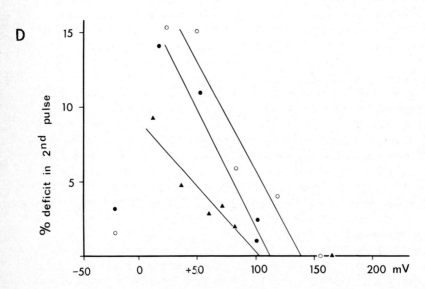

Fig. 4. A: Currents 1 and 2 are evoked by $+65$ mV steps each of 30 msec duration from a holding potential of -45 mV. Simultaneous sampling of K^+ signals is shown in B (n = 25). The superimposed averaged trajectory of single pulses is indicated by heavier dots. C: Ratios of pulse 2 to pulse 1 outward current integrals ($Q_{2:1}$) and corresponding ratios of potassium fluxes ($FK_{2:1}$) of this cell are plotted as a function of pulse duration. Durations of pulse 1 and 2 were kept equal and were varied in unison. The difference between K^+ flux and charge transfer persists while shortening the pulse durations to 50 msec. D: Sampled deficit in outward current of second pulse ($FK_{2:1} - Q_{2:1}$) is plotted against absolute voltages of pairs of clamp pulses. Different symbols denote experiments on different cells.

amplitude than 35 mV (to about −10 mV), it declined (Fig. 4) to spurious values during tests with pulses of more than + 100 mV. An inversal of the deficit in the net outward current could not be found. Replacement of Ca^{2+} in the medium by Co^{2+} abolishes both the deficit in the current trajectories of subsequent pulses and the deficit in net outward current obtained by determining $FK_{n:1} - Q_{n:1}$ (*Heyer* and *Lux* in prep.).

Discussion

These findings suggest that the long-term suppression of net outward current by a prior conditioning pulse consists of two components: One component is a true time- and voltage-dependent partial inactivation of the delayed potassium system, similar in some respects to the inactivation of the sodium system described by *Hodgkin* and *Huxley* (1952). It is manifested as a reduction of extracellular K^+ accumulation during subsequent pulses relative to accumulation of the first pulse. Since it develops during periods between pulses with durations approximating that of the action potential (Fig. 4), the inactivation of outward current described here (*Lux* and *Eckert* 1974) appears more likely to have functional importance under physiological conditions than any reduction in outward current which may be produced by the extracellular accumulation of K^+ with prolonged depolarizing pulses (*Alving* 1969, *Eaton* 1972). The second component contributing to depression of delayed outward current during subsequent pulses appears to be a previously unrecognized inward current which partially shortcircuits the outward potassium current. Its existence is supported by independent lines of evidence. The reduction in K^+ efflux during following pulses, resulting from a true K^+ inactivation, is smaller than the reduction in net outward current. In addition, the rising phase of subsequent pulses exhibits slower kinetics than that of the leading pulse, constant with a superimposed delayed inward current. The inward current component disappears when Ca^{2+} in the bulk solution is replaced by Co^{2+}. The results of other Ca^{2+} substitution also suggest that the inferred inward current is largely carried by calcium ions.

In principle, it should be feasible to compare directly the amount of transferred charge during the outward current of a single voltage pulse and the amount of excess potassium which appears in the bulk solution (*Neher* and *Lux* 1973). However, due to deviations from supposed radial geometry and to inhomogeneity of membrane current density the achieved resolution was found insufficient in the range of the differences between net and overall charge displacement as inferred from the present measurements.

At small voltages below the threshold of the outward current a small, slowly or noninactivating inward current was recently observed in this type of cell (*Eckert* and *Lux* 1974). From ion substitution experiments the current was seen to be carried primarily by Ca^{2+}. With usual assumptions about its voltage dependence it can be followed that at higher voltages this slow inward current during a single pulse will be only a few percent of the net outward current. However, at low voltages and in the absence of an outward current a facilitation of this inward current component by a prior pulse was not observed. It is not clear as yet whether the presence of a potassium current is conditional for the appearance of an augmented inward current during subsequent pulses. It is nevertheless interesting that the activation of this inward current, unlike the other membrane current components, is facilitated by a preceding depolarization. A probably related observation with the use of aequorin as Ca^{2+} indicator was made on bursting pacemaker cells in *Aplysia* during consecutive spikes which evoked a disproportionate intracellular accumulation of Ca ions

(*Stinnacre* and *Tauc* 1974). Both inactivation of the K⁺ system and activation of a slow inward current may explain the progressive increase in spike duration and overshoot and the positive shift in afterpotentials which occur during the train of spikes comprising a 'burst'.

References

Alving, B. O.: Differences Between Pacemaker and Nonpacemaker Neurons of *Aplysia* on Voltage Clamping. J. gen. Physiol. 54 (1969) 512–531

Baker, P. F., Hodgkin, A. L., Ridgway, E. B.: Depolarization and Calcium Entry in Squid Giant Axons. J. Physiol. Lond. 218 (1971) 709–755

Baker, P. F., Meves, H., Ridgway, E. B.: Phasic Entry of Calcium in Response to Depolarization of Giant Axons of *Loligo forbesi*. J. Physiol. Lond. 216 (1971) 70 P

Eckert, R., Lux, H. D.: A Non-Inactivating Inward Current Recorded During Small Depolarizing Voltage Steps in Snail Neurons. Brain Res. 83 (1975) 486–489

Eaton, D. C.: Potassium Ion Accumulation Near a Pacemaking Cell of *Aplysia*. J. Physiol. Lond. 224 (1972) 421–440

Gainer, H.: Patterns of Protein Synthesis in Individual Identified Molluscan Neurons. Brain Res. 39 (1972) 369–385

Geduldig, D., Gruener, R.: Voltage Clamp on the *Aplysia* Giant Neurone: Early Sodium and Calcium Currents. J. Physiol. Lond. 211 (1970) 217–244

Gola, M.: Evolution de la forme des potentiels d'action par stimulations répétitives. Analyse par la méthode du voltage imposé (Neurones d'*Helix pomatia*). Pflügers Arch. 346 (1974) 121–140

Hodgkin, A. L., Huxley, A. F.: Currents Carried by Sodium and Potassium Ions through the Membrane of the Giant Axon of *Loligo*. J. Physiol. Lond. 116 (1952) 449–472

Hodgkin, A. L., Keynes, R. D.: The Potassium Permeability of a Giant Nerve Fibre. J. Physiol. Lond. 128 (1955b) 61–88

Kerkut, G. A., Meech, R. W.: The Internal Chloride Concentration of H and D Cells in the Snail Brain. Comp. Biochem. Physiol. 19 (1966) 819–832

Khuri, R. N., Agulian, S. K., Kalloghlian, A.: Intracellular Potassium in Cells of the Distal Tubule. Pflügers Arch. 335 (1972) 297–308

Krishtal, O. A., Magura, I. S.: Calcium Ions as Inward Current Carriers in Mollusc Neurones. Comp. Biochem. Physiol. 35 (1970) 857–866

Leicht, R., Meves, H., Wellhöner, H. H.: Slow Changes of Membrane Permeability in Giant Neurones of *Helix pomatia*. Pflügers Arch. 323 (1971) 63–79

Lux, H. D., Eckert, R.: Inferred Slow Inward Current in Snail Neurones. Nature 250 (1974) 574–576

Lux, H. D., Neher, E.: A G_K Inactivation Phenomenon in a Nerve Cell Membrane. IV. Biophys. Congr. Moscow 3 (1972) 222–223

Lux, H. D., Neher, E.: The Equilibration Time Course of $[K^+]_0$ in Cat Cortex. Exp. Brain Res. 17 (1973) 190–205

Meves, H.: The Ionic Requirements for the Production of Action Potentials in *Helix pomatia* Neurones. Pflügers Arch. 303 (1968) 215–241

Meves, H., Vogel, W.: Calcium Inward Currents in Internally Perfused Giant Axons. J. Physiol. Lond. 235 (1973) 225–265

Neher, E., Lux, H. D.: Voltage Clamp on *Helix pomatia* Neuronal Membrane; Current Measurement Over a Limited Area of the Soma Surface. Pflügers Arch. 311 (1969) 272–277

Neher, E., Lux, H. D.: Properties of Somatic Membrane Patches of Snail Neurons Under Voltage Clamp. Pflügers Arch. 322 (1971) 35–38

Neher, E., Lux, H. D.: Rapid Changes of Potassium Concentration at the Outer Surface of Exposed Single Neurons During Membrane Current Flow. J. gen. Physiol. 61 (1973) 385–399

Stinnacre, J., Tauc, L.: Ca-Influx in Active *Aplysia* Neurones Detected by Injected Aequorin. Nature New Biol. 242 (1973) 113—115

Walker, J. L., Jr.: Ion Specific Liquid Ion Exchanger Microelectrodes. Analyt. Chem. 43 (1971) 89–92-A

Vyskočil, F., Kříž, N.: Modifications of Single- and Double-Barrel Potassium Specific Microelectrodes for Physiological Experiments. Pflügers Arch. 337 (1972) 265–276

A. Lehmenkühler, E.-J. Speckmann and H. Caspers

Cortical Spreading Depression in Relation to Potassium Activity, Oxygen Tension, Local Flow and Carbon Dioxide Tension

(3 Figures)

The aim of the present investigations was to study some elementary mechanisms underlying cortical spreading depression (SD). For this purpose the relations between changes in K+ activity and in cortical DC potentials were examined. Furthermore, the time relationship between bioelectric events and changes in local metabolism was analyzed. Experimental results reported in the literature indicate that oxidative metabolism increases considerably during the development of an SD wave (*Mayevsky* and *Chance* 1974; *Mayevsky, Zeuthen* and *Chance* 1974; *Rosenthal* and *Somjen* 1973). These results were obtained by combining optical and electrical recording techniques. Since the cortical areas from which the two signals were sampled differed in size, exact timing of bioelectric and metabolic processes proved difficult. Therefore, the problem was reassessed using an arrangement of combined microelectrodes which allowed registration of tissue pO_2, K+ activity and DC potentials at the same site. Finally, the role played by neuronal action potentials in producing SD and associated reactions of tissue pO_2 was examined.

Methods

The experiments were performed on rats anaesthetized with phenobarbital and artificially ventilated with the body temperature kept constant at 37 ($\pm$ 1) °C. Repetitive waves of SD activity were elicited by applying a KCl crystal to the surface of the cerebral cortex. Both the conventional EEG and the DC potential recorded from frontoparietal and occipital surface areas as well as from deeper layers of the cortex served to indicate SD waves. A common reference electrode was placed in the nasal bone of the animal. Aside from the bioelectric events extracellular K+ activity in the cortex was determined by double-barrelled micropipettes with a tip diameter of about 2 µ. A liquid ion exchanger was used according to the techniques described by *Lux* and *Neher* (1973). In addition, tissue pO_2 (*Speckmann* and *Caspers* 1970), arterial pO_2 and pCO_2 were measured. In the majority of the experiments the tips of the microelectrodes recording the DC potential, K+ activity and tissue pO_2 were arranged together. Furthermore, regional and local flow were determined. These measurements were performed continuously with a thermo-electric method and controlled by means of H_2-clearance curves according to the methods developed by *Lübbers, Kessler, Knaust, McDowall* and *Wodick* (1966).

Results

In a first series of experiments the time relationship between changes of cortial DC potentials and of K+ activity during repetitive SD was examined. In accordance with earlier reports

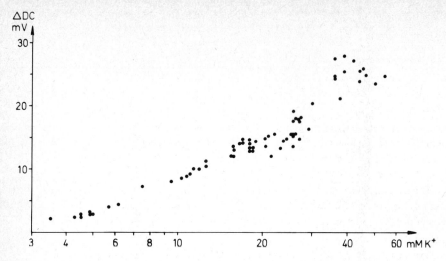

Fig. 1. Relation between amplitudes of maximum negative DC shifts (Δ DC) and extracellular K^+ activity at DC recording site during repetitive waves of cortical spreading depression.

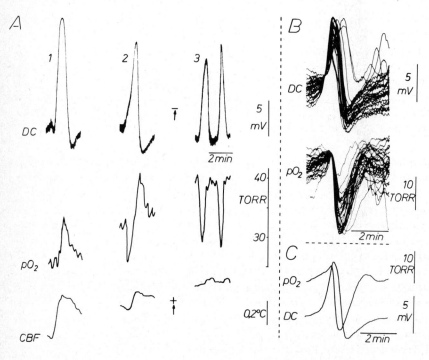

Fig. 2. Cortical DC shifts (DC), changes in tissue pO_2 and in cortical blood flow (CBF) in the course of repeated waves of spreading depression. A: Single recordings selected from the first (1), intermediate (2) and last parts (3) of a series of spreading depression. B: Superposition of thirty simultaneously recorded DC and pO_2 reactions during SD. C: DC and pO_2 curves averaged from the tracings in B.

(*Mayevsky, Zeuthen* and *Chance* 1974; *Vyskočil, Kříž* and *Bureš* 1972) DC shifts and variations of K^+ activity during SD showed a similar time course. In addition, there was a close relationship between the peak amplitude of DC shifts and changes in K^+ activity during SD. In Figure 1 the potassium activity is plotted logarithmically against the amplitudes of the corresponding DC deflections. The graph reveals an almost linear relationship between the two values at least in a medium range. A tenfold change in K^+ activity was associated with a DC shift of about 20 mV.

In order to obtain more information about the metabolic situation during SD, tissue pO_2 and DC potentials were recorded with microelectrodes at the same site. The results of a typical experiment are displayed in Figure 2. The tracings in A represent recordings of the DC potential, of local pO_2 and of cortical blood flow. At the beginning of a series of repetitive SD waves, tissue pO_2 increased (A 1). Subsequently, this pattern of pO_2 reaction became biphasic (A 2). Finally, monophasic decreases in pO_2 were observed (A 3). The changes in phasic responses of the pO_2 during each SD wave were associated with an increase of the mean level of the oxygen pressure.

The mechanisms responsible for the different reactions of cortical pO_2 were analyzed by simultaneous measurements of local cortical blood flow (CBF). At the onset of SD activity each SD wave was accompanied by a considerable phasic increase in cortical flow (Fig. 2, A 1). Due to these increases the mean level of local flow rose progressively to a maximum while the phasic responses of CBF were concomitantly reduced (Figs. 2, A 2 and A 3). The rise of the flow level was also demonstrated by applying the H_2-clearance technique. This behavior of local flow was obviously responsible for the pattern of pO_2 reactions during SD waves.

To study the oxygen consumption during SD, the DC and pO_2 reactions occurring in late phases of a series were evaluated. In Figure 2B thirty original tracings of DC shifts and pO_2 reactions during SD were superimposed graphically and evaluated by a special method of pattern recognition and correlation analysis (*Knoll, Speckmann* and *Caspers* 1974). Figure 2 C represents the averaged time course of these curves. The evaluation reveals that the fall in pO_2 always started in the *ascending* slope of the negative DC shift, preferentially in its last part.

The question arose whether the decrease in pO_2 during SD shifts is attributable to an increase in neuronal spike activity frequently found in the early rising phase of the negative DC deviation. To study this problem, SD shifts were elicited after neuronal action potentials had been extinguished by critical hypoxia. Figure 3A shows continuous recordings of the conventional EEG, of the cortical DC potential and of tissue pO_2. Ligation of both common carotid arteries and simultaneous ventilation with 6% O_2 in N_2 led to a disappearance of EEG activity, to the typical negative DC shift and to a drop of tissue pO_2. Together with the extinction of the EEG also multiple neuronal spike activity recorded by a semi-microelectrode was suppressed. Reopening of the arteries and reventilation of the animal with 100% oxygen induced a steep reincrease of cortical tissue pO_2 and a deflection of the cortical DC potential in the positive direction. The application of a KCl crystal elicited periodically recurring DC shifts associated with typical decreases of pO_2 even in this phase, in which EEG activity was abolished. These experiments show that the occurrence of SD itself as well as the increase of oxygen consumption is independent of neuronal spike activity in the cerebral cortex.

Previous investigations showed that an increase in pCO_2 exerts a hyperpolarizing effect on cortical neurons and leads to an interruption of drug-induced seizure activity (*Caspers* and *Speckmann* 1972; *Speckmann* and *Caspers* 1974). These observations raised the question as to

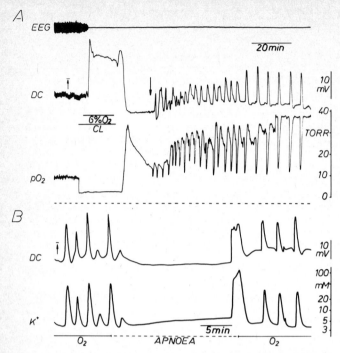

Fig. 3. A: Changes of conventional EEG, of cortical DC potential (DC) and of cortical tissue pO_2 evoked by clamping both common carotid arteries (CL) and breathing 6% O_2 in N_2 (horizontal bars). Before and after this period 100% O_2 was applied. After extinction of the EEG the application of a KCl crystal is still effective to induce series of SD waves associated with a decrease in pO_2. B: Suppression of intracortical DC shifts and of corresponding changes in K^+ activity within a series of SD waves by hyperoxic hypercapnia (apnoea).

whether SD waves and the corresponding variations of K^+ activity could also be influenced by an increase in pCO_2 in the cerebral cortex. In Figure 3B the DC potential measured in a deeper cortical layer as well as K^+ activity are displayed. During a period of apnoea which is characterized by an increase of arterial pCO_2 and by a rise of tissue pO_2, SD waves were abolished and K^+ activity showed a slight continuous rise (see also *Lehmenkühler, Caspers* and *Speckmann* 1973). Comparing DC and K^+ reactions under SD and hypercapnic conditions, it becomes apparent that an increase in K^+ activity can be accompanied by DC shifts in negative as well as in positive directions.

References

Caspers, H., Speckmann, E.-J.: Cerebral pO_2, pCO_2 and pH: Changes During Convulsive Activity and their Significance for Spontaneous Arrest of Seizures. Epilepsia 13 (1972) 699–725

Knoll, O., Speckmann, E.-J., Caspers, H.: Ein Verfahren zur Korrelierung verschiedener biolektrischer Vorgänge mit definierten Potentialmustern im EEG. Z. EEG-EMG 5 (1974) 199–205

Lehmenkühler, A., Caspers, H., Speckmann, E.-J.: Actions of CO_2 on Cortical Spreading Depression. Pflügers Arch. 339 (1973) R 83

Lübbers, D. W., Kessler, M., Knaust, K.,

McDowall, D.G., Wodick, R.: Die Verwendung von Wasserstoff und Sauerstoff zur Messung der lokalen Gewebedurchblutung in situ mit der Platinelektrode. Pflügers Arch. 289 (1966) R 99

Lux, H. D., Neher, E.: The Equilibration Time Course of $(K^+)_0$ in Cat Cortex. Exp. Brain Res. 17 (1973) 190–205

Mayevsky, A., Chance, B.: Repetitive Patterns of Metabolic Changes During Cortical Spreading Depression of the Awake Rat. Brain Res. 65 (1974) 529–533

Mayevsky, A., Zeuthen, T., Chance, B.: Measurements of Extracellular Potassium, ECoG and Pyridine Nucleotide Levels During Cortical Spreading Depression in Rats. Brain Res. 76 (1974) 347–349

Rosenthal, M., Somjen, G.: Spreading Depression, Sustained Potential Shifts and Metabolic Activity of Cerebral Cortex of Cats. J. Neurophysiol. 36 (1973) 739–749

Speckmann, E.-J., Caspers, H.: Messung des Sauerstoffdrucks mit Platinmikroelektroden im Zentralnervensystem. Pflügers Arch. 318 (1970) 78–84

Speckmann, E.-J., Caspers, H.: The Effect of O_2- and CO_2-tensions in the Nervous Tissue on Neuronal Activity and DC Potentials. In: Handbook of Electroencephalography and Clinical Neurophysiology, Vol. 2 Part C, pp. 71–89, ed. by A. Remond. Elsevier, Amsterdam 1974

Vyskočil, F., Kříž, N., Bureš, J.: Potassium-Selective Microelectrodes Used for Measuring the Extracellular Brain Potassium During Spreading Depression and Anoxic Depolarization in Rat. Brain Res. 39 (1972) 255–259

E. Betz

Pial Vascular Smooth Muscle Reactions During Perivascular Microperfusion with Artificial CSF of Varying Ionic Composition

(3 Figures)

The cerebrospinal fluid (CSF) is the normal extracellular fluid compartment for the smooth muscle cells of pial vessels. Since the pial arteries do not have vasa vasorum and since the smooth muscle cells of these vessels can be easily reached by artificial extracellularly applied substances, they are excellent models for vascular smooth muscle research. In order to study the effects of ionic variations in the perivascular space on the pial vascular smooth muscles, we exposed the subarachnoidal vessels in cats and perfused the perivascular space with mock CSF. We took series of photos for documentation of the effects of variation in the ionic composition of the perivascular fluid. Systemic blood pressure, respiration and blood electrolytes were kept constant during the experiments.

The H^+ ions are assumed to be the most important ions for the physiological regulation of the smooth muscle tone in cerebral vessels. Local perivascular application of CSF with pH of 6.9 causes dilatation, and artificial CSF with pH of more than 7.4 leads to constriction of the pial arterioles (*Betz* 1973). By means of electrical stimulation of a single vessel it is possible to

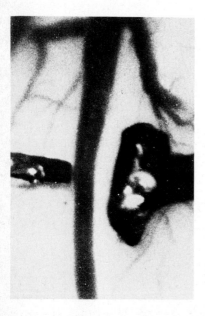 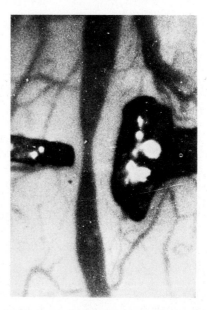

Fig. 1. The effect of local electrical stimulation of a pial artery $\varnothing \approx 150\,\mu m$ with DC. The picture was taken 3 min after the stimulus, which lasted 10 s. (Stimulation with rectangular impulses 1 ms, 5 mA per impulse during continous perfusion of the perivascular space with normal CSF).

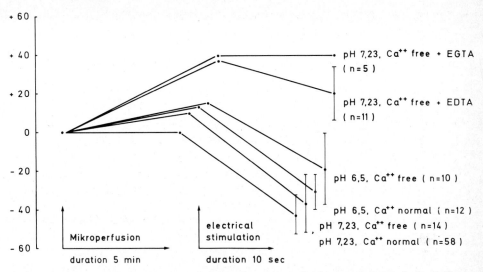

Fig. 2. Effect of microperfusion of the perivascular space and subsequent electrical stimulation on pial vascular diameters. Prior to the stimulation the vicinity of the single vessel was perfused for 5 min with mock CSF with different pH, different Ca^{++} contents or with CSF containing EDTA or EGTA. Subsequent stimulation with DC impulses (see text). The measurements of the effects were made 0–5 s after the end of the electrical stimulus. They are represented as mean values with standard errors.

study the variation of its reactivity. For this purpose a microelectrode, 100 μm in diameter, was placed at one side of the pial artery and the reference electrode (indifferent electrode) was attached to the other side of the artery or arteriole. The reference electrode had dimensions of $0.2 \cdot 0.2 \cdot 2.0$ mm. For stimulation we used series of DC impulses (3 or 5 mA for 1 msec, 10 Imp/sec). The total stimulation period lasted 10 s. Figure 1 gives an example of the type of the most frequently occurring vascular constriction after stimulation. Besides the spindleshaped form of a constriction we observed unilateral constrictions appearing only at the site at which the current electrode touched the arterial wall, or we observed constrictions propagated up or down along the stimulated vessel.

In Figure 2 it can be seen that the degree of the constriction after stimulation is not significantly different in acidotic CSF and normal CSF. If, however, the perivascular space was previously perfused with EDTA or EGTA the dilated vessels did not respond to the stimulation, indicating that pial vascular constriction is Ca^{++}-dependent.

Is there a connection between the normally occurring dilatation during acidosis and the perivascular Ca^{++}? Under normal vascular conditions a decrease of pH in the perivascular space always leads to a dilatation of all the observed vessels. If the Ca concentration in the acidotic perivascular CSF is increased, less vessels dilate, indicating an interaction between H^+ ions and Ca^{++} on the smooth muscles. Increase of Ca^{++} in acid CSF leads to pial vascular constriction. It has been assumed that an inhibition of a Ca^{++}-activated ATP-ase by hydrogen ions is the reason for this interaction of H^+ and Ca^{++} (*Gruen* 1972).

Another ion which influences the vascular smooth muscle tone is potassium. If the potassium concentration in the CSF with normal pH is zero, the pial arteries respond with constriction. If potassium is higher than normal, the arteries and arterioles dilate. This can be seen in Figure 3. High concentrations of potassium of more than 50 meq/l K^+ in the perivascular CSF again reduce the diameters of the pial vessels, so that in our experiments the diameters were about the same as before the experiments, when K^+ was at 50–60 meq/l in the CSF. At 100 meq/l K^+ in CSF the pial vessels again showed marked constrictions.

Electrical stimulation in the various states always resulted in constriction. However, the degree of the finally obtained vascular diameter depended on the concentration of K^+ in the CSF.

If the mock CSF contained no potassium the perivascular constriction after stimulation was more pronounced than in normal CSF. Strong constrictory effects were also seen during application of very high K^+ concentrations. During moderate increases the finally obtained constrictory state, however, did not exceed the initial arterial diameter.

It has been demonstrated by *Böning* (1973) that the effects of H^+ and K^+ on the pial vascular smooth muscles interact. In the state of pial vascular constriction because of perivascular alkalosis, addition of K^+ ions caused a reduction of the degree of the constriction. We do not yet know the site of this effect. However, it can be concluded from experiments of *Keatinge* (1968), *Speden* (1970) and *Siegel* et al. (1974) that the K^+ ions act on the constriction via the

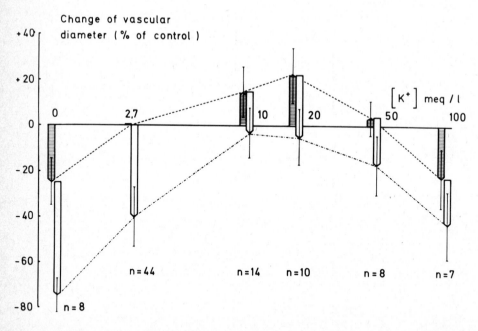

Fig. 3. The effect of microperfusion of the pial perivascular space with mock CSF containing different concentrations of K^+. The dark columns represent the changes of pial arterial diameters during the perfusion. 5 min after the start of the perfusion the vessels were locally stimulated with DC impulses. In every case the response of the vessels was a constriction. As can be seen from the response at 50 or 60 meq/l K^+ in the CSF, the vessels constrict only to their initial diameters, whereas in low or very high CSF-K^+ the degree of the constriction is considerably greater.

variation of the membrane potential, whereas it is not clear whether the above-mentioned ions exert their effects on the interaction of myosine and actine also in the smooth muscle mainly via action potentials.

References

Betz, E., Enzenroß, H. G., Vlahov, V.: Interaction of H^+ and Ca^{++} in the Regulation of Local Pial Vascular Resistance, *Pflügers* Arch. 343 (1973) 79–88

Böning, U.: Der Einfluß von extrazellulären Kalium-, Calcium- und Magnesiumionen auf den Durchmesser pialer Arteriolen. Diss., Tübingen 1973.

Grün G.: Ca^{++}-Antagonismus – ein neu erkanntes Prinzip der Relaxation glatter Muskelzellen. Habil.-Schrift, Freiburg 1972

Keatinge, W. R.: Sodium Flux and Electrical Activity of Arterial Smooth Muscle. J. Physiol. (London) 194 (1968) 183

Siegel, G., Jäger, R., Nolte, J., Bertsche, O., Roedel, H., Schröter, R.: Ionic Concentrations and Membrane Potential in Cerebral and Extracerebral Arteries. In: Pathology of Cerebral Microcirculation, p. 96–120, ed. by. J. Cervós-Navarro. *de Gruyter*, Berlin 1974

Speden, R. N.: Excitation of Vascular Smooth Muscle. In: Smooth Muscle, p. 558, ed. by E. Bülbring, A. F. Brading, A. W. Lones, T. Tomitas. Arnold, London 1970

D. Heuser and E. Betz

Measurements of Potassium Ion and Hydrogen Ion Activities by Means of Microelectrodes in Brain Vascular Smooth Muscle

(5 Figures)

The hypothesis that the pH in the brain plays the major role in the regulation of cerebral vascular resistance has been modified by recent studies (*Knabe* and *Betz* 1972; *Kuschinsky, Wahl, Bosse* and *Thurau*; 1972). They showed that the activity of several cations in the perivascular fluid of cerebral vessels evidently influences their tone and contractility. *Betz* (1974) showed that aside from the important H^+ ions, K^+ and Ca^{++} ions are also involved in the ionic mechanism participating in vascular smooth muscle contraction. These conclusions

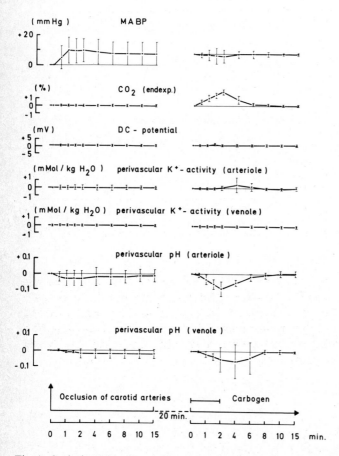

Fig. 1. Occlusion of both carotid arteries and subsequent inhalation of 4.1% CO_2 in O_2 (n = 22).

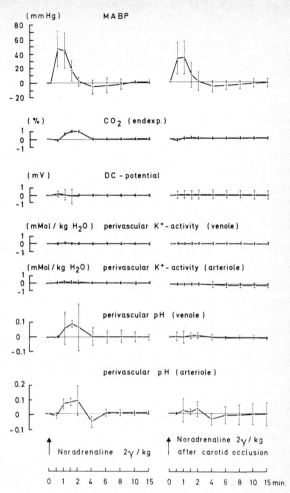

Fig. 2. Influence of Norepinephrine (2 µg/kg i.v.) on perivascular pH and potassium, before and after occlusion of carotid arteries (n = 22).

were drawn from experiments in which the ionic composition of the perivascular fluid was changed by means of the micropipette technique. However, no answer was found as to whether the changes of perivascular ion activities induced by certain pharmacological and patho-physiological stimuli are of a magnitude that could influence cerebral cortical vascular resistance. As a contribution to this discussion we measured perivascular H^+ and K^+ activities continuously by means of microelectrodes under various experimental contitions.

Methods

We performed experiments on 22 cats under Pentobarbital anaesthesia (25 mg/kg). The animals were relaxed with Flaxedil® and artificially ventilated.
Blood gases were controlled by means of a micro-system, body temperature was kept within normal values, end-tidal CO_2 concentration was recorded by means of an URAS, MABP

with a Statham transducer via a PVC catheter in the abdominal aorta. Pneumothorax and drainage of CSF minimized brain movements occurring synchronously with respiratory movements. Occlusion of the carotid arteries was performed with special occluders. Vasoactive substances were given intravenously. The skull was opened with a hand drill in order to avoid heating the cortical surface. We removed the dura within an area of 4–6 mm² and cut off parts of the arachnoidea along a small artery and a venole using a sharp needle. This was done under control by a stereo microscope. For continuous measurements of K^+ ions, H^+ ions and DC potential we used microelectrodes. The K^+ probes with tip diameters between 1 and 3 μm were made according to a technique similar to that of *Walker*. They were filled with Corning Potassium Exchange Resin No. 477317 and calibrated in several KCl concentrations in 150 mmolar NaCl solution. The pH electrodes had tip diameters between 20 and 100 μm and were manufactured as recently described (*Heuser, Astrup, Lassen* and *Betz*, 1975). As reference electrodes we used a micropipette filled with 150 mmolar NaCl. The signals from the sensitive sites were recorded differentially against the reference cell with neutralized capacitance DC preamplifiers with an input resistance of about 10^{14} ohms.

DC-potential was recorded from the reference electrode against ground. By means of micromanipulators we placed the electrodes directly on the surface of a single pial arteriole or

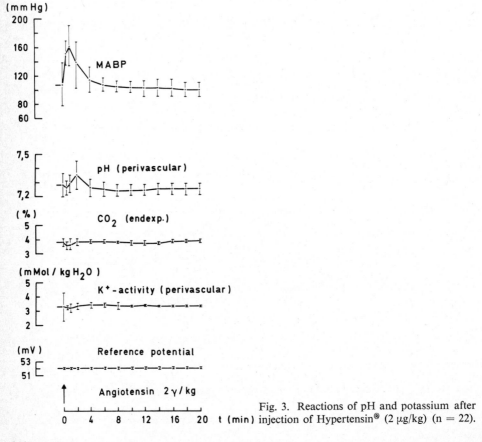

Fig. 3. Reactions of pH and potassium after injection of Hypertensin® (2 μg/kg) (n = 22).

Results

In the beginning of the experiments we found a perivascular pH difference of 0.15 ($\pm$ 0.2) units between arteries and veins in 20 measurements. The veins were more acid.

The difference tended to increase in the course of the experiment within the following hours. In Figure 1 reactions of all parameters after occlusion of both carotid arteries and subsequent inhalation of 4.1% CO_2 are demonstrated: occlusion causes a perivascular acidosis comparable to that of breathing 4.1% CO_2 during 2 minutes. We could not observe significant increases in potassium activity, though in some experiments K^+ activity was raised less than 1 mmol. The DC potential also was not significantly changed.

In Figure 2 one can see the reactions of the measured parameters after injection of Norepinephrine (2 µg/kg). There are polyphasic perivascular pH-reactions, often starting with a small transient acidosis. The reactions of arteries and veins are similar and the type of reaction is not very much affected by occlusion of the carotid arteries.

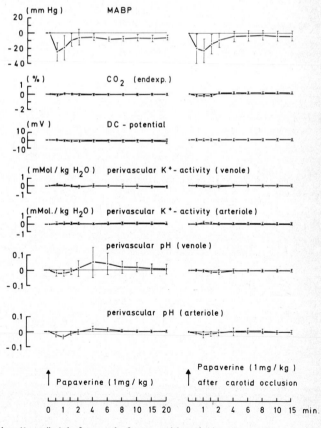

Fig. 4. Reactions after papaverine (1 mg/kg) before and after carotid occlusion (n = 15).

Table 1a. Occlusion of carotid arteries   min

	init. val	½	1	1½	2	4	6	8	10	15
MABP	116	9.6	19.7	18.4	18.7	17	15	15	15	15
	±26	15	17	16	15	16	15	15	15	15
end-exp. CO_2	3.3	0	0	0	0	0	0	0	0	0
	±0.2	0	0	0	0	0	0	0	0	0
DC-pot.	10.6	0	0	0	0	0	0	0	0	0
	±4.1	0	0	0	0	0	0	0	0	0
K^+-act. art.	2.65	0.025	0.025	0	0.025	0.03	0.03	0.07	0.07	0.07
	±0.97	0.27	0.27	0.3	0.3	0.3	0.3	0.17	0.17	0.17
K^+-act. ven.	2.7	0	0	0	0.06	0	0	0	0	0
	±1.2	0	0	0	0.1	0	0	0	0	0
pH art.	7.22	0	−0.02	−0.03	−0.03	−0.02	−0.02	−0.02	−0.01	−0.01
	±0.15	0	0.03	0.05	0.05	0.06	0.05	0.04	0.04	0.04
pH ven.	7.17	0	0	−0.01	0.01	−0.02	−0.02	−0.02	−0.02	−0.02
	±0.1	0	0.01	0.01	0.03	0.03	0.03	0.03	0.03	0.03

Table 1b. Inhalation of 4.1% in CO_2 in O_2 min

	init. val	½	1	1½	2	4	6	8	10	15
MABP	131.5	−1	−0.6	1.7	3.2	1.3	−1.1	−0.3	−0.7	−0.7
	±28.5	3	7.3	10	9	4.4	5.4	2.9	15	1.5
end-exp. CO_2	4.1	0.5	1.1	1.4	1.8	0.7	0.2	0.1	0	0
	±0.4	0.3	0.4	0.3	0.3	0.5	0.2	0.1	0	0
DC-pot.	8.8	0	−0.04	0.1	−0.08	−0.4	0	0	0	0
	±7.7	0	0.1	1.6	2.1	0.7	0	0	0	0
K^+-act. art.	3.27	0	0	0.02	0.17	0.47	0.22	0.02	0.05	0.1
	±0.8	0	0	0.05	0.28	1.01	0.5	0.12	0.17	0.27
K^+-act. ven.	3.14	0	0.05	0.06	0.07	0	0	0	0	0
	±0.6	0	0.07	0.1	0.1	0	0	0	0	0
pH art.	7.20	−0.01	−0.04	−0.08	−0.1	−0.06	−0.02	−0.01	0.006	0.01
	±0.17	0.01	0.03	0.03	0.05	0.02	0.02	0.03	0.02	0.02
pH ven.	7.15	0	−0.03	−0.04	−0.06	−0.08	−0.04	0	0	0
	±0.13	0	0.02	0.02	0.08	0.1	0.01	0.01	0.01	0

Measurements of Potassium Ion and Hydrogen Ion Activities

Table 2a. Norepinephrine (2 μg/kg)   min

	init. val	½	1	1½	2	4	6	8	10	15
MABP	118	46.3	44.6	16	1.3	—6,7	—5.4	—3.0	—0.9	—0.4
	±23	24	24	14	6	9.5	9.6	7.1	1.9	2.5
end-exp. CO_2	3.9	—0.17	0.7	0.12	0.1	0.03	0.03	0.03	0.03	0.03
	± 0.5	0.09	0.21	0.13	0.09	0.06	0.06	0.1	0.1	0.1
DC-pot.	8.14	0.2	0.1	0.1	0	0	0	0	0	0
	± 5.3	0.4	0.7	1.0	0	0	0	0	0	0
K^+-act. art.	3.3	0.08	0.13	0.14	0.16	0.06	0.03	0.01	0	0
	± 0.91	0.11	0.23	0.25	0.28	0.12	0.06	0.04	0	0
K^+-act. ven.	3.4	0.06	0.1	0.05	0	0.025	0	0	0	0
	± 1.2	0.1	0.2	0.16	0.05	0.08	0.08	0.08	0.08	0.08
pH art.	7.28	0.025	0.06		0.08	—0.06	0	0	0	0
	± 0.10	0.02	0.3		0.01	0.02	0.02	0.02	0.01	0.01
pH ven.	7.26	0	0.06	0.09	0.07	0	0	0	0.01	0
	± 0.09	0.01	0.11	0.2	0.17	0.06	0.04	0.05	0.06	0.04

Table 2b. Norepinephrine (2 ug/kg) after carotid occlusion   min

	init. val	½	1	1½	2	4	6	8	10	15
MABP	133.8	32.6	34.0	9.2	0.8	6.2	—4.5	—3.1	—1.3	—0.8
	±31.8	22	23	17	13	8	6.2	5.8	5.8	5.9
end-exp. CO_2	3.7	— 0.2	0.1	0.09	0.05	0	0	0	0	0
	± 0.28	0.11	0.19	0.1	0.08	0	0	0	0	0
DC-pot.	8.2	0.1	0.15	0.15	0.15	0.15	0.18	0.18	0.15	0.15
	± 9.5	0.5	0.5	0.6	0.6	0.6	0.5	0.5	0.6	0.6
K^+-act. art.	3.35	0.01	—0.05	0.02	0.01	—0.08	—0.12	—0.12	—0.11	—0.1
	± 1.1	0.1	0.19	0.2	0.18	0.38	0.35	0.35	0.35	0.35
K^+-act. ven.	3.37	0.01	0.05	0.04	0.01	0.11	0.12	0.12	0.08	0.03
	± 1.0	0.08	0.27	0.2	0.16	0.18	0.19	0.19	0.13	0.12
pH art.	7.20	— 0.001	0.02	0.01	0.03	—0.04	—0.01	—0.008	—0.004	0.001
	± 0.14	0	0.08	0.02	0.05	0.04	0.06	0.07	0.07	0.08
pH ven.	7.13	0.005	0.001	0.01	6.01	—0.002	—0.006	—0.002	—0.003	—0.01
	± 0.11	0.03	0.04	0.02	0.02	0.02	0.01	0.01	0.01	0.01

Injection of Hypertensin® (2 µg/kg) causes an initial slight acidosis, a subsequent alkalosis and thereafter again an acidosis. There is a very small but not significant increase in perivascular potassium activity. End-tidal CO_2 did not change and MABP reacted as expected.

Injection of papaverine (1 mg/kg i.v.) is also followed by a biphasic shift of perivascular pH (Fig. 4) with an initial acidosis, but the time course is more prolonged than after injection of vasoconstrictive drugs. The reactions are slightly influenced by carotid occlusion. Release of the carotid occlusion had no effect on perivascular ion activities, but MABP decreased by about 17 mmHg.

In Figure 5 the reactions of the measured parameters after inhalation of nitrogen for two minutes are demonstrated. There is an initial increase, then a distinct decrease of MAP and a subsequent return to normal values after about 25 min. End-tidal CO_2 is 0.5% below the initial value and a marked acidosis occurs (from 7.2 to 6.86). Perivascular K^+ activity is

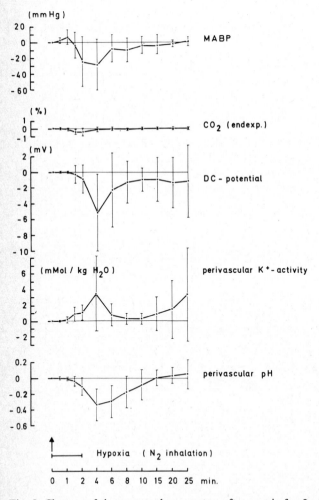

Fig. 5. Changes of the measured parameters after anoxia for 2 minutes (n = 16)

Table 3a. Papaverine (1 mg/kg)

	init. val		1/2	1	1¹/₂	2	4	6	8	10	15	20	25	30 min
MABP	123.6		−25	−19	−9.3	−5.8	−5.7	−7.7	−7.4	−6.5	−6.8	−6.3		
	±26.3		11.8	15.7	9.2	10.3	4.4	2.9	2.9	3.5	4.3	3.0		
end.-exp. CO₂	3.74		−0.1	0.05	0.03	0	0	0	0	0	0	0		
	±0.44		0.09	0.07	0.04	0	0	0	0	0	0	0		
DC-pot.	7.85		−0.2	−1.0	−1.2	0.6	−0.7	−0.1	−0.1	−0.1	−1.4	−2.2		
	±5.33		0.7	1.3	1.2	1.9	1.7	1.4	1.4	1.8	2.5	3.3		
K⁺-act. art.	3.2		0.01	0.19	0.13	0.03	0.02	0.1	0.14	0.14	0.14	0.14		
	±0.8		0.17	0.2	0.18	0.2	0.18	0.2	0.21	0.21	0.21	0.21		
K⁺-act. ven.	3.1		0.08	0.05	0.05	0.1	0.06	−0.06	−6.05	0.11	−0.03	−0.05		
	±1.06		0.17	0.17	0.10	0.2	0.2	0.12	0.27	0.25	0.17	0.19		
pH art.	7.31		−0.01	−0.03	−0.01	−0.005	0.02	0.02	0	0	0	0		
	±0.15		0.01	0.01	0.01	0.015	0.02	0.01	0.01	0.01	0.01	0.01		
pH ven.	7.25		−0.02	−0.02	−0.01	0.01	0.05	0.04	0.02	0.02	0.01	0.01		
	±0.09		0.02	0.02	0.02	0.05	0.1	0.07	0.04	0.03	0.03	0.03		

Table 3b. Papaverine (1 mg/kg) after carotid occlusion   min

	init. val	1/2	1	1½	2	4	6	8	10	15
MABP	133	—21	—24	—15	—10	—5	—4	—3	—4	—4
	± 3.10	19	16	13	10	7	7	6	6	6
end.-exp. CO_2	3.9	— 0.15	— 0.05	— 0.01	0	0	0	0	0	0
	± 0.58	0.12	0.12	0.05	0	0	0	0	0	0
DC-pot.	10.1	— 0.3	— 0.7	— 0.1	— 0.15	0	—0.03	—0.03	—0.03	—0.03
	± 7.4	0.79	2.0	1.6	1.3	1.2	1.0	1.0	1.0	1.0
K^+-act. art.	3.1	0.03	0.22	0.13	0.07	0.05	0.11	0.12	0.12	0.12
	± 0.9	0.2	0.2	0.2	0.2	0.16	0.21	0.29	0.29	0.29
K^+-act. ven.	3.7	0.05	0.07	0.1	0.07	0	—0.03	—0.03	—0.05	—0.07
	± 1.1	0.14	0.08	0.11	0.08	0.11	0.11	0.11	0.14	0.10
pH art.	7.20	— 0.01	— 0.03	— 0.02	— 0.01	0.01	0.02	0.02	0.01	0.01
	± 0.15	0.01	0.04	0.03	0.04	0.05	0.04	0.03	0.02	0.02
pH ven.	7.16	— 0.004	— 0.01	— 0.02	— 0.02	0.005	0.006	0.008	0.009	0.01
	± 0.11	0.01	0.01	0.02	0.03	0.02	0.02	0.02	0.02	0.02

raised from about 3 to 7 mmol, DC potential changed by 6 mV. The time course of K^+ and DC changes is obviously homologous. On the average potassium and DC did not return completely to the initial values.

Discussion

The presented results give evidence that changes of the pial perivascular K^+ activity do not play an important role during the action of the drugs mentioned above. The polyphasic perivascular pH reactions are not easy to explain. The slight initial acidosis after injection of vasoconstrictory agents could be interpreted in relation to intracellular pH shifts during liberation of Ca^{++} ions be means of energy-dependent processes (*Kulbertus* 1964). We still have no explanation for the papaverine-induced initial acidosis, but the consecutive alkalotic shift could be interpreted as a washout effect of CO_2. The distinct reactions of pH and potassium after anoxia give evidence that under these conditions there could be an interference of H^+ and K^+ ions which governs the regulation of cerebral vascular resistance. According to the results of *Kuschinsky* et al. (3), changes of K^+ activity predominate over those of perivascular pH.

Table 4. Anoxia (N_2 inhalation)

	init. val	1/2	1	1½	2	4	6	8	10	15	20	25 min
MABP	130.7 ±35	3.3 3.4	7.0 9.2	−4.7 15.9	−25.7 31.8	29.9 32	−8.9 17.2	−10.2 15.5	−4.2 11.7	−4.0 9.9	−2.0 3.9	4.2 6.0
end.-exp. CO_2	4.02 ± 0.5	−0.09 0.08	−0.18 0.17	−0.4 0.3	−0.4 0.5	−0.16 0.28	−0.07 0.27	−0.1 0.2	−0.01 0.15	0 0.16	0.04 0.14	
DC-pot.	10.5 ± 7	0	0	−0.3 1.1	−0.9 1.8	−5.2 4.9	−2.3 4.7	−1.04 2.7	−0.95 1.5	−1.4 2.8	−1.25 3.2	4.6
K^+-act. art.	3.2 ± 0.81	0.03 0.06	0.21 0.43	0.9 0.9	1.2 0.9	3.5 4.7	0.86 1.4	0.35 0.7	0.3 0.7	0.9 2.1	1.59 3.7	3.4 6.0
pH art.												
K^+-act. ven.	7.20 0.1	0 0	−0.01 0.02	−0.04 0.06	0.11 0.11	−0.34 0.2	−0.30 0.2	−0.18 0.2	−0.09 0.19	0 0.1	0.02 0.1	0.05 0.18
pH ven.												

References

Betz, E.: Pial Vascular Smooth Muscle Reaction During Perivascular Microperfusion with Artificial CSF of Varying Ionic Composition. Int. Workshop on Ion-Selective Electrode and on Enzyme Electrodes in Biology and in Medicine, Schloß Reisensburg near Ulm, Germany 1974

Heuser, D., Astrup, J., Lassen, N. A., Betz, E.: Brain Carbonic Acid Acidosis after Acetazolamid. Acta Physiol. Scand. 93 (1975) 385–390

Knabe, U., Betz, E.: The Effect of Varying Extracellular K^+, Mg^{++} and Ca^{++} on the Diameter of Pial Arterioles. In: Vascular Smooth Muscle, pp. 83–85, ed. by *E. Betz.* Springer, Berlin 1972

Kulbertus, H.: Etudes des échanges d'ions H^+ au cours de la contraction des parois carotidiennes soumises à differents agents vasomoteurs. Angiologica 1 (1964) 275–283

Kuschinsky, W., Wahl, M., Bosse, O., Thurau, K.: Perivascular Potassium and pH as Determinants of Local Pial Arterial Diameter in Cats. A microapplication study. Circulat. Res. 31 (1972) 240–247

J. Höper, M. Kessler, W. Simon

Measurements with Ion-selective Surface Electrodes (pK, pNa, pCa, pH) During No-flow Anoxia

(3 Figures)

Electrically neutral ion-carrier ligands for potassium (10), sodium and calcium (1, 2, 12) enabled us to construct new ion-selective electrodes (6) which allow measurements on organ surfaces and thus do not entail extensive tissue lesion. As pointed out by *Kessler* (8), it seems unlikely that real interstitial measurements of ion activity can be performed with microelectrodes. However in organs which have "leaky junctions" instead of tight junctions such as the kidney, the gall bladder and the liver the interstitial cation activities can be monitored with ion selective surface electrodes.

A typical liquid membrane surface electrode is shown in Figure 1. A silver wire is inserted into a small PVC (Trovidur®) cylinder and fixed with a glue. At the tip the silver is coated with a ligand-impregnated PVC membrane. By means of a treatment of the silver surface with tretraphenylborate and with the addition of tetraphenylborate to the membrane phase (12) one is able to produce solid contacted electrodes of high EMF stability. The drift of the ion-selective electrode is about 0.1 mV/hour. For the experimental conditions given 90% of the final signal amplitude is attained within 5 to 10 seconds. As reference a silver chloride electrode was used. The absolute potential of this measuring circuit amounts to about 50 mV in 100 millimolar solutions. The slope of the K^+- and Na^+-electrodes lies within a range of 54–58 mV per logarithmic activity unit and that of the Ca^{++}-electrode to 24–26 mV per logarithmic activity unit. The selectivity corresponds to the recently reported values (1, 2). For pH measurements glass electrodes are used.

Using potassium and sodium sensitive surface electrodes we investigated the extracellular ionic alterations of the isolated and hemoglobin-free perfused rat liver (5, 9) before and after perfusion stop.

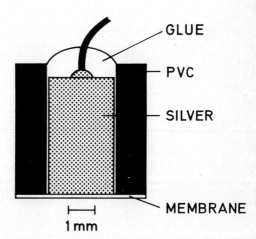

Fig. 1. Scheme of an ion-selective surface electrode. The diameter can be varied. The smallest electrodes are 1 mm in diameter.

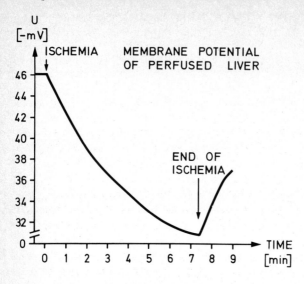

Fig. 2. Changes in extracellular ion activity during no-flow anoxia. Sodium and potassium ion activities measured with ion-selective surface electrodes. pH with glass-needle electrodes.

When no-flow anoxia is induced in the isolated organ (5, 9) typical ionic disturbances are observed (Fig. 2): A biphasic increase in hydrogen ion activity occurs immediately. The first rapid increase is mainly due to the increase in PCO_2 while the second phase is caused by the more gradual accumulation of lactic acid produced by anaerobic glycolysis. The acidosis presumably causes a depolarisation of the cell membranes which induces the subsequent decrease in extracellular sodium activity and the increase in potassium activity (4).

Simultaneous measurements of the DC membrane potential with intracellular microelectrodes showed a resting potential of —46 mV which decreases in 7 min of no-flow anoxia to —31 mV (Fig. 3).

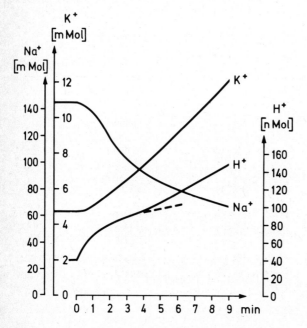

Fig. 3. Measured change in DC-membrane potential during no-flow anoxia.

Measurements with Ion-selective Surface Electrodes (pK, pNa, pCa, pH) During No-flow Anoxia

By applying the Nernst equation as modified by *Hodgkin* and *Horrowicz* (3) and inserting our measured values, we calculated the ratio of the permeability coefficient of sodium to that of potassium.
The calculation was made according to the equation

$$E = -\frac{RT}{F} \ln \frac{[K^+]_i + \alpha [Na^+]_i}{[K^+]_e + \alpha [Na^+]_e}.$$

The potassium and sodium content of perfused liver was determined by flame photometry. The volume of the extracellular space needed for the calculation of intracellular concentration was measured with inulin (13) after freeze stopping to avoid loss of medium. The mean value of extracellular space was $34.3 \pm 2.1\%$ (n = 10) of the total volume.

We calculated an intracellular concentration of 124 mM for potassium and 36 mM for sodium. For extracellular sodium and potassium activities we used our measured values.

By inserting these values in the equation and assuming that the activity coefficient for intracellular sodium is 0.2 as given by *Lev* (10), an α of 0.11 results under normal conditions (Table 1). This α is in good agreement with data of other authors for non-excitable cells (see 14). After 7 min of no flow anoxia when the measured extracellular sodium activity is decreased to 75.2 mM and the potassium activity is increased to 10.35 mM, an α of 0.37 results. This indicates that the change in membrane potential is caused not only by the potassium ions but also by the sodium ions.

Table 1. Values for intra- und extracellular ion activities. The extracellular values are measured with ionselective electrodes. Intracellular ion activities are calculated. The upper values in each square are before, the lower values are after 7 min of no-flow anoxia.

	i	e
K^+ [mMol/l]	124.0	4.7
	123.0	10.3
Na^+ [mMol/l]	7.2	145.2
	14.2	75.2

α (RP) = 0.11
α (7 min) = 0.37

The influx of sodium is not fully compensated by an efflux of potassium. The available quantities of Mg^{++}-, Ca^{++}- and NH_4^+-ions do not suffice to make up the deficiency, so it seems that electrically neutral molecules must also be transferred across the cell membrane to compensate for the extracellular decrease in osmotic pressure.

In apparent support of this concept, electron microscopic investigations of liver tissue show vacuolisation of the cell plasma (6). We assume that these vacuoles may transfer osmotic active organic molecules, e. g. glucose.

Summarizing we can say that during no-flow anoxia (ischemia) accumulation of CO_2 and lactate causes an intra- and extracellular acidosis. The increase in hydrogen ion activity causes a protonation of the plasma membrane which may induce a partial depolarization of the liver cells as observed in our experiments.

References

Ammann, D., Pretsch, E., Simon, W.: Darstellung von neutralen, lipophilen Liganden für Membranelektroden mit Selektivität für Erdalkali-Ionen. Helv. chim. Acta 56 (1973) 1780

Ammann, D., Pretsch, E., Simon, W.: A Sodium Ion-selective Electrode Based on a Neutral Carrier. Anal. Letters 7 (1) (1974) 23

Hodgkin, A. L., Horrowicz, P.: The Influence of Potassium and Chloride Ions on the Membrane Potential of Single Muscle Fibres. J. Physiol. 148 (1959) 127

Höper, J., Kessler, M., Starlinger, H.: Preservation of ATP in the Perfused Liver. In: Oxygen Transport to Tissue, p. 371, ed. by H. J. Bicher and D. F. Bruley. Plenum Press, New York 1973

Kessler, M.: Normale und kritische Sauerstoffversorgung der Leber bei Normo- und Hypothermie. Habil.-Schrift, Marburg/Lahn 1968

Kessler, M., Höper, J., Schäfer, D., Starlinger, H.: Sauerstofftransport im Gewebe. In: Mikrozirkulation, p. 36–52, ed. by S. W. Ahnefeld, C. Burri, W. Dick, M. Halmágyi. Springer, Berlin 1974

Kessler, M., Höper, J., Simon, W.: Methodology and Application of a Multiple Ion Selective Surface Electrode (pH, pK, pNa, pCl) for Tissue Measurements. 58th Annual FASEB Meeting 1974, Feder. Proc. Vol. 33, No. 3 (1974) 279

Kessler, M., Hájek, K., Simon, W.: Four-Barrelled Microelectrode for the Measurements of Potassium, Sodium and Calciumion Activity. This symposium pp. 136

Lang, H.: Die Regulation der Lactat- und Pyruvatkonzentration durch die hämoglobinfreie perfundierte Leber bei unterschiedlichen CO_2- und O_2-Drucken. Diss., Bochum 1970

Lev, A.: Determination of Activity and Activity Coefficients of Potassium and Sodium Ions in Frog Muscle Fiber. Nature 201 (1964) 1132

Pioda, L. A. R., Stankova, W., Simon, W.: Highly Selective Potassium Ion Responsive Liquid-Membrane Electrode. Anal. Letters 2 (12) (1969) 665

Simon, W., Pretsch, E., Ammann, D., Morf, W. E., Güggi, M., Bissig, R., Kessler, M.: Recent Development in the Field of Ion-Selective Electrodes. Pure and Applied Chem. 44 (1975) in press

Williams, J. A., Woodbury, D. M.: Determination of Extracellular Space and Intracellular Electrolytes in Rat Liver in Vivo. J. Physiol. 212 (1971) 85

Williams, J. A.: Origin of Transmembrane Potentials in Non-Excitable cells. J. theor. Biol. 28 (1970) 287

M. Kessler, J. Höper, B. Krumme and H. Starlinger

Disturbance and Compensation of Cellular Cation Activity During Anoxia and in Shock

(6 Figures)

Introduction

Since many years it is well known that immediately after the beginning of ischemia (no-flow anoxia) a fast decrease in energy-rich phosphates occurs (3, 4, 28). The cellular mechanisms which induce this depletion of ATP within a priod of only a few minutes have not yet been explained completely.

The measurements of cation activities in the tissues with recently developed ion-selective electrodes enabled us to monitor directly and continuously the kinetics of cellular ionic disturbanes induced by "normal-flow anoxia" or by "no-flow anoxia" (17). The results of these ionic measurements were applied to develop a new hypothesis which could explain the fast depletion of energy-rich phosphates during ischemia.

Fig. 1. Microscope photometer for fluorometric measurements in perfused organs.

Methods

As a model for our investigations we used the isolated and hemoglobin free perfused rat liver (10, 21). The ionic measurements were performed with both ion selective microelectrodes (23, 15) and surface electrodes (17, 8). NADH fluorescence was monitored with a microscope photometer (see also 23, 24) developed recently (Fig. 1) and tissue Po_2 with multiwire Po_2 electrodes (10, 11, 13, 14).

Results

Figure 2 shows the ATP content of the isolated liver after one hour of normal-flow anoxia (A) respectively after 10 minutes of no-flow anoxia (B). The results indicate clearly that in normal flow anoxia the ATP content remains relatively high whereas in ischemia the well-known fast depletion of ATP occurs, also in the isolated perfused organ.

The intracellular pH during normal-flow anoxia was investigated with the DMO-method and with the Thomas type pH-electrode (12, 18) as shown in Figure 3. The intracellular pH remained constant during 1 hour of anoxia, although the extracellular pH was varied from 6.9 to 7.9 and in spite of the fact that anaerobic glycolysis induced an intracellular production of hydrogen ions. When the extracellular pH was further decreased from 6.9 to 6.7 by addition of HCl to the perfusate, the intracellular pH decreased to 6.9.

It is remarkable that the perfused liver is able to maintain an almost constant intracellular pH under the conditions of anaerobic glycolysis and pronounced extracellular acidosis. This observation strongly supports the evidence that a mechanism may exist which is able to induce an active transport of hydrogen ions across the plasma membrane and thus control the intracellular hydrogen ion activity (see also 16).

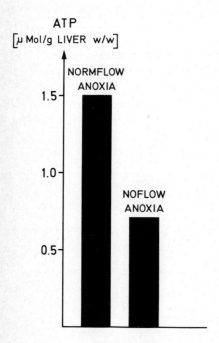

Fig. 2. ATP content of isolated perfused rat liver. The left column represents values after 1 hour of anoxic perfusion (normal-flow anoxia) and the right one those measured after 10 min perfusion stop (no-flow anoxia).

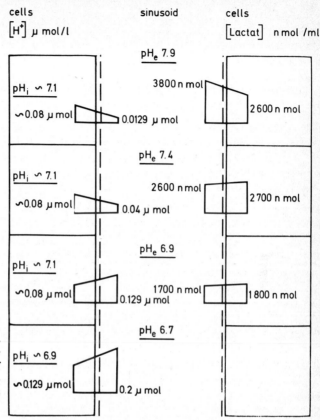

Fig. 3. Intracellular pH of perfused liver during normal flow anoxia and during variation of pH in the perfusate. The intra- and extracellular lactate levels are shown also.

In contrast to the regulation of intracellular pH observed in anoxic perfusion, the intracellular pH decreases to 6.4 or less already after five minutes of ischemia. As described by *Höper* (8) a shift of sodium and potassium develops rapidly after a perfusion stop. It is very puzzling that these ionic disturbances are initiated so early. Figur 4 shows an ischemic reaction in which

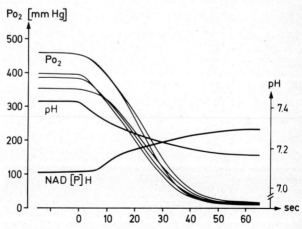

Fig. 4. Tissue values of local P_{O_2}, NAD(P)H and pH during an ischemic reaction.

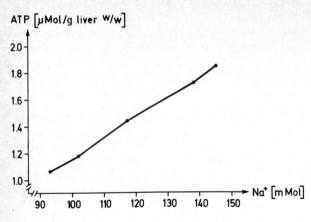

Fig. 5. Relationship between depletion of ATP and sodium influx after the beginning of no-flow anoxia.

oxygen tension of tissue, NAD(P)H fluorescence and pH were recorded simultaneously. As is to be seen, the decrease in pH starts very early when oxygen tension in tissue is still high. Since the intracellular lactate level increases more gradually than CO_2, presumably the initial decrease in pH is caused mainly by CO_2 which accumulates during the period of decreasing tissue oxygen tension.

When we correlate the depletion of ATP to the decrease in extracellular sodium ion activity, an almost linear relationship results (Fig. 5). This suggests that the ischemic decrease of ATP is caused by an activation of the sodium-potassium pump, induced mainly by the influx of sodium into the cell.

Furthermore, during ischemia an increase of calcium activity from 1.24 mM to 2.5 mM is observed. This increase cannot be explained only by the dissociation of the extracellular, protein-bound calcium. Presumably an additional efflux of intracellular calcium is induced by an active transport of calcium.

Discussion

In Figure 6 the different results of our investigations are summarized in a hypothetical scheme: During a period of about 30–40 seconds after the perfusion stop the accumulation of CO_2 causes a fast increase in intra- and extracellular hydrogen ion activity. These hydrogen ions presumably induce the depolarization of the plasma membrane and the subsequent shift of sodium and potassium ions.

An activation of the sodium-potassium pump as well as a direct stimulation of the NaK-ATPase by the sodium influx (29) may occur thus causing the depletion of ATP.

It is still unknown whether or not the formation of vacuoles observed during ischemic reactions (17) also requires energy.

The extracellular acidosis causes a dissociation of the protein-bound calcium. Furthermore, the extracellular calcium activity is increased by an efflux of intracellular calcium. There is evidence that the observed increase in extracellular calcium activity inhibits the further depolarization of the plasma membrane. Concomitantly the ionic shift is slowed down significantly thus protecting the cellular integrity. In this context it may be meaningful that several investigators found an increase in membrane permeability for potassium and sodium when intracellular calcium was raised (6, 9, 19, 20, 22, 26).

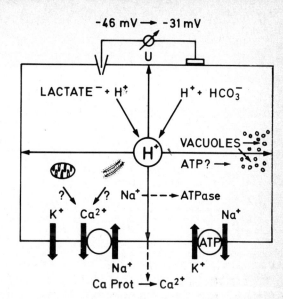

Fig. 6. Hypothetical scheme of cellular reactions occuring during no-flow anoxia.

The assumption that an increase in extracellular calcium activity may decrease the permeability of the plasma membrane for cations during no-flow anoxia is supported by recent observations. Investigations in the isolated perfused liver showed that the potassium efflux which is induced by noflow anoxia is strongly increased when the extracellular calcium is bound by EGTA.

After a period of about 20–30 minutes of ischemia, a sudden decrease in extracellular calcium activity is observed. This might be the instant when cellular necrosis is initiated (see also (5)). In a very careful investigation performed in the dog brain *Schindler* (1974) was able to demonstrate that during extreme hypercapnia with blood pH values of 6.4 a complete depletion of energy-rich phosphates does not occur. This shows that when severe acidosis develops in the tissues the cells are able to maintain their integrity as long as oxygen is available. However, the fact that the creatine phosphate of brain is decreased shows that both the energy requirement as well as the energy production may be disturbed when the activity of protons is so drastically increased.

References

1. *Blaustein, M. P.:* The Interrelationship Between Sodium Calcium Fluxes Across Cell Membranes. Rev. Physiol. Biochem. Pharmacol. 70 (1974) 33
2. *Borle, A. B.:* Calcium Metabolism at the Cellular Levee. Fed. Proc. 32 (1973) 1944
3. *Bretschneider, H. J.:* Überlebenszeit und Wiederbelebenszeit des Herzens bei Normo- und Hypothermie. Verh. dtsch. Ges. Kreisl.-Forsch. 30 (1964/65) 11
4. *Deuticke, B., Gerlach, E.:* Abbau freier Nucleotide in Herz, Skelettmuskel, Gehirn und Leber der Ratte bei Sauerstoffmangel. Pflügers Arch. 292 (1966) 239
5. *Fleckenstein, A., Janke, J., Döring, H. J., Leder, O.:* Myocardial Fiber Necrosis Due to Intracellular Calcium Overload – a New Principle in Cardial Pathophysiology. In: Myocardial Biology, p. 563, ed. by *N. S. Dhalla.* Urban & Schwarzenberg, München 1974
6. *Gardos, C.:* The Role of Calcium in the Perma-

bility of Human Erythrocytes. Acta physiol. Acad. Sci. hung. 15 (1959) 121
7. *Höper, J., Kessler, M., Starlinger, H.:* Preservation of ATP in the Perfused Liver. In: Oxygen Transport to Tissue, Vol. A, p. 371, ed. by *H. I. Bicher* and *D. F. Bruley.* Plenum Press, New York 1973
8. *Höper, J., Kessler, M., Simon, W.:* Measurements with Ion-Selective Surface Electrodes (pK, pNa, pCa, pH) During No-Flow Anoxia
9. *Jansen, J. K. S., Nicholls, J. G.:* Conductance Changes, an Electrogenic Pump and the Hyper-Polarization of Leech Neurones Following Impulses. J. Physiol. (Lond.) 229 (1973) 635
10. *Kessler, M.:* Normale und kritische Sauerstoffversorgung der Leber bei Normo- und Hypothermie. Habil.-Schrift, Marburg 1968
11. *Kessler, M.:* Normal and Critical O_2 Supply of the Liver. In: Oxygen Transport in Blood and Tissue, p. 242, ed. by *D. W. Lübbers, U. C. Luft, G. Thews.* Thieme, Stuttgart 1968
12. *Kessler, M.:* Methodology and Application of pH-sensitive Microelectrodes for Intracellular Measurements. Int. Symposium on Microchemical Techniques, Pennsylvania State Univ. 1973, paper No. 28
13. *Kessler, M., Grunewald, W.:* Possibilities of Measuring Oxygen Pressure Field in Tissue by Multiwire Platinum Electrodes. Progr. Resp. Res. 3 (1969) 147–152
14. *Kessler, M., Lübbers, D. W.:* Aufbau und Anwendungsmöglichkeiten verschiedener Po_2-Elektroden. *Pflügers* Arch. 291, (1966) R 82
15. *Kessler, M., Hájek, K., Simon, W.:* Four Barrelled Microelectrode for the Measurement of Potassium, Sodium and Calcium
16. *Kessler, M., Höper, J., Schäfer, D., Starlinger, H.:* Sauerstofftransport im Gewebe. In: Microcirculation, pp. 36–52, ed. by *S. W. Ahnefeld, C. Burri, W. Dick, M. Halmágyi.* Springer, Berlin 1974
17. *Kessler, M., Höper, J., Simon, W.:* Measurements with Ion-Selective Surface Electrodes (pK, pNa, pCa, pH) During No-Flow Anoxia
18. *Kessler, M., Schmeling, D.:* Methodology of pH Sensitive Microelectrodes for Intracellular Measurements. Pflügers Arch. Suppl. 343 (1973) 128
19. *Kregenow, F. M., Hoffmann, J. F.:* Some Kinetic and Metabolic Characteristics of Calcium Induced Potassium Transport in Human Red Cells. J. gen. Physiol. 60 (1972) 406
20. *Krnjevic, K., Lisieviicz, A.:* Injection of Cylcium Ions into Spinal Motoneurones. J. Physiol. (Lond.) 225 (1972) 363
21. *Lang, H.:* Die Regulation der Lactat- und Pyruvatkonzentrationen durch die hämoglobinfrei perfundierte Leber bei unterschiedliehen CO_2- und O_2-Drucken. Diss., Bochum 1970
22. *Meech, R :* Intracellular Calcium Injection Causes Increased Potassium Conductance in Aplysia Neurones. Comp. Biochem. Physiol. 42A (1972) 493
23. *Neher, E., Lux, H. D.:* Messung extrazellulärer Kalium-Anreicherung während eines Voltage-Clamp-Pulses. Pflügers Arch. Suppl. 332 (1972) R 88
24. *Rahmer, H., Kessler, M.:* Influence of Hemoglobin Concentration in Perfusate and in Blood on Fluorescence of Pyridine Nucleotides (NADH and NADPH) of Rat Liver. In: Oxygen Transport to Tissue, Vol. A, p. 377, ed. by *H. I. Bicher* and *D. F. Bruley.* Plenum Press, New York 1973
25. *Rahmer, H., Kessler, M.:* Die korrigierte Messung der NAD(P)H-Fluoreszenz der Rattenleber in vivo am Modell des hämorrhagischen Schocks. Langenbecks Arch. Suppl. Chir. Forum, (1974) 297–300
26. *Romero, P. J., Whittam, R.:* The Control by Internal Calcium of Membrane Permeability to Sodium and Potassium. J. Physiol. (Lond.) 214 (1971) 481
27. *Schindler, U.:* Stoffwechseluntersuchungen am Cortex ceribri der Katze unter Hypercapnie. Diss., Tübingen 1974
28. *Schmahl, F. W., Betz, E., Dettinger, E., Hohorst, H. J.:* Energiestoffwechsel der Großhirnrinde und Elektroencephalogramm bei Sauerstoffmangel. Pflügers Arch. 292 (1966) 46
29. *Skou, J. C.:* The Influence of Some Cations in the Adenosine Triphosphotase from Peripheral Nerves. Biochem. Biophys. Acta 23 (1967) 394

L. Mela, C. W. Goodwin and L. D. Miller

Mitochondrial Calcium Transport Activity Following Acute and Chronic Changes of Tissue Oxygenation

(4 Figures)

Adaptive responses of mitochondrial enzyme concentrations have been reported to occur during high-altitude acclimation of animals (*Shertzer* and *Cascarano* 1972). Cytochrome oxidase concentration was shown to decrease in heart, liver and kidney mitochondria while the concentrations of dehydrogenases increased (*Reynafarje* and *Green* 1960; *Shertzer* and *Cascarano* 1972; *Tappan* et al. 1957). Our previous studies have indicated that in addition to the changes in cytochrome b c and $(a + a_3)$ concentrations, chronic hypoxia also induced changes in the capacity of mitochondria to perform respiratory reactions linked to oxidative phosphorylation (*Park* et al. 1973). Similar activity changes also occur after acute changes of oxygen availability even in the absence of any alterations of respiratory enzyme concentrations (*Mela* et al. 1973).

The data reported here indicate that also other mitochondrial energy-linked functions, such as Ca^{++} accumulation, show altered capacity after changes of oxygenation both in adult and newborn animals.

Experimental Methods

Acute hypoxia was induced in normoxic male rats by replacing breathing air with 9% oxygen at 1 Atm. (*Mela* et al. 1973). Chronic hypoxia was induced in dogs via an inferior vena cava to right pulmonary vein anastomosis to induce a constant hypoxic level PaO_2 40 mmHg (*Park* et al. 1973). Newborn puppies, who at birth are severely hypoxic, PaO_2 13 mmHg, were studied the first weeks of life under normoxic conditions.

Mitochondria were isolated from rat livers or dog and puppy hearts by conventional methods. Oxygen utilization of the mitochondrial suspensions was determined with a Clark oxygen electrode. Ca^{++} transport activity was measured in a dual wavelength spectrophotometer using murexide as a calcium indicator at 540–507 nm (*Mela* and *Chance* 1968). Cytochrome c and $(a + a_3)$ concentrations of the mitochondrial preparations were determined from the redox changes from fully oxidized to fully reduced level in a dual wavelength spectrophotometer. Cytochrome c was measured at 550–540 nm and cytochrome $(a + a_3)$ at 445–460 nm. Extinction coefficients of 19 $mM^{-1} cm^{-1}$ for cytochrome c and 160 $mM^{-1} cm^{-1}$ for $(a + a_3)$ were used for calculations.

Results and Discussion

Figure 1 indicates rising State 3 respiratory rates and rising Ca^{++} transport rates in rat liver mitochondria after increasing periods of acute hypoxia. Both traces reach a maximum after 40–50 minutes of hypoxia. Thus, both of the energy-linked functions, the oxidative

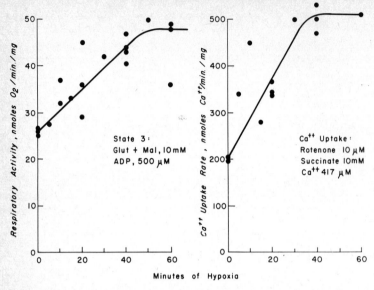

Fig. 1. Measured in rat liver mitochondria. Respiration and Ca^{++} uptake isolated after 0–60 min of in vivo hypoxia, FIO_2 9%.

phosphorylation as well as Ca^{++} transport, show increased capacity after in vivo hypoxia. This phenomenon however, as we well know, does not occur during in vitro hypoxia. It is thus possible that hypoxia removes an inhibitory factor capable of controlling the mitochondrial capacity to perform energy-linked functions in vivo. This hypothetical factor presumably is a cytoplasmic one. The sites of interaction of the factor with the mitochondrial enzyme system are at the level of membrane transport of substrates, cations and adenine nucleotices and/or the dehydrogenase activity. Further experiments to specifically study the point of control are under way.

Figure 2 represents another example of mitochondrial response to changing levels of in vivo oxygenation. The PaO_2 of newborn puppies exposed to air rises rapidly during the first days

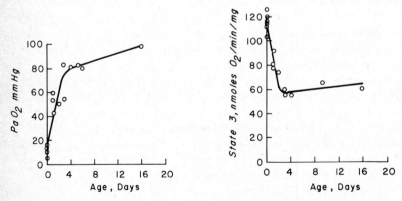

Fig. 2. PaO_2 and cardiac mitochondrial respiration in newborn puppies. Zero time indicates time at birth before breathing air.

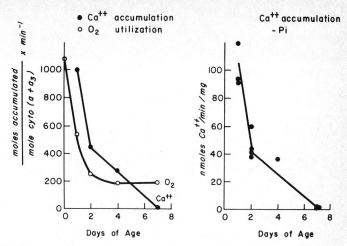

Fig. 3. Heart mitochondrial Ca^{++} accumulation and respiration in newborn puppies zero to nine days of age.

of life. Simultaneously the heart mitochondrial respiratory capacity decreases, a phenomenon opposite to the one induced by lowering the PaO_2. The correlation between the rising PaO_2 and lowering mitochondrial respiration in State 3 is linear. Thus, we can state that *in vivo* the amount of oxygen available to tissues regulates the amount of mitochondrial respiratory activity.

As can be seen in Figure 3, Ca^{++} transport activity of newborn heart mitochondria responds to increased oxygenation in a similar fashion during the first days of life. Mitochondrial cytochrome concentrations change during the same time period. By day seven, cytochrome c increases from a fetal (term) value of 0.25 nmoles/mg protein to the adult level of 0.65 nmoles/mg protein. Cytochrome oxidase rises from an extremely low fetal (term) value of 0.065 nmoles/mg protein to 0.14 nmoles/mg at day seven. This is still considerably below the adult level in dog heart mitochondria, 0.21 nmoles/mg protein.

Taking into account the changes in the cytochrome $(a + a_3)$ concentration, the activity of both State 3 and Ca^{++} transport per cytochrome decreases dramatically during the first seven days of life. During the next few weeks both the State 3 and Ca^{++} transport activities of heart mitochondria slowly reach adult levels.

When adult dogs were made chronically hypoxic, a response different from State 3 respiration was observed in the heart mitochondrial Ca^{++} transport activity during the acute phase of hypoxia, as seen in Figure 4. As we have reported earlier (4) State 3 respiratory capacity increases during the first days of hypoxia and stabilizes after about five days at a new level, 40% above control. Ca^{++} transport activity, however, decreases sharply during the first days of hypoxia, and by the end of the first week, returns to normal control level.

At the present time, it is difficult to understand the exact physiological meaning of the mitochondrial respiratory and Ca^{++} transport capacity changes induced by *in vivo* hypoxia. They might play an important role in the metabolic adaption of tissue to new levels of oxygenation.

The interesting question of how and through what kind of a cellular or intramitochondrial control mechanism the mitochondrial adaptation to changing oxygen concentrations occurs,

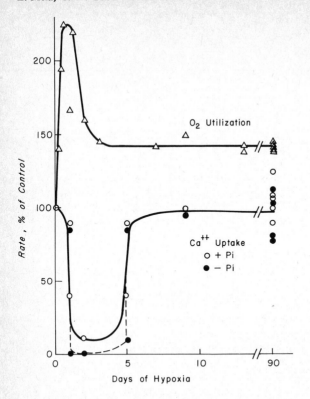

Fig. 4. Dog heart mitochondria during chronic hypoxia.

remains unanswered. An answer to this question would provide us invaluable information of the still poorly understood *in vivo* control of mitochondrial enzymatic activity.

Acknowledgements

These studies were supported by USPHS grant GM-19867 and Career Development Award GM-50318 to Leena Mela. The expert technical assistance of Mrs. *Agnes Knee* and Mr. *Joseph Campbell* are greatly appreciated.

References

Mela, L., Chance, B.: Spectrophotometric Measurements of the Kinetics of Ca^{2+} and Mn^{2+} Accumulation in Mitochondria. Biochemistry 7 (1968) 4059.

Mela, L., Miller, L. D., Bacalzo, L. V., Olofsson, K., White, R. R.: Role of Intracellular Variations of Lysosomal Enzyme Activity and Oxygen Tension in Mitochondrial Impairment in Endotoxemia and Hemorrhage in the Rat. Ann. Surgery 178 (1973) 727.

Park, C. D., Mela, L., Wharton, R., Reilly, J., Fishbein, P., Aberdeen, E.: Cardiac Mitochondrial Activity in Acute and Chronic Cyanosis. J. Surg. Res. 14 (1973) 139.

Reynafarje, B., Green, J.: Pyridinenucleotide-Cytochrome c Reductases in Rats Exposed to Low Oxygen Tensions. Proc. Soc. Exp. Biol. Med. 103 (1960) 224.

Shertzer, H. G., Cascarano, J.: Mitochondrial Alterations in Heart. Liver and Kidney of Altitude-Acclimated Rats. Amer. J. Physiol. 223 (1972) 632.

Tappan, D. V., Reynafarje, B., Potter, V. R., Hurtado, A.: Alterations in Enzymes and Metabolites Resulting from Adaption to Low Oxygen Tensions. Amer. J. Physiol. 190 (1957) 93.

T. Zeuthen and C. Monge

Electrical Potentials and Ion Activities in the Epithelial Cell Layer of the Rabbit Ileum in vivo*)

(1 Figure)

Theories about the intra- and pericellular environment of the epithelial cells of the intestine during transport of sodium and water from the lumen of the gut into the blood have been developed primarily from observations on the changes occurring in the solutions bathing the luminal and serosal surfaces of the organ. Furthermore the reported values of the intracellular electrical potential as recorded with microelectrodes are conflicting: e.g., *Field* and *Curran* (unpublished observations, quoted by *Schultz* and *Curran* 1968) reported potentials of —10 to —15 mV in the rabbit ileum *in vitro* as opposed to *Rose* and *Schultz* (1971), who reported —36 mV in the same preparation. We therefore decided to determine *in vivo* the profile simultaneously of electrical potential and ion activities for potassium (K^+) and chloride (Cl^-) across both the mucus and epithelial cell layers and to correlate these profiles with the microanatomy of the tissues by means of iontophoretic staining through the electrode.

In the ileum of the rabbit the route of transport from the lumen is first across the mucus layer, which consists of mucopolysaccharides coating the inner surface of the intestine, then across the epithelial cell layer and finally into the capillary system or lymphatic drainage. In the terminal ileum of the rabbit the epithelial cells are about 40 μm long and have a diameter of roughly 5 μm. They are closely packed in a columnar fashion with their mucosal ends facing the lumen of the intestine and their serosal ends abutting the basement membrane facing the underlying capillaries, lymphatics and connective tissue. When Krebs' or similar solutions are placed in the lumen it is well established (see *Edmonds* 1970) that the solution is transported into the underlying tissues. During this transport the sodium activity in the gut lumen remains constant, the chloride activity decreases and the bicarbonate activity increases.

Methods

Microelectrodes. Three types of double-barrelled microelectrodes were used. In all three, one of the barrels was used to record the electrical potential, and the other was used either for iontophoretic staining of the recording site or for determining potassium or chloride activity when filled with the appropriate liquid ion changer (Corning 477317 potassium exchanger or 477315 chloride exchanger). Staining was done either by lithium carmine (*Villegas* 1962) or by Procion Yellow MHRS (*Stretton* and *Kravitz* 1968). The ion-selective microelectrodes were prepared as described by *Zeuthen* et al. (1974). The reference barrel was filled with either 2M KCl or 2M NaCl and in a few experiments with M NH_4NO_3. The impedance of the reference barrel was about $7-10 \times 10^6$ ohm when measured in 2M KCl and had a tip potential that varied less than 3 mV in different isotonic solutions of a strength of 150 mM mixed from NaCl and KCl. The electrodes had a tip diameter of less than 0.6 μm.

---

*) This study was initiated at A.R.C. Institute of Animal Physiology, Babraham, and continued at the Department of Pathology, University of Bristol.

Surgical procedure. In White New Zealand rabbits anaesthetized with sodium pentobarbitone an incision 5 cm long was made in the midline of the abdominal wall into the peritoneal cavity. With the rabbit lying on one side a 20 cm loop of the terminal ileum was taken out and supported on a Perspex® plate warmed to 38 °C. A slit 6 mm long was made through the upper side of the serosa of the intestine so that a perspex tube of 5 mm outer diameter with a wall thickness of 1 mm could be passed vertically through the slit and pressed against the opposite mucosal side of the gut, the serosal side of which was resting on the perspex plate. The position of the tube was adjusted so that the blood supply to the tissue was maintained and observed through the heated Perspex® plate. In this position the seal between the tube and the tissue prevented leakage of the mucosal solution (usually 145 mM NaCl, 5 mM KCl buffered to pH 7.8 with phosphate buffer, or Krebs' solution) which was placed in the tube. Microelectrodes were advanced into the tissue via the tube. If peristaltic movements disturbed the microelectrode recordings, incisions were made in the muscle layers. During a 3-hour period the tissue looked healthy, as determined from direct microscopic and subsequent histological examination. Net absorbtion of sodium and water by this tissue *in vivo* was determined essentially as described by *Visscher* et al. (1944).

Results and discussion

In about half of 92 animals the electrode passed first through an identifiable layer of up to 50 µm thickness, probably the mucus layer, before it reached the cells. In this layer potassium activity increased towards the tissue, i.e., it had the activity of the luminal solution (5 mM) in its outer parts and typically increased to 25–30 mM adjacent to the cell surface. The electrical potential was a few mV positive in this layer. This gradient of potassium in the mucus layer shows that potassium would be transported into the lumen when the intra-luminal potassium activity was low. However naturally occurring fluids in the terminal ileum contain 20–25 mM potassium (*Edmonds* 1970), which corresponds to the values we found at the cell surfaces. When the tip of an electrode was advanced from the lumen through the layer of epithelial cells and into the underlying tissues a series of increasingly negative steps in potential were recorded with the reference barrel (Fig. 1, lower trace). This was finally followed by an abrupt shift towards small negative values. The electrode was advanced at a rate of about 2 µm/sec until a sudden negative step was recorded when the advance was stopped for 10–20 seconds to record the potential. When Procion Yellow was injected after every second or third change in potential it stained cells which appeared in subsequent histological preparations of the epithelial layer at intervals separated by at least one cell diameter. Thus each negative step corresponded to the intracellular electrical potential of an epithelial cell. Stain deposited after the potential had returned to near zero values was found to be located in the tissues below the basement membrane. Similar step-wise profiles of decreasingly negative potential were recorded when the electrode was advanced from the serosal side, through the epithelium into the lumen. If the solution at the lumen side was stirred the potentials which were recorded remained unchanged. When sulphate ions were substituted for chloride ions in the mucosal solution, or potassium activity was increased to 50 mM and sodium activity decreased to 100 mM, step-wise profiles were still obtained.

With 2 M KCl in the reference barrel the first negative potential recorded as the electrode was advanced into the epithelial cells from the luminal side was -5.0 mV ± 0.29 ($\pm$ S.E.M. 9 recordings from 6 animals) and the final negative potential of -34.4 mV ± 2.7 was recorded

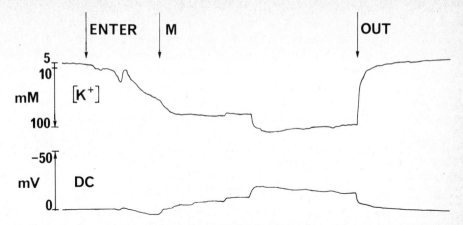

Fig. 1. The potassium activity (K^+) and the electrical potential (DC) as recorded by a microelectrode advanced into the rabbit ileum epithelial cells. The electrode was advanced across the mucus layer between ENTER and M and continued at a speed of about 2 µm/sec. until a stepwise negative increase in potential was recorded. At each step the advance was arrested for 10–20 seconds. Each plateau of electrical potential is recorded intracellularly as ascertained by iontophoretic staining. In this experiment the electrode was pulled OUT into the lumen when 4 cells had been impaled. The last cell impaled was held for about 60 sec.

just before the electrode entered the underlying tissues. With 2M NaCl in the barrel the first negative potential was —6.6 mV ± 1.0 (7 recordings from 4 animals) and the final was —43.9 mV ± 3.8. The differences in potentials obtained with the different filling solutions were not significant: Student's t-test; $p < 0.1$ in the luminal end of the cell, $p < 0.06$ in the distal end. When the tips of the electrodes were deep to the epithelial cell layer both types of probe recorded —6.5 mV ± 4.9 (23 recordings from 6 animals). In a few experiments where the reference barrel was filled with $M NH_4NO_3$ we obtained potentials similar to those above. The intracellular potassium activity recorded with ion-selective double-barrelled electrodes also depended on the depth of penetration: each stepwise decrease in electrical potential was associated with a stepwise increase in the potassium activity (Fig. 1, upper trace). If the reference barrel was filled with 2M KCl we measured a potassium activity of 40–50 mM (about 30 mM if the preparation had no mucus layer) in the mucosal end of the cell. In the distal end of the cell we obtained a maximum average of 160 mM (sometimes as high as 200 mM). In the underlying tissues we obtained a mean of 11 mM. If the reference barrel was filled with 2M NaCl similar results were obtained, although the recorded potassium activity tended to decline after its initial stepwise increase. Finally in two animals the chloride activity and electrical potential were recorded across the epithelial cell layers. Each increasingly negative step in potential was associated with a stepwise decrease in chloride activity: a maximum of roughly 80 mM was obtained intracellularly close to the lumen and a minimum of about 10 mM was obtained near the basement membrane. Extracellularly, deep to the basement membrane a hyperosmolarity of roughly 300 mM was obtained. In these experiments the reference barrel was filled with 2M KCl.

These findings can be explained if there exists intracellularly along the length of each epithelial cell a gradient of electrical potential, of potassium activity and of chloride activity: the cells are depolarized at the end facing the lumen, with a low membrane pontential, low potassium activity and probably a high chloride activity, while they become more polarized towards their serosal ends. If these measurements of chloride activity are confirmed by other experiments with solutions in the reference barrel which do not contain chloride, or with electrodes of a smaller tip diameter, then the extracellular environment around the serosal end of the cells is hypertonic to that of blood serum.

The gradient of electrical potential along the length of one individual cell can now be reconstructed, it was constant: -0.85 mV/μm (U.C.L.* $= -0.95$ mV/μm, L.C.L.** $= 0.73$ mV/μm, n $= 94$). In those preparations in which there was a pronounced mucopolysaccharide layer the gradient of chemical potential of the potassium ion was about equal but opposite in magnitude to the gradient of electrical potential, which means that there is no net force acting on the potassium ions here. However towards the distal end of the cell, over a distance of about 10 μm, the gradient of the chemical potential of potassium increased to about 1.5 mV/μm close to the basement membrane. This means that a net force is exerted here on the potassium ions and the net flux of potassium towards the lumen can be calculated from the *Nernst-Planck* equation to be $2.6 \cdot 10^{-7} \frac{\text{Mol}}{\text{sec cm}^2}$ (*Bockris* and *Reddy* 1972) if the intracellular diffussion constant for potassium is the same as that in water. Since there was no net flux of the ions in the luminal end of the cell and as the luminal potassium activity increases at a rate several orders of magnitude smaller than would result from the flux above, the flux of potassium in the distal end of the cell must proceed across the cell wall into the extracellular space and back into the underlying tissues.

In preparations with no mucus layer there tended to be a net electrochemical gradient for potassium at the luminal as well as at the serosal end with movement of potassium into the lumen.

The nature of the intracellular gradients can be explained tentatively as a diffusion potential. If, for example, a 150 mM solution of KCl is separated from a solution of 100 mM NaCl plus 50 mM KCl by a membrane in which (a) potassium moves freely, (b) sodium mobility is only one fifth of its value in water and (c) most of the negative charges are fixed, a diffusion potential of 30 mV will develop inside the membrane, the potassium-rich side being negative, as estimated from the equations of *Planck* (1890).

Our values for the net fluxes of sodium from the lumen into the blood $3.0 \cdot 10^{-9} \frac{\text{Mol}}{\text{sec cm}^2} \pm 0.3$, n $= 4$ and of water $2.0 \cdot 10^{-8} \frac{1}{\text{sec cm}^2} \pm 0.2$, n $= 4$ are about 4 times as large as those obtained from the ileum *in vitro* (*Schultz* and *Zalusky* 1964). Even so our data of the intracellular environment are compatible with the data from some *in vitro* studies. An intracellular potential of -36 mV was recorded by *Rose* and *Schultz* (1971) from the rabbit ileum *in vitro*, while potentials of -10 to -15 mV have also been observed in this preparation by *Field* and *Curran* (unpublished observations, quoted by *Schultz* and *Curran* 1968). *Lyon* and *Sheerin* (1971) obtained about -10 mV in the jejunum and ileum of the rat, which is similar to the value *Barry* and *Eggenton* (1972) found in rat jejunum. In neither of these studies however were attempts made to correlate the recording site with the microanatomy of the tissue. It is therefore possible that *Rose* and *Schultz* systematically recorded closer to the serosal end of the cell than the other workers. *Schultz* et al. (1966) reported an

---

*) U.C.L.: Upper confidence limit.
**) L.C.L.: Lower confidence limit.

average intracellular potassium concentration of 140 mM and *Frizzell* et al. (1973) reported an average chloride activity of about 67 mM; these values are slightly higher than the averaged values we obtained from our study. *Lee* and *Armstrong* (1972) obtained a potassium activity of 85 mM in the bullfrog small intestine, but a membrane potential of —40 mV.
The existence of hyperosmolar areas is in agreement with current theories on water transport (*Curran* 1960; *Diamond* 1964). The possibility of an intracellular gradient of ion activities in the epithelium of the gut has been hypothesized by *Lindemann* and *Pring* (1969) in relation to models of standing gradients of solutes in the extracellular spaces (*Diamond* and *Bossert* 1967).
If the potential of the lumen was taken as zero mV we recorded on the average a slightly negative potential behind the epithelial cell layer where as most absorbing epithelial tissues (see review by *Schultz* and *Curran* 1968) have a slightly positive potential when recorded between two large electrodes placed in solutions bathing the lumen and the serosal side of the intestine. However the two measurements are not directly comparable since we did not attempt to determine whether the potentials recorded by the microelectrodes were recorded close to or inside capillaries, lymphatics or other tissues. Thus the mean value of the potentials inside the capillaries might still be positive.

References

Barry, J. C., Eggenton, J.: Membrane Potentials of Epithelial Cells in the Rat Small Intestine. J. Physiol. 227 (1972) 201–216

Bockris, J. O. M., Reddy, A. K. N.: Modern Electrochemistry, Vol. 1, pp. 622. Plenum Press, New York 1970.

Curran, P. F.: Na, Cl and Water Transport by Rat Ileum *In Vitro*. J. Gen. Physiol. 43 (1960) 1137–1148

Diamond, J. M.: The Mechanism of Isotonic Water Transport. J. Gen. Physiol. 48 (1964) 15–42

Diamond, J M., Bossert, W. H.: Standing Gradient Osmotic Flow. A Mechanism for Coupling of Water and Solute Transport in Epithelia. J. Gen. Physiol. 50 (1967) 2061–2083

Edmonds, C. J.: Water and Ionic Transport by Intestine and Gall Bladder. In: Membranes and Ion Transport, Vol. 2, pp. 79–110, ed. by E. E. Bittar. Wiley-Interscience, New York 1970

Frizzell, R. A., Nellans, H. N., Rose, R. C., Markscheild-Kaspi, L., Schultz, S. G.: Intracellular Cl-Concentration and Influxes Across the Brush Border of Rabbit Ileum Amer. J. Physiol. 224 (1973) 328–337

Lee, C. O., Armstrong, W. McD.: Activities of Sodium and Potassium Ions in Epithelial Cells of Small Intestine. Science 175 (1972) 1261–1263.

Lindemann, B., Pring, A.: A Model of Water Absorbing Epithelial Cells with Variable Cellular Volume and Variable Width of the Lateral Intercellular Gaps. Pflügers Archiv. 307 (1969) R 55–R 56

Lyon, I., Sheerin, H. E.: Studies on Transmural Potential *In Vitro* in Relation of Intestinal Absorbtion. IV. The Effect of Sugars on Electrical Potential Profiles in Jejunum and Ileum. Biochim. biophys. Acta 249 (1971) 1–14

Planck, M.: Über die Potentialdifferenz zwischen zwei verdünnten Lösungen binärer Electrolyte. Ann. Physik (3) 40 (1890) 561–576

Rose, R. C., Schultz, S. G.: Studies on the Electrical Potential Profile Across Rabbits' Ileum. J. Gen. Physiol. 57 (1971) 639–663

Schultz, S. G., Curran, P. F.: Intestinal Absorbtion of Sodium Chloride and Water. In: Handbook of Physiology, Alimentary Canal, Vol. III, pp. 1245–1275, ed. by C. F. Code 1968

Schultz, S. G., Fuisz, R. E., Curran, P. F.: Amino Acid and Sugar Transport in Rabbit Ileum. J. Gen. Physiol. 49 (1966) 849–866

Schultz, S. G., Zalusky, R.: Ion Transport in Isolated Rabbit Ileum. I. Short Circuit Current and Sodium Flux. J. Gen. Physiol. 47 (1964) 567–584

Stretton, A. O. W., Kravitz, E. A.: Neural Geometry: Determination with a Technique of Intracellular Dye Injection, Science 162 (1968) 132–134

Villegas, L.: Cellular Location of the Electrical Potential Difference in Frog Gastric Mucosa. Biochim. biophys. Acta 64 (1962) 359–367

Visscher, M. B., Fletcher, E. S., Carr, C. W., Gregor, H. P., Bushey, M. S., Barker, D. E.: Isotope Tracer Studies on the Movement of Water and Ions Between Intestinal Lumen and Blood. Amer. J. Physiol. 142 (1944) 550–575

Zeuthen, T., Hiam, R. C., Silver, I. A.: Recording of Ion Activities in the Brain. In: Ion Selective Microelectrodes, pp. 202, ed. by *H. Berman* and *N. Herbert.* Plenum Press, New York 1974

J. L. Walker and R. O. Ladle

Intracellular Chloride Activity in Heart Muscle

(1 Figure)

Intracellular chloride activity (a_{Cl}^i) has been measured in frog ventricle with chloride microelectrodes. The fabrication and calibration of the electrodes have been described previously (*Walker*, 1971). A strip of frog ventricle approximately 1 mm wide and 1 cm long is pinned to the bottom of a tissue bath with the endocardial surface up. The cells are then penetrated with microelectrodes attached to micromanipulators.

To calculate a_{Cl}^i it is necessary to make measurements with two electrodes: a chloride electrode and a 3 M KCl filled electrode. The reason for this can best be seen by examination of Equation 1.

$$\Delta E = E_M - \frac{nRT}{F} \log_e \frac{a_{Cl}^i}{a_{Cl}^0} \tag{1}$$

ΔE(mv) is the difference in potential between the inside and outside of the cell as measured with the chloride electrode. E_M(mv) is the membrane potential, in this case the diastolic membrane potential. R, T and F have their usual meaning. n (dimensionless) is an empirical constant that is necessary because the slope of the electrodes is frequently less than *Nernstian*. The value of n is between 0.95 and 1.0. Before a_{Cl}^i can be calculated, equation 2, E_M must be measured with a 3 M KCl filled micropipette.

$$a_{Cl}^i = a_{Cl}^0 \exp \frac{(\Delta E - E_M) F}{nRT} \tag{2}$$

Frog ventricular cells are so small, 5–7 μm diameter, that it is not possible to simultaneously record from the same cell with both electrodes. Our solution to this problem is to make about five measurements with each of the electrodes and use the mean values of ΔE and E_M to calculate a_{Cl}^i. The results of twelve such experiments are presented in Table 1. The entries are mean values and the standard errors of the means.

Table 1.

ΔE (mv)	E_M (mv)	a_{Cl}^i (mM/L)	E_{Cl} (mv)
-50.4 ± 0.53	-90.4 ± 0.61	17.6 ± 0.57	-41.9 ± 0.81

The value of a_{Cl}^i is too high for passive distribution of chloride at the diastolic membrane potential as it is in skeletal muscle. The high value of a_{Cl}^i raises the question of whether this is an apparent rather than a true value due to the contribution to the electrode potential of other anions inside the cells. This concern arises because the chloride electrode is not very selective with respect to other anions, including organic anions. As a control experiment, a_{Cl}^i was measured in the same piece of ventricle with both a liquid ion exchanger chloride electrode and a Ag-AgCl microelectrode of the design described by *Neild* and *Thomas* (1973). The results of this experiment are shown in Table II. Values for a^i_{Cl} are not included because no E_M measurements were made. It is sufficient, since the slopes of the two electrodes were the same, that the two ΔE values are not significantly different.

Table 2.

	Number of Penetrations	Electrode Slope (mv)	Average ΔE (mv)	Standard Error (mv)
Liquid Ion Exchanger Cl⁻	7	56	−57.4	1,29
Ag-AgCl	5	56	−59.8	2.58

Having established that the liquid ion exchanger chloride micro-electrode does indeed read the true value of a^i_{Cl} in this preparation, the next step was to test the hypothesis of *Hutter* and *Noble* (1961). Their hypothesis states that chloride is passively distributed with the chloride equilibrium potential, E_{Cl}, equal to the mean membrane potential during the cardiac cycle. The hypothesis can be tested in two ways.

The mean membrane potential can be changed and if the hypothesis is correct, a^i_{Cl} should change until E_{Cl} becomes equal to the new mean membrane potential. This experiment was done by allowing the ventricle strips to remain quiescent in which case the mean membrane potential becomes equal to the resting membrane potential, E_M. In six preparations which were quiescent for up to 18 hours, there were no significant changes from the control values of a^i_{Cl}. The results of these experiments are presented in Table III. The entries for E_M and a^i_{Cl} are the mean values and the standard errors of the means. The entries in the time column are the

Table 3.

	Time (hr)	E_M (mv)	a^i_{Cl} (mM/L)
Control		−94.6 ± 2.3	16.5 ± 1.8
	3.5	−98.4 ± 2.9	13.5 ± 1.8
	9	−97.8 ± 3.4	21.2 ± 3.2
Control		−85.6 ± 1.3	24.1 ± 2.2
	3	−89.9 ± 2.1	22.3 ± 2.7
Control		−91.6 ± 1.1	20.0 ± 1.6
	3	−96.9 ± 1.2	16.3 ± 1.1
Control		−86.7 ± 1.9	18.4 ± 2.3
	3	−88.6 ± 1.9	17.6 ± 1.5
Control		−90.2 ± 2.2	17.1 ± 2.0
	3	−94.1 ± 2.3	17.6 ± 1.9
Control		−92.6 ± 1.9	16.8 ± 1.4
	18	−93.0 ± 1.7	17.5 ± 1.6

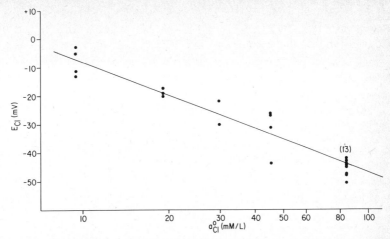

Fig. 1. Each point represents an experiment in which the chloride in the Ringer's solution was replaced by glucuronate. The points for $a_{Cl}^0 = 90$ mM/L are the controls (100% Cl$^-$) for the points for which $a_{Cl}^0 < 100\%$. The number 13 above the control point denotes the fact there are 13 control values, one for each experiment. The values of E_{Cl} were calculated from the known values of a_{Cl}^0 and the measured values of a_{Cl}^i. The line is the best least squares fit to the points, it has a slope of -38.5 mv and a correlation coefficient of 0.97. The ordinate is E_{Cl} (mv) and the abscissa is a_{Cl}^0 on a logarithmic scale.

lengths of time that the preparations had been quiescent when the measurements were made. The second way in which the hypothesis can be tested is to change E_{Cl} by altering a_{Cl}^0, and if the hypothesis is correct, a_{Cl}^i should change so as to return E_{Cl} to its original value. a_{Cl}^0 was changed by replacing chloride with glucuronate which, at least in frog skeletal muscle, is an impermeant anion (*Woodbury* 1973). The selectivity coefficient for glucuronate with respect to chloride has not been determined, but when 90% of the chloride in frog Ringer's is replaced with glucuronate, the chloride electrode measures the same chloride activity as a Ag-AgCl electrode. These experiments were carried out with quiescent preparations to test whether the results presented in Table III could be explained on the basis of low resting chloride permeability.

Within one hour after a_{Cl}^0 is reduced, a_{Cl}^i reaches a new steady state value and remains there for up to 24 hours. When a_{Cl}^0 is restored to normal, a_{Cl}^i returns to the control value within one hour. There is no significant change in E_M when chloride is replaced with up to 90% glucuronate. Figure 1 shows the results of these experiments. E_{Cl}, calculated from the measured a_{Cl}^i, is plotted as a logarithmic function of a_{Cl}^0. The line, fit to the points by a linear regression, has a slope of -38.5 mv and a correlation coefficient of -0.97. If the hypothesis of *Hutter* and *Noble* is correct, the slope of the line fitting the points should be zero. The -38.5 mv slope is significantly different from zero at the 0.005 level.

Our conclusion is that the *Hutter* and *Noble* hypothesis is invalid for frog ventricle and that chloride is actively transported into these cells.

Acknowledgement

This work was supported by a grant-in-aid from the American Heart Association.

References

Hutter, O. F., Noble, D.: Anion Conductance of Cardiac Muscle. J. Physiol. (London) 157 (1961) 328–334

Neild, T. O., Thomas, R. C.: New Design for a Chloride-Sensitive Microelectrode. J. Physiol. (London) 231 (1973) 7P–8P

Walker, J. L.: Ion Specific Liquid Ion Exchanger Microelectrodes. Anal. Chem. 43 (1971) 89A–92A

Woodbury, J. W.: Personal Communication 1973

J. F. White*) and J. A. M. Hinke

Use of the Sodium Microelectrode to Define Sodium Efflux and the Behavior of the Sodium Pump in the Frog Sartorius**)

(9 Figures)

About two years ago, we began to examine the behavior of the Na pump in the frog sartorius muscle fiber by means of the Na-sensitive microelectrode, much like *Thomas* (1972) did in the snail neutrone. In addition, we combined Na microelectrode measurements with isotope efflux studies in order to better identify that fraction of total Na efflux from the whole muscle which belongs to the membrane pump. One of our initial aims was to confirm whether the pump rate in the frog sartorius could be related to the cube of the internal Na concentration (*Keynes* and *Swan* 1959; *Mullins* and *Frumento* 1963).

The Thomas-type microelectrode design (Fig. 1) was selected because it could measure intracellular Na activity (^{a}Na)$_i$ after a small penetration of the 1–2 μm tip beyond the membrane. The degree of membrane damage by this microelectrode is comparable to that produced by a conventional open-tipped microelectrode in the frog sartorius fiber. Figure 2 provides some representative voltage recordings from the Na microelectrode during penetration into the fiber. In A, B, and C, one can see that the response time of the electrode after the initial penetration varied from 1 to 3 minutes; in B, one can also see that the intrafiber potential varied sometimes with increasing increments of penetration. This latter behavior is probably not a property of the Na microelectrode since we observed similar behavior with open-tipped (3 M KCl) micropipettes. In about half the experiments to be reported, the Na microelectrode

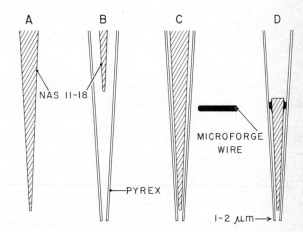

Fig. 1. Diagrams showing construction of Na microelectrode with recessed Na-sensitive glass from the 1–2 μm open tip.

---

*) Holder of a Public Health Service postdoctoral fellowship.
**) This research was supported by the Medical Research Council of Canada (MRC 1039).

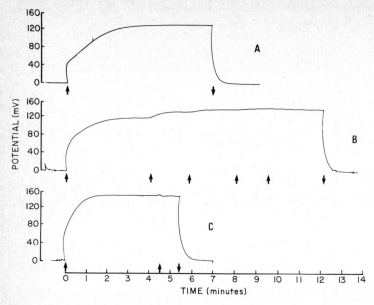

Fig. 2. Sample voltage recordings (A, B and C) from a Na microelectrode inside a muscle fiber (referred to a bath reference electrode via a saturated KCl bridge). At each upward arrow, the microelectrode was advanced slightly into the fiber; at each downward arrow, the microelectrode was removed from the fiber.

was referred to a 3 M KCl micropipette (containing a Ag-AgCl wire) rather than to a large conventional reference bath electrode. As shown in Figure 3, if such a reference micropipette is inserted into the same fiber containing the Na-sensitive microelectrode one obtains a direct measure of the intrafiber Na$^+$ activity.

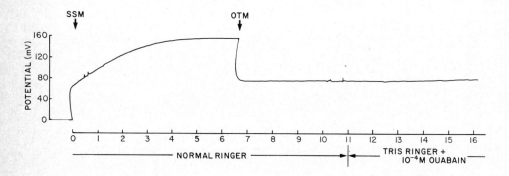

Fig. 3. Sample voltage recording from a Na microelectrode inside a fiber and referred to a conventional open-tipped micropipette electrode. At SSM, the Na-sensitive microelectrode was advanced into the fiber; at OTM, the open-tipped micropipette was advanced into the same fiber. The change in voltage is a measure of membrane potential.

Sodium Activity and Sodium Efflux

Figure 4 is a typical example of how the free intrafiber Na activity (^{a}Na)$_i$ decreases when the external bath is changed from normal Ringer's solution to one in which all the Na has been replaced by TrisCl (tris-hydroxymethyl aminomethane). Obviously, the change in (^{a}Na)$_i$ with time cannot be described by a single exponential function.

The two curves in Figure 5 summarize the mean changes in total muscle Na (curve A, ^{22}Na loss) and in intrafiber Na$^+$ (curve B, Na microelectrode) following replacement of external Na$^+$. The most important observation to make here is that relatively large amounts of intrafiber Na are being lost quite early, in a time period usually reserved for extracellular Na$^+$ only.

Following convention, we have described curves A and B by the method of the sum of exponential functions. The results of such an analysis are summarized in Table 1. Observe first that curve A requires three rate constants ($k_1 = 0.13$, $k_2 = 0.03$ and $k_3 = 0.006$ min^{-1}). Observe secondly that the rate constants for curve B are similar in magnitude to the largest and smallest rate constant for curve A.

When curve B is subtracted from curve A, the resultant can be satisfactorily defined by two exponential functions with rate constants (Table 1) similar in magnitude to the fast and intermediate rate constants of curve A. It would seem quite reasonable to define the movements from the extracellular space as the difference between curves A and B. If so, then we must identify two extracellular spaces: one which contains about 14% of the muscle water and releases Na rather quickly ($k_i = 0.15$ min^{-1}) and another which contains about 7.5% of

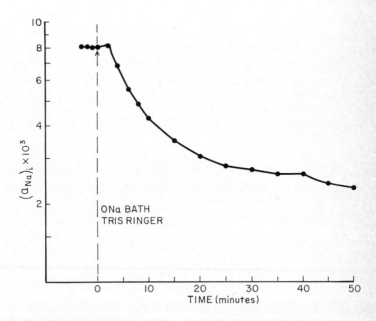

Fig. 4. Typical example of the decrease in intrafiber Na$^+$ activity (^{a}Na)$_i$ when the Na$^+$ ion in the bath Ringer's solution is replaced by tris (hydroxymethyl) aminomethane. The logarithmic scale is used on the ordinate. Comparable results were obtained when choline$^+$ or Li$^+$ was used to replace bath Na$^+$.

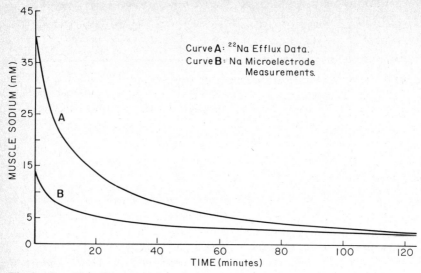

Fig. 5. Mean Na⁺ efflux curves from frog muscle following Na⁺ bath replacement at t = o. Curve A represents mean Na⁺ loss (as measured by ^{22}Na⁺ loss) from 4 isotope-loaded whole muscle preparations. Curve B represents mean intrafiber Na⁺ loss (as measured by the Na⁺ microelectrode) from 4 fibers from different muscles. Total muscle Na⁺ in Curve A was measured by atomic absorption spectroscopy on companion muscles. Total intrafiber Na⁺ in Curve B (at t = o) was made equal to $\frac{.75 \, (^aNa)_i}{.5}$ at t = o), where .75 is the fraction of muscle water located inside the fibers and .5 is the apparent intrafiber activity coefficient found experimentally.

the muscle water and releases Na rather slowly ($k_2 = 0.03$ min^{-1}). It is worth mentioning that the ^{14}C-sorbitol space determined on companion muscles to those used for the ^{22}Na flux study was found to be about 22% of total muscle water. This estimation of extracellular space agrees rather well with the one obtained (14%+7.5%) from curve analysis.

Table 1. Analysis of Na efflux curves
$(Na)_F = A_1 e^{-k \cdot 1 \cdot t} + A_2 e^{-k \cdot 2 \cdot t} + A_3 e^{-k \cdot 3 \cdot t}$

Curve	Initial content*) (mmoles)			Rate constant (min^{-1})		
	A_1	A_2	A_3	k_1	k_2	k_3
1. ^{22}Na vs t Isotope Efflux	27	10	3	0.13	0.03	0.004
2. (Na)**) vs t Na Electrode	10	–	5	0.15	–	0.006
3. 1–2 Extracellular	16	10	–	0.15	0.03	–

*) At t = o, $(Na)_F = 40$ mmoles and $(Na)_i = 15$ mmoles per total fiber water. All A and k values were obtained graphically, then corrected according to *Huxley* (1960).

**) $(Na)_i = \frac{.75 \, (^aNa)_i}{.5}$ where intrafiber water equals 75% of total muscle water and the apparent activity coefficient equals 0.5 (obtained experimentally).

Use of the Sodium Microelectrode to Define Sodium Efflux

Our analysis (Table 1) of curve B (Fig. 5) indicates that intrafiber Na+ leaves through the membrane by two pathways, one faster than the other. However, the same mean experimental data can be fitted equally well (Fig. 6, curve A) by a curve which satisfied the cubic relation, $\frac{\delta (^aNa)_i}{\delta t} = k(^aNa)_i^3$. If this latter function is preferred, one must conclude there is only one major exit path for internal Na+. Whether the first or the second analysis is preferred does not alter the important experimental observation that much of the internal Na leaves the fiber rapidly (50% in 10 minutes). *Thomas* (1972) also observed internal Na in the snail neurone to decrease at a rapid rate (50% in 10 minutes) under comparable experimental conditions, except his efflux curve was nicely fitted by one exponential term.

Effect of Ouabain on Intrafiber Na

Curve B in Figure 6 shows how ouabain (10^{-4}M) altered the decrease in intrafiber Na following external Na replacement, and Table II summarizes the results of curve fitting by the two methods, the sum of exponential functions and the single power function method. In the first case, notice that the two exponentials are still present with the same rate constants ($k_1^1 = 0.14$ and $k_3^1 = 0.005$ min^{-1}), but with altered initial capacities. This analysis identifies the Na pump with the fast rate constant and indicates that ouabain probably made 50% of the membrane carrier ineffective. Put another way, instead of 66% of the internal Na leaving at a fast rate, we now see only 33% leaving at the fast rate.

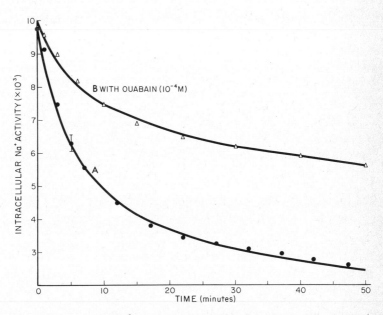

Fig. 6. Effect of ouabain (10^{-4} M) on the decrease of intrafiber Na+ activity ($^aNa)_i$ following Na+ bath replacement at t = o. Each experimental point at each time period contains data from 4 fibers (normalized so that $(^aNa)_i = 0.010$ at t = o). The bar through point on curve A at t = 5 min equals 2 × S.E. Curve A results from the integration of $\frac{\delta(^aNa)_i}{\delta t} = k\,(^aNa)_i^3$ where $k = 1.6 \times 10^{-3}$; Curve B results from the integration of $\frac{\delta(^aNa)_i}{\delta t} = k^1\,(Na)_i^6$ where $K' = 6.7 \times 10^{-7}$.

Table 2. Two methods of analysis of $(Na)_i$ vs t data with and without ouabain (10^{-4} M)

A. *Sum of Exponentials:*
  1. No Ouabain.
     $(Na)_i = 10e^{-.15t} + 5e^{-.006t}$

  2. With Ouabain.
     $(Na)_i = 5e^{-.11t} + 10e^{-.006t}$

B. *Power Rule:*
$$\frac{\delta(Na)_i}{\delta t} = k(Na)_i^n$$

	n	k
1. No Ouabain	3	1.6×10^{-3}
2. With Ouabain	6	6.7×10^{-7}

In the second case, notice that a power of 6 was required instead of a power of 3 before a single power function could be found to fit the data (curve B, Fig. 6). This finding forces one to speculate that ouabain might invoke a change in the Na carrier such that it must transport twice as many Na^+ ions, but at a greatly reduced rate (since k decreases from 1.6×10^{-3} to 6.7×10^{-7}). Since we are not fond of such an explanation for the action of ouabain, we are inclined to prefer the first description of internal Na efflux as the sum of two exponentials each representing a separate pathway through the fiber membrane.

Figure 7 demonstrates what happens to $(^aNa)_i$ when ouabain is added to the normal Ringer bath solution. After about a 2-minute delay, the $(^aNa)_i$ level began to rise at a steady rate of about 0.11 mM/min, from a starting $(^aNa)_i$ value of 4 mM. From the analysis in Table 2, we

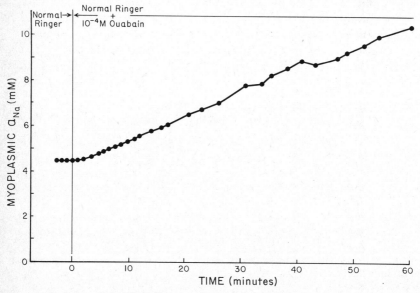

Fig. 7. Effect of ouabain (10^{-4} M) on intrafiber Na^+ activity $(_aNa)_i$ when a fiber is bathed in Ringer's solution. All the data are from one fiber.

can expect the initial Na efflux of this ouabain-poisoned fiber to be about 0.21 mM/min, hence the initial Na influx must be 0.32 mM/min. Since the external [Na] is 116 mM and has an acticvity coefficient of 0.77, we calculate the influx rate constant (k_1) from $\frac{\delta (^a Na)_i}{\delta t} =$ $-k_o (^a Na)_i + k_i (^a Na)_o$ to be about 0.004 min^{-1}. Notice that this rate constant is similar in magnitude to the slow efflux rate observed earlier when external Na was removed. Probably, this slow rate represents the leakage pathway for Na across the membrane and may be independent of the action of the pump.

Effect of K$^+$ on Intrafiber Na

Figure 8 shows that the Na activity in two fibers was little altered after doubling the external K$^+$ in the external Ringer solution. Figure 9 shows, however, that the Na activity always increases, sometimes quite rapidly, when the external K$^+$ is reduced to zero (solid line zones). Furthermore, this zero-K$^+$-induced increase in Na activity was arrested and sometimes reversed by re-exposure to normal K$^+$ (2.5 mM). *Thomas* (1972) observed a similar relation between external K$^+$ and internal Na$^+$ in the snail neurone.

Summary

(1) Sodium loss from the fiber interior into a zero-Na bath can be described either by the sum of two exponential terms,

$$(Na)_i = A_1 e^{-k_1 t} + A_3 e^{-k_3 t}$$

or by the integrated cubic function,

$$(Na)_i = \frac{A_T}{\sqrt{2 A_T^2 \, kt + 1}}$$

Following the administration of ouabain, sodium loss can still be described by the first equation, but the cubic rule, the derivative of the second equation, is no longer applicable. (2) The two rate constants for the first equation ($k_i = 0.15$ and $K_3 = 0.005$ min^{-1}) are

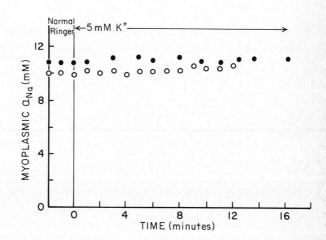

Fig. 8. Effect of doubling bath K$^+$ in Ringer's solution on intrafiber Na$^+$ activity. Results from two fibers are shown.

J. F. White and J. A. M. Hinke

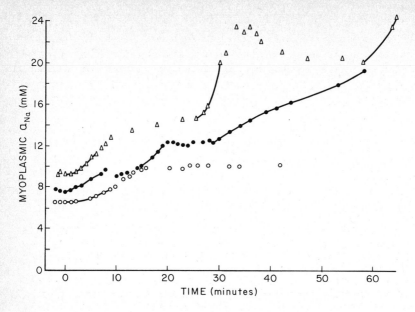

Fig. 9. Changes in intrafiber Na$^+$ activity when the bath K$^+$ content is alternated between zero (solid line zones) and normal (2.5 mM). Results from three fibers are shown.

unaltered by ouabain. However, ouabain does reduce the A_1 term by 50% indicating that the fast efflux component belongs to the Na pump. Since the rate constant for Na influx is similar to k_3, we identify the slow efflux component as a passive leakage channel.
(3) Sodium loss from the extracellular spaces of the whole frog sartorius muscle can also be described by two first order rate constants ($k_i = 0.15$ and $k_2 = 0.03$ min^{-1}), one of which is similar to the rate constant for the Na pump.

Summary

Sodium-sensitive glass microelectrodes with 1–2 μm tips were constructed according to *Thomas* (1972) and were used to measure the free Na$^+$ inside the fibers of the frog sartorius muscle. The efflux of intrafiber Na$^+$ into zero sodium Ringer solutions was compared with the efflux of total Na$^+$ from the whole muscle (^{22}Na isotope-loaded washout method). Sodium loss from the fiber interior can be described either by the sum of two exponential terms,

$$(Na)_i = A_1 e^{-k_1 t} + A_3 e^{-k_3 t} \tag{1}$$

or by the integrated cubic function,

$$(Na)_i = \frac{A_T}{\sqrt{2 A_T^2 \, kt + 1}} \tag{2}$$

Following the administration of ouabain (10^{-4}M), sodium loss can still be described by the first equation, but the second equation is no longer applicable. Rather than alter the rate constants in Equation 1 ($k_1 = 0.15$ and $k_2 = 0.005$ min^{-1}), ouabain reduced the A_1 term by 50%, thus indicating that the fast efflux component in Equation 1 belongs to the Na pump.

The rate constant for Na influx in the presence of ouabain was similar to k_3, which suggests that the slow efflux component in Equation 1 is a passive leakage channel. Finally, sodium loss from the extracellular space(s) of the whole muscle (defined as the difference between whole muscle $^{22}Na^+$ efflux and intrafiber Na^+ efflux) can also be described by two rate constants as in Equation 1 ($k_1 = 0.15$ and $k_2 = 0.03$ min^{-1}), one of which is similar to the rate constant for the Na pump.

Acknowledgements

We are grateful to Mrs. Juanna Li for technical assistance.

References

Huxley, A. F.: Mineral Metabolism; Appendix 2. pp 163–166, ed. by *C. L. Comar* and *F. Bronner*. Academic Press, New York 1960

Keynes, R. D., Swan, R. C.: The Effect of External Sodium Concentration on the Sodium Fluxes in Frog Skeletal Muscle. J. Physiol. (London) 147 (1959) 591–625

Mullins, L. J., Frumento, A. S.: The Concentration Dependence of Sodium Efflux from Muscle. J. gen. Physiol 46 (1963) 629–654

Thomas, R. C.: Intracellular Sodium Activity and the Sodium Pump in Snail Neurones. J. Physiol. (London) 220 (1972) 55–71

Raja N. Khuri

Intracellular Potassium in Single Cells of Renal Tubules

(5 Figures)

I. Intracellular Physiology

A. *Intracellular Potassium*

The true internal environment is the cytoplasmic aqueous solution each cell contains within its membrane. Reliable estimation of the ionic composition of the intracellular environment of single cells is an essential prerequisite for our understanding of such basic phenomena as muscle contraction, nerve conduction, membrane potentials, enzyme activity, and active and passive transport.

Potassium is the major determinant of the electrophysiological properties of cells, the renal tubular cells being no exception (*Windhager* and *Giebisch* 1965). The kidney is the homeostatic or regulatory organ responsible for maintaining the organism in a state of

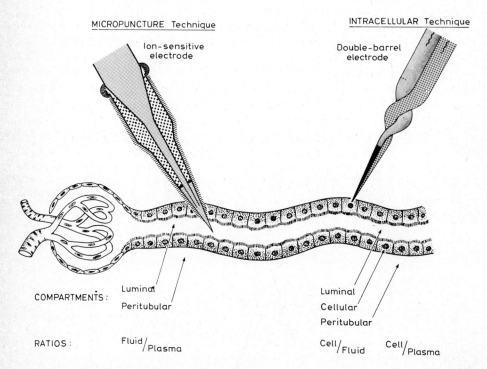

Fig. 1. The renal micropuncture technique and the intracellular technique have both been adopted in using ion-selective microelectrodes.

potassium balance. Almost all the filtered K+ is reabsorbed by the proximal tubule and the variable amount of K+ that is eliminated in the urine is derived from distal tubular secretion. The primary determinant of the K+ secretory and, therefore, excretory rates is the concentration of K+ within the regulatory distal tubular cells. The other factors that can affect distal tubular K+ secretion are the passive K+ permeability and the electrical potential difference across the luminal cell membrane of the distal epithelium.

B. *Intracellular Technique*

The renal tubular epithelium is essentially a three-compartment system: interstitial, intracellular and luminal. These three aqueous phases are separated by two lipid plasma membranes, the outer peritubular and the inner luminal cell membrane. The micropuncture technique (Fig. 1) ignores the cellular compartment of the tubular epithelium. It deals with fluid/plasma transepithelial ratios, thus considering the system as a two-compartment system. The intracellular technique using ion-selective microelectrodes determines the ionic activity of renal tubular cells as well as the electrical gradients across the two cell boundaries. It yields simultaneous electrical and chemical transcellular ratios, e.g., cell/plasma and cell/fluid K+ concentration ratios. Thus with the intracellular microelectrode technique the renal epithelium is treated as a three-compartment system.

The intracellular electro-chemical technique employs double-barrelled ion-selective liquid ion-exchanger micro-electrodes to determine directly the electrical potential differences across the individual cell membranes simultaneously with the intracellular ionic activities in single cellular elements. This yields the gradients of electrical potential and ionic activities across individual cell membranes. With the electrochemical driving forces for the movement of a given ion both directly and accurately quantitated, sound conclusions may be reached on the mechanisms of transport involved.

Double-barrelled K+-seletive liquid ion-exchanger micro-electrodes (Fig. 2) were used. The reference barrel is the electrical sensor; the other barrel the chemical (K+) sensor. The leads of the two barrels were connected to an electrometer and this gave the K+ signal alone (lower tracing of Fig. 3). The leads of the reference barrel and an external salt bridge were connected to another electrometer and the PD between them represented the membrane PD (upper tracing of Fig. 3). As the double-barrelled microelectrode impales the peritubular membrane, there is an abrupt rise in the membrane potential of about — 65 mV associated with a more gradual rise of the K+ potential. This is because the K+ sensor has a slower response time than the electrical cell. Figure 3 shows an average deflection of the K+ potential of 58 mV in going from an external [K+] of 4.3 mM to an intracellular [K+] of 54 mM for a proximal tubule

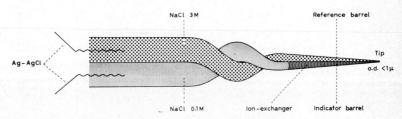

Fig. 2. Double-barrelled, K+-selective liquid ion-exchange microelectrodes.

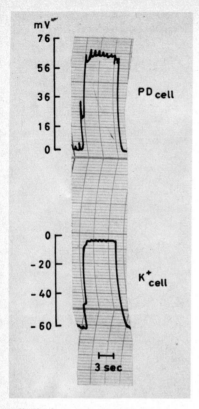

Fig. 3. Simultaneous recording of peritubular membrane PD (upper) and potassium potential (lower) by the two barrels of the double-barrelled K⁺ microelectrode.

cell. If, after recording the cellular potentials, the double-barrelled electrode was advanced by means of the hydraulic micro-drive a few micra into the lumen of a late proximal tubule two stable potentials are recorded: a transepithelial PD of practically nil and a tubular fluid K^+ potential slightly below that of the covering Ringer.

II. Proximal Tubule

Sixty-two electrometric determinations in vivo in the proximal tubule of the rat kidney yielded a mean intracellular K^+ concentration of 54.4 ± 2.5 mM. This accounts for just over 1/3 of the total K^+ content of proximal tubule cells and yields a mean ionic activity coefficient of about 0.4. In Necturus proximal tubule (*Khuri* et al. 1972a) 3/4 of the total K^+ content is electrometrically active. The observed discrepancy between the active electrometric intracellular $[K^+]$ and the total intracellular K^+ content is reminiscent of some analogous discrepancies. Several studies on kidney cortex slices have demonstrated that only a fraction of the total K^+ content is diffusible (chemical leaching of tissue) or exchangeable (with tracer isotopes). The diffusible fraction was estimated at 2/3 (*Mudge* 1951; *Foulkes* 1962; *Kamm* and *Strope* 1973). Kidney cortex K^+ was found to be exchangeable at two different rates, a rapid one and a very slow one (*Mudge* 1953; *Whittam, Davies* 1954). But the relationship between active electrometric concentration, chemical diffusibility and isotopic exchangeability remain speculative. However, these discrepancies may be accounted for by one of two possibilities: first, a non-uniform distribution of K^+ with different intracellular compartments or compartmentalisation of kidney cell K^+, and second, protein binding of intracellular K^+.

Intracellular Potassium in Single Cells of Renal Tubules

Table 1. Second half of a rat proximal tubule. Concentration ratios, electrical potential differences and calculated K$^+$ equilibrium potentials across the tubular epithelium (transepithelial) and the inner (luminal) and outer (peritubular) cell boundaries. Mean values ± standard error are given. All values given are derived from purely electrometric determinations.

Transepithelial				
$(TF)_k$ mM	$(P)_k$ mM	$(TF/P)_k$	E_k mV	E_m mV
3.7	4.3	0.85	±4.2	—0.7
±0.1	±0.1	±0.02	±0.2	±0.2

Luminal				
$(Cell)_k$ mM	$(TF)_k$ mM	$(Cell/TF)E_k$	E_k mV	E_m mV
54.4	3.7	14.7	70.1	67.8
±2.5	±0.1	±0.7	±1.2	±1.0

Peritubular				
$(Cell)_k$ mM	$(P)_k$ mM	$(Cell/P)_k$	E_k mV	E_m mV
54.4	4.3	12.7	66.1	68.5
±2.5	±0.1	±0.6	±1.2	±1.0

As summarized in Table 1 and Figure 4 (*Khuri* et al. 1974) the calculated K$^+$ equilibrium potential (E_K) is not significantly different from the measured membrane electrical PD (E_m) across both the inner (luminal) and the outer (peritubular) cell boundaries. Thus potassium ion exhibits an electro-chemical equilibrium distribution across the two individual cell mem-

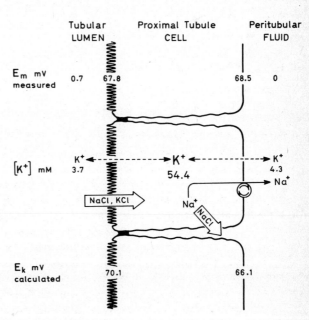

Fig. 4. A schematic representation of K$^+$ and Na$^+$ transport in the 3-compartment system of the proximal convoluted tubule of the rat kidney.

branes. The implication of this equilibrium distribution in the presence of a luminal active K+ pump (reabsorptive) is obviated. However, since Cl− ions cannot cross from lumen-to-cell passively (*Khuri, Agulian, Bogharian* 1974), the active mechanism for Cl− transport in the luminal membrane may be a neutral pump that transports neutral chloride salts (NaCl and KCl). For the peritubular cell membrane the need to postulate the presence of an active K+ influx component of the peritubular cation Na-K exchange pump is also obviated. Figure 4 shows an Na+ pump in the peritubular cell membrane since Na+ extrusion is not necessarily directly linked with K+ uptake into the cell. It is our thesis that the Na+ pump is at least partly electrogenic and that most of the K+ uptake across the peritubular cell membrane is by passive diffusion.

III. Distal Tubule

Since the intracellular [K+] of distal tubule cells is the major determinant of K+ secretory, and therefore, excretory rates, a study was undertaken of the various states that are associated with changes in K+ excretory rates. These states include changes in potassium balance, changes in acid-base balance and the effects of adrenal mineralocorticoids. The results are summarized in Table 2. (*Khuri, Agulian, Kalloghlian* 1972a).

A. *Normal Potassium Balance*

A mean intracellular effective [K+] of 46.5 ± 1.6 mM was obtained in 94 electrometric determinations. Other investigators (*Malnic, Klose, Giebisch*, 1964), who assumed a high value for cellular [K+], had to postulate the presence of a K+ reabsorptive pump in the luminal membrane to account for the K+ concentration in the tubular fluid which was always much below the predicted value. The measured low distal cell [K+] obviates the need to postulate the presence of a luminal K+ active reabsorptive pump. Since more than 95% of mammalian kidney cortex slices are proximal tubules, any correlation between total K+ content of whole cortical slices and the electrometric active [K+] of distal tubule cells is

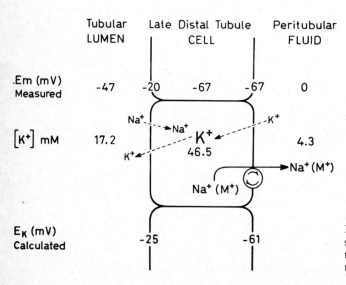

Fig. 5. A schematic representation of K+ and Na+ transport in a late distal tubule cell of the rat kidney.

irrelevant. The electrometric finding of a low K⁺ is consistent with the observation of a low radiokinetic K⁺ exchangeability in the distal tubule of the Amphuima kidney (*Wiederholt* et al. 1971).

Figure 5 is a schematic representation of K⁺ and Na⁺ transport in a late distal tubule cell under normal conditions. At the peritubular cell boundary the measured E_m of -67.0 ± 1.0 mV is significantly ($P < .001$) more negative than the calculated E_K of -60.7 ± 0.9 mV. Therefore, K⁺ ion can move passively from peritubular fluid into the cell. The need to postulate the presence of an active K⁺ influx component of the peritubular cation pump is obviated. At the luminal cell membrane E_m of -20.0 ± 1.0 mV is significantly ($P < .005$) less negative than E_K of -24.5 ± 0.9 mV, thus indicating the presence of a residual net electrochemical driving force favoring K⁺ passive secretion from cell-to-lumen. Thus the distal tubule K⁺ secretory process consists of two passive diffusion steps in series: (1) from interstitium-to-cell and (2) from cell-to-lumen.

B. *Changes in Potassium Balance*

As shown in Table 2 the effective intracellular [K⁺] increased by 14 mM (30%) with chronic K⁺ loading and decreased by 10 mM (22%) with chronic K⁺ depletion. The luminal membrane PD is an electrical force that opposes the leakage of K⁺ from the cell. The depolarization of the luminal E_m in the high K⁺ rats and its hyperpolarization in the low K⁺ rats respectively promote and oppose the passive diffusion of K⁺ from cell-to-lumen. Thus the chemical and electrical factors summate to promote or oppose K⁺ secretion by passive diffusion in K⁺ excess and K⁺ depletion states.

Table 2. Late distal tubule potassium concentration ratios, electrical potential differences, and calculated K⁺ equilibrium potentials across the different boundaries of the distal tubular epithelium under different metabolic conditions.

Parameter	Normal	High K⁺	Low K⁺	Acidosis metabolic	Alkalosis metabolic
[K⁺] plasma	4.3	5.2	2.9	4.8	3.4
$(TF/P)_{k^+}$	4	9	1.5	2	7
E_k (calculated)	−35	−55	−10	−18	−49
E_m transepithelial	−47	−62	−19	−14	−52
[K⁺] cell	46.5	60	36	39	52
[K⁺]$_{TF}$	17.2	47	4.4	9.6	23,8
$(Cell/TF)_{k^+}$	2.7	1.3	8.2	4.1	2.1
E_k (calculated)	−26	−7	−54	−36	−19
E_m luminal	−20	−6	−47	−35	−15
[K⁺] cell	45.6	60	36	39	52
$(Cell/P)_{k^+}$	10.8	11.5	12.4	8.1	15.3
E_k (calculated)	−61	−63	−65	−54	−69
E_m peritubular	−67	−68	−66	−49	−67
K⁺ excretory rate ($\mu M \cdot min^{-1} \cdot kg^{-1}$)	2.12	18.28	0.23	0.73	38.3

C. Changes in Acid-Base Balance

The inverse relationship between potassium excretion and urinary acidification is well known (*Berliner* et al. 1951). The effective distal tubule cell [K^+] decreased by 7.8 mM (17%) with metabolic acidosis and increased by 5 mM (11%) with metabolic alkalosis. Again, by comparison with normal rats the luminal membrane PD (E_m) became hyperpolarized with acidosis and depolarized with alkalosis.

Thus the two kaliuretic states of K^+ loading and metabolic alkalosis are characterized by an increase in the effective intracellular source [K^+] and a decrease (depolarization) of the luminal membrane PD. In contrast, the two kaliupenic states of K^+ depletion and metabolic acidosis are characterized by a decrease in cell [K^+] and an increase (hyperpolarization) of the luminal membrane PD. In brief, the summation of the observed chemical and electrical driving forces between the cellular and luminal compartments could quantitatively account for the different rates of passive entry of K^+ into the lumen of the distal tubule in metabolic states which are associated with changes in the secretion, and therefore, excretion of potassium.

D. Effect of Aldosterone

It is known that in the absence of aldosterone distal tubular K^+ secretion is diminished. Table 3 (*Wiederholt, Agulian, Khuri* 1974) is a summary of the effect of aldosterone on distal tubular cells effective [K^+] and the peritubular membrane PD as recorded by the simultaneous electrochemical technique. In adrenalectomized rats the effective intracellular [K^+] decreased significantly by 8 mM (17%) and the peritubular PD decreased to 48.9 ± 0.8 mV. Only after 5 days of aldosterone replacement therapy were both the [K^+] and the E_m normalized. The reduction in intracellular [K^+] and peritubular PD after adrenalectomy is similar to that observed with metabolic acidosis (Table 2). Metabolic acidosis can decrease the diffusible K^+ in renal cortical cells (*Foulkes* 1962; *Kamm* and *Strope* 1973). Therefore, an indirect effect of the lack of adrenal steroids via acidosis on intracellular

Table 3. Summary of distal tubule intracellular effective potassium concentration (electrometric) and peritubular membrane potential difference in rats. Mean ± S.E.

Group	K-intracellular (mMol/l)	PD-peritubular (mV)	n	K-intracellular P	PD-peritubular P
Control rats	47.2 ± 1.8	68.0 ± 2.1	25	–	–
Control rats	46.5 ± 1.6	67.4 ± 1.0	94	n. s.	n. s.
Adrenalectomized	39.1 ± 1.4	48.9 ± 0.8	75	< 0.001	< 0.001
Adren. + aldo (1–2 days)	39.2 ± 1.4	48.9 ± 1.2	48	< 0.001	< 0.001
Adren. + aldo. (5 days)	45.2 ± 0.9	63.9 ± 0.8	53	n. s.	n. s.

a) Data from *Khuri* et al (7).
b) The difference between adrenalectomized rats + aldosterone (5 days) and adrenalectomized rats without hormone substitution were significant at the level of $P < 0.005$ for K-intracellular and of $P < 0.001$ for PD-peritubular.

[K$^+$] cannot be excluded. The diminished peritubular membrane PD after adrenalectomy could be due to a reduction in the active extrusion of Na$^+$ across that membrane by an electrogenic, potential-generating Na$^+$-pump. Following the stimulation by aldosterone of an at least partly electrogenic peritubular cationic pump it would take several days for the increased K$^+$ uptake (both passive and active) to fill up the large pool of intracellular K$^+$ and restore the normal level of intracellular [K$^+$]. This recovery time is unduely prolonged perhaps on account of the aldosterone-induced, increased K$^+$-passive permeability of the luminal membrane resulting in augmented cell-to-lumen K$^+$ efflux (*Wiederholt* et al. 1973). In brief, the studies presented on the effective intracellular K$^+$ of the distal tubule cells under a variety of metabolic states demonstrate the physiological significance of this effective [K$^+$] as the primary determinant of the process of K$^+$ secretion.

Acknowledgement

This work was supported in part by the Libanese National Research Council.

References

Berliner, R. W., Kennedy, T. J., Orloff, J.: Relationship Between Acidification of the Urine and Potassium Metabolism. Amer. J. Med. 11 (1951) 274–282

Foulkes, E. C.: Exchange and Diffusion of Potassium in Kidney Cortex Slices. Amer. J. Physiol. 203 (1962) 655–661

Kamm, D. E., Strope, G. L.: Glutamine and Glutamate Metabolism in Renal Cortex from Potassium-depleted Rats. Amer. J. Physiol. 224 (1973) 1241–1248

Khuri, R. N., Agulian, S. K., Bogharian, K.: Electrochemical Potentials of Potassium in Proximal Renal Tubule of Rat. Pflügers Arch. 346 (1974) 319–326

Khuri, R. N., Agulian, S. K., Kalloghlian, A.: Intracellular Potassium in Cells of the Distal Tubule. Pflügers Arch. 335 (1972a) 297–308

Khuri, R. N., Hajjar, J. J., Agulian, S., Bogharian, K., Kalloghlian, A., Bizri, H.: Intracellular Potassium in Cells of the Proximal Tubule of Necturus Maculosus. Pflügers Arch. 338 (1972b) 73–80

Malnic, G., Klose, R. M., Giebisch, G.: Micropuncture Study of Renal Potassium Excretion in the Rat. Amer. J. Physiol. 206 (1974) 674–686

Mudge, G. H.: Electrolyte and Water Metabolism of Rabbit Kidney Slices: Effect of Metabolic Inhibitors. Amer. J. Physiol. 167 (1951) 206–223

Mudge, G. H.: Electrolyte Metabolism of Rabbit-Kidney Slices: Studies with Radioactive Potassium and Sodium. Amer. J. Physiol. 173 (1953) 511–522

Whittam, R., Davies, R. E.: Relations Between Metabolism and the Rate of Turnover of Sodium and Potassium in Guinea Pig Kidney-Cortex-Slices. Biochem. J. 56 (1954) 445–453

Wiederholt, M., Agulin, S. K., Khuri, R. N.: Intracellular Potassium in Distal Tubule of the Adrenalectomized and Aldosterone Treated Rat. Pflügers Arch. 347 (1974) 117–123

Wiederholt, M., Schoormans, W., Fischer, F., Behn, C.: Mechanism of Action of Aldosterone on Potassium Transfer in the Rat Kidney. Pflügers Arch. 345 (1973) 159–178

Wiederholt, M., Sullivan, W. J., Giebisch, G.: Potassium and Sodium Transport Across Single Distal Tubules of Amphiuma. J. Gen. Physiol. 57 (1971) 495–525

Windhager, E. E., Giebisch, G.: Electrophysiology of the Nephron. Physiol. Rev. 45 (1965) 214–244

Hj. Hirche, C. Steinhagen, I. Hosselmann, J. Manthey and U. Bovenkamp

The Interstitial pH of the Isolated Skeletal Muscle of the Dog at Rest and During Exercise*)

(3 Figures)

Gebert and *Friedman* (1973) recently described interstitial pH changes measured with H^+-sensitive microelectrodes in working gastrocnemius muscles of the rat. Dependent on the stimulation frequencies, they observed an initial increase of pH followed by a long-lasting pH decrease. The aim of the present study was to measure pH changes of the interstitial fluid of isolated gastrocnemii of dogs with H^+-sensitive microelectrodes during and after exercise performed under various metabolic acid-base conditions.

Materials and Methods: 10 mongrel dogs (31.4 ± 2.5 kg body weight) were anesthetized by pentobarbital (25 mg/kg iv). Expiratory pCO_2 was recorded by an IRAS apparatus (*Hartmann* u. *Braun*, Frankfurt). The animals were ventilated artificially to maintain an end-tidal pCO_2 of about 40 Torr. The left gastrocnemius muscle (average wet weight 140.6 ± 9 g) was isolated as described recently in detail (*Hirche* et al. 1971, 1974). All branches of the left femoral artery and vein not supplying the gastrocnemius were dissected and ligated. The femur was fixed and the Achilles tendon was freeloaded with weights of 5–15 kg. Shortening of the muscle could be measured using a linear displacement transducer. The muscle surface was kept wet and prevented from drying and cooling by an insulating cover. The sciatic nerve was cut and the distal end was attached to a bipolar electrode. For indirect stimulation square wave impulses, 0.5 msec in duration and at a frequency of 100 sec were used. Trains of these impulses which lasted 0.2 sec were given at intervals of 0.5 s. This means that the muscles were tetanically stimulated 86 times per minute. The arterial influx was measured electromagnetically. Arterial and venous pH were measured continuously by flow-through electrodes (Fa. Ingold, Frankfurt, type 402–611) as well as blood temperature (thermistors) and venous oxygen saturation. Oxygen consumption and lactate (LA) release from the muscle tissue were measured according to the Fick principle. The interstitial pH was measured using bulb-type buffer-filled microelectrodes made of Corning 0150 glass according to *Gebert* (1972). The diameter and the length of the H^+-sensitive tip were both about 100 μ. The internal resistance of these electrodes was 10^{10}–10^{11} ohms. As reference electrodes KCl/Agar-filled microcapillaries 50 μ in diameter connected via a KCl/Agar bridge with conventional Ag/AgCl-electrodes were used. The potential measurements were carried out by DC amplifiers (Keithley-Elektrometer 602) with an input impedance of 10^{14} ohms shunted by 20 pf. The electrodes were calibrated at muscle temperature before and after the experiments, which lasted 5 to 7 hours. The drift was about 0–2 mV during this time. The microelectrodes were inserted between parallel muscle fibers to prevent breaking and dislocation of the electrodes by the muscle contraction. The tips of H^+-sensitive and reference electrodes were placed 5–7 mm under the muscle surface. Interstitial pH was always measured in two different parts of the muscle. Differences of H^+ activity did not exceed 2 to 3 nmol/l. The calibration and the

---

*) Supported by the Bundesinstitut für Sportwissenschaft.

The Interstitial pH of the Isolated Skeletal Muscle of the Dog

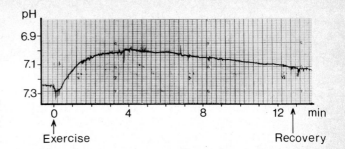

Fig. 1. Original recording of interstitial pH changes measured with a microelectrode.

experiments were performed within a large *Faraday* cage. Acid base parameters were measured according to the *Astrup* method using nomogramms according to *Siggaard-Andersen*. Metabolic acidosis was induced by the infusion of L-arginine-hydrochloride and metabolic alcalosis was induced by the infusion of 8.4% $NaHCO_3$ solutions.

Results and Discussion

A typical original recording of an interstitial pH measurement is demonstrated in Figure 1. Disturbances due to changes of capacity during blood sampling are recognizable as spikes. Neither electrical stimulation nor muscle contraction interfered. Values of blood flow, oxygen consumption, force and LA release at rest, during exercise, and recovery shown in Figure 2

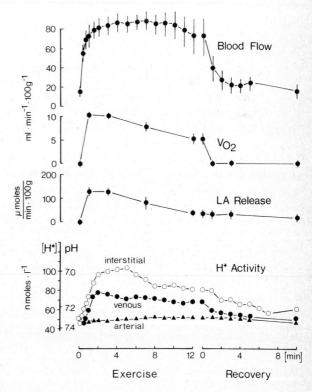

Fig. 2. From top to bottom: blood flow, oxygen consumption ($\dot{V}O_2$), lactate (LA) release, interstitial, venous, and arterial H^+ activity of the gastrocnemius muscle of the dog during exercise and recovery. (Mean values $\pm$ SE, N = 7).
See text for further explanation.

373

are consistent with the data which were reported recently from our laboratory (*Hirche* et al. 1971). During the first 10–15 sec of exercise interstitial H$^+$ activity always decreased transiently by about 5 nmoles/l and increased thereafter reaching maximal values after 2–3 minutes as is shown in Figure 2. In experiments with metabolic acidosis (BE = $-$ 13 meq/l) H$^+$ activity decreased transiently by about 30 nmol/l while in experiments with alkalosis (BE = + 8.3 meq/l) H$^+$ activity decreased only by about 2–3 nmol/l. The time course of this early transitory interstitial "alkalosis" corresponds with the known intracellular "alkalosis" which is due to the rapid creatine phosphate breakdown splitting at the onset of exercise (*Lipmann* and *Meyerhof*, 1930, *Manthey* et al. 1973). Therefore we suggest that rapid H$^+$ ion exchange mechanisms across muscle cell membrane can occur.

In all experiments H$^+$ activity in the interstitial fluid was higher than in venous blood. While in experiments with alkalosis H$^+$ activity differences between interstitial space and venous blood of only 20 nmoles/l were measured this gradient reached values up to 90 nmoles/l in metabolic acidosis. In experiments with acidosis interstitial H$^+$ activity decreased only very slowly during recovery and was elevated even 20 min after the end of exercise. Consistent with recent observations (*Hirche* et al. 1974) maximal LA release during exercise in acidosis was only 90 meq · 100 g^{-1} · min^{-1} while in alkalosis values of 300 meq · 100 g^{-1} · min^{-1} were reached.

In Figure 3 interstitial H$^+$ activities are plotted against venous LA concentrations measured during exercise. Thereby venous LA concentrations are considered to correspond to interstitial LA concentrations. The different slopes of the three regression lines demonstrate that the same LA concentration in acidosis increases interstitial H$^+$ activity much more than in alkalosis. The high H$^+$ activity in the interstitial space during exercise in acidosis supports the view that H$^+$ ions and LA$^-$ can permeate only very slowly across the muscle cell membrane in metabolic acidosis (*Hirche* et al. 1974). The high H$^+$ activity gradients between interstitial space and venous blood observed in metabolic acidosis during exercise and recovery cannot be explained from these experiments. These observations suggest that H$^+$ ions and LA$^-$ permeability within the interstitial space is impeded by severe metabolic acidosis.

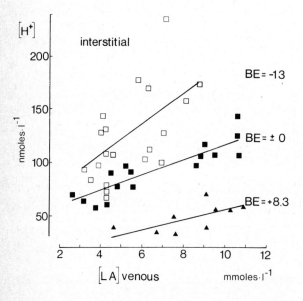

Fig. 3. Interstitial H$^+$ activity of isolated dog gastrocnemius muscles plotted against venous LA concentration measured during exercise in experiments with metabolic acidosis (N = 9), normal acid base balance (N = 7) and metabolic alkalosis (N = 4). The following regression lines were calculated: Acidosis: y = 14.3 × + 50.2, r = 0.58 (P < 0.02), Normal: Y = 7.2 × + 45.4, r = 0.92 (P < 0.001), Alkalosis: y = 4.8 × + 8.0, r = 0.73 (P < 0.05). BE = base excess in meq per liter blood.
See text for further explanation.

Conclusion

It is concluded that the bulb-type buffer-filled H$^+$-sensitive microelectrode (Corning glass 0150) is a useful tool to measure rapid changes and absolute values of the interstitial pH of isolated working skeletal muscles.

References

Gebert, G.: Die Messung von Na$^+$-, K$^+$- und H$^+$-Aktivitäten in Gewebe mit Glasmikroelektroden. Ärztl. Forsch. 26 (1972) 379–385

Gebert, G., Friedman, S. M.: An Implantable Glass Electrode Used for pH Measurement in Working Skeletal Muscle. J. appl. Physiol. 34 (1973) 122–124

Hirche, Hj., Hombach, V., Langohr, H. D., Wacker, U., Busse, J.: Lactic Acid Permeation Rate in Working Gastrocnemius of Dogs During Metabolic Alkalosis and Acidosis Pflügers Arch. 356 (1975) 209–222

Hirche, Hj., Wacker, U., Langohr, H. D.: Lactic Acid Formation in the Working Gastrocnemius of the Dog. Int. Z. angew. Physiol. 30 (1971) 52–64

Lipmann, F., Meyerhof, O.: Über die Reaktionsänderung des tätigen Muskels. Biochem. Z. 227 (1930) 84–109

Manthey, J., Busse, J., Bovenkamp, U., Hirche, Hj., Hombach, V.: Lactacid and Alactacid Oxygen Deficit in the Isolated Gastrocnemius of the Dog. Pflügers Arch. 339 (1973) R 50

G. Gebert

Tissue Electrolytes and Blood Flow in Skeletal Muscle

(5 Figures)

The tone of vacular smooth muscle cells is influenced by the ionic environment. An extracellular acidosis usually dilates the vascular bed of skeletal muscle (e. g., *Emerson, Parker* and *Jelks* 1974), a slight to moderate increase in interstitial K^+ has a similar effect at least during the first minutes (e.g., *Brace* 1974), and enhanced Na^+ lowers the vascular resistance through the change in osmolality (*Mellander, Johansson, Gray, Jonsson, Lundvall* and *Ljung* 1967). On the other hand, it is well known that working skeletal muscle cells release H^+ ions as well as K^+ ions, thereby increasing the interstitial K^+ and H^+ activity. Furthermore, the exercising skeletal muscle takes up interstitial water (*Lundvall* 1972), augmenting the concentration of all solutes including Na^+ in the interstitial fluid. These findings lead to the conclusion that changes in interstitial H^+, K^+ and Na^+ activity are involved in the local regulation of skeletal muscle blood flow in adaptation to the metabolic demand. To obtain further information on this problem, the local changes in interstitial ion activity had to be measured simultaneously with the blood flow during muscular work. This could be done by means of ion-selective electrodes.

Methods

Two types of miniature glass electrodes were constructed for the measurement of H^+, K^+ and Na^+ activity. Bulb-type electrodes were used for Na^+ and K^+ determinations (NAS_{11-18} glass from Corning, N. J., and pMe glass from Schott & Gen., Mainz). The H^+ measurements were

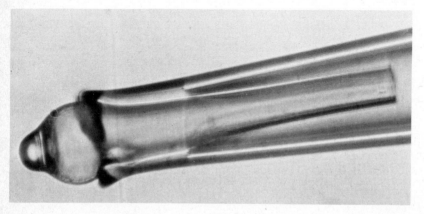

Fig. 1. Tip of an ion-selective microelectrode (bulb type). Tip diameter about 30 µm. The glass-glass seal is clearly visible.

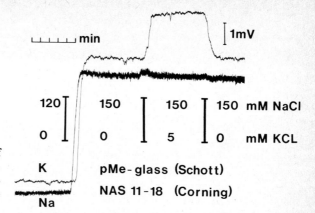

Fig. 2. Simultaneous calibration of Na^+ and K^+ electrodes at 22°C. Slope of the Na^+ electrode 56.8 mV, of the K^+ electrode for Na^+ 58.8mV, selectivity K^+/Na^+ 2.6.

performed using metal-connected electrodes constructed from Corning 0150 glass. Details of the construction are given by *Gebert* (1972) and *Gebert* and *Friedman* (1973). The tip diameter of the bulb-type electrodes varied from 15 to 150 μ, and the sensitive part of the metal-connected electrodes had an outer diameter of 100–300 μ. An example of bulb-type electrodes is given in Figure 1.

The experiments were performed in anesthetized rabbits (m. quadriceps femoris) and rats (m. gastocnemius). In the experiments in rabbits the blood flow of the femoral artery was measured using a Statham electromagnetic flowmeter, and the changes in ion activity during muscular exercise were determined additionally in the venous blood using a bypass between the femoral and the jugular vein.

While the potential changes of the H^+ and the Na^+ electrodes could be considered directly as changes in the corresponding ion activities (with the exception of a transient response of some NAS_{11-18} electrodes to H^+ and K^+), the electrodes made of pMe glass had a selectivity for K^+ over Na^+ of only 1–10. Therefore the shifts in potassium activity had to be calculated from the potential changes of both the K^+ and the Na^+ electrodes. A calibration curve is shown in Figure 2.

Results

The potassium and the sodium activities rose during muscular exercise in the interstitium of skeletal muscle as well as in the venous blood. The changes, especially of potassium acitivity, closely paralleled the functional hyperemia at least after a brief frequent stimulation (10–40 Hz for 15 sec). An example is given in Figure 3.

After a brief tetanus the potassium activity increased by up to 5 mM (interstitial space) or by 0.5–2.5 mM (venous blood), respectively. The sodium activity rose by 1–5 mM without a significant difference between plasma and interstitial values.

Different results were obtained when a stimulation of 0.2 to 3 pulses per second for 2–3 minutes was used. The potential changes of the Na^+ and the K^+ electrodes were comparable to the change in blood flow also in these experiments (Fig. 4), but the maximum increase in K^+ is now only 0.2 to 0.8 mM while the rise in Na^+ amounts to 1–4 mM.

The H^+ activity did not rise when a stimulation frequency below 1 pulse per second was chosen. On the contrary, a more or less pronounced interstitial alkalosis (0.01–0.15 pH units)

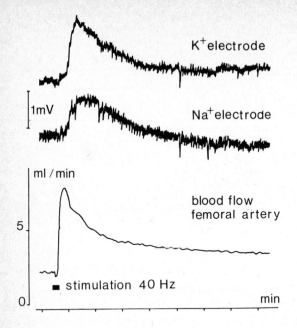

Fig. 3. Potential changes of two ion-sensitive electrodes rinsed by rabbit femoral venous blood. Supramaximal stimulation of the whole leg musculature. Maximal increase in K$^+$ 1.8 mM, in Na$^+$ 4.9 mM

occurred. Only with higher frequencies the response became biphasic, and after 15–30 seconds (depending on the frequency) the initial alkalosis turned into an acidosis of up to 0.25 pH units (Fig. 5).

Discussion

The observed maximum shifts in interstitial potassium activity after a brief frequent stimulation are sufficient to elicit a considerable vasodilation in vivo (*Kjellmer* 1965) as well as

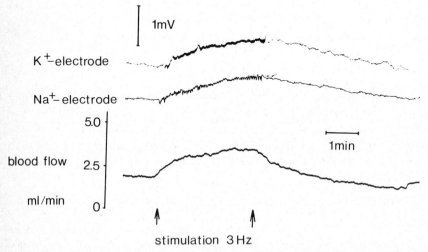

Fig. 4. Potential changes of electrodes inserted into the m. quadriceps of an anesthetized rabbit. Lower curve: blood flow in the femoral artery. Stimulation of the whole leg. Maximal increase in K$^+$ 0.5 mM, in Na$^+$ 3.5 mM.

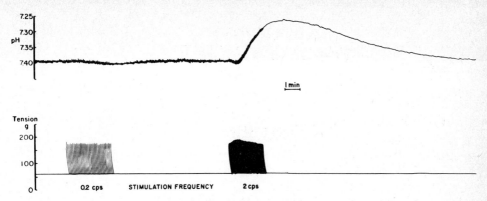

Fig. 5. Interstitial pH of rat gastrocnemius muscle measured by an inserted glass microelectrode. Supramaximal stimulation of the distal end of the sciatic nerve.

in vitro (*Konold, Gebert* and *Brecht* 1968). Therefore in that type of work K^+ can be considered as an important factor in the induction of functional hyperemia. In the experiments in which the muscles were stimulated for several minutes using low frequencies, the rise in K^+ was much smaller, indicating that potassium plays a minor role when the exercise hyperemia has to be maintained over a longer period. This agrees with the findings of *Chen, Brace, Scott, Anderson* and *Haddy* (1972) in the perfused dog gracilis muscle.

The interstitial sodium activity rose during muscular exercise only by about 4% compared to the maximum increase in K^+ activity by more than 100%. But even this small increase, the time course of which is also close to that of the functional hyperemia, indicates a rise in interstitial osmolality by up to 10 mosmole/kg. Taking into account the further increase due to potassium etc., the osmolality might be an important factor in the exercise hyperemia as postulated by *Mellander* et al. (1967).

In contrast to the pK and pNa, the interstitial pH of working skeletal muscle did not change in the same way as the blood flow. Initially a slight alkalosis was observed, and only in experiments with stimulation frequencies above 1 pulse per second an interstitial acidosis developed after 10–30 seconds. Considering the rapid onset of functional hyperemia (*Chen* et al. 1972), the interstitial pH cannot play a decisive role, at least not within the first minute of exercise.

In the case of higher contraction frequencies, the falling pH might contribute to the local vasodilation, though its effect seems to be relatively small (*Emerson* et al. 1974).

Summarizing the described results and considerations, neither K^+ nor H^+ seem to be of crucial importance for the regulation of blood flow in working skeletal muscle, though the changes in their interstitial activity contribute more or less to the fall in flow resistance during muscular exercise.

References

Brace, R. A.: Time Course and Mechanisms of the Acute Effects of Hypokalemia and Hyperkalemia on Vascular Resistance. Proc. Soc. exp. Biol. Med. 145 (1974) 1389–1394

Chen, W.-T., Brace, R. A., Scott, J. B., Anderson, D. K., Haddy, F. J.: The Mechanism of the Vasodilator Action of Potassium. Proc. Soc. exp. Biol. Med. 140 (1972) 820–824

Emerson, Jr., T. E., Parker, J. L., Jelks, G. W.: Effects of Local Acidosis on Vascular Resistance in Dog Skeletal Muscle. Proc. Soc. exp. Biol. Med. 145 (1974) 273–276

Gebert, G.: Die Messung von Na^+-, K^+- und H^+-Aktivitäten im Gewebe mit Glasmikroelektroden. Ärztl. Forsch. 26 (1972) 379–385

Gebert, G., Friedman, S. M.: An Implantable Glass Electrode Used for pH Measurement in Working Skeletal Muscle. J. appl. Physiol. 34 (1973) 122–124

Kjellmer, I.: The Potassium Ion as a Vasodilator During Muscular Exercise. Acta physiol. scand 63 (1965) 460–468

Konold, P., Gebert, G., Brecht, K.: The Effect of Potassium on the Tone of Isolated Arteries. Pflüg. Arch. ges. Physiol. 301 (1968) 285–291

Lundvall, J.: Tissue Hyperosmolality as a Mediator of Vasodilatation and Transcapillary Fluid Flux in Exercising Skeletal Muscle. Acta physiol. scand. 379 (1972) 1–142

Mellander, S., Johansson, B., Gray, S., Jonsson, O., Lundvall, J., Ljung, B.: The Effect of Hyperosmolarity on Intact and Isolated Vascular Smooth Muscle. Possible Role in Exercise Hyperemia. Angiology 4 (1967) 310–322

L. Coleman

Recent Advances in Areas of Clinical Utility and Instrumentation for the Determination of Ionized Calcium

Implications for the Future of Chemical-Sensing Electrodes in Clinical Medicine

(7 Figures)

I. Introduction

Quantitative differentiation between the active and inactive fractions of many species is a major obstacle to the definitive analysis of biological fluids. Instrumentation for measuring free calcium in serum by ion-selective electrodes demonstrates the diagnostic value of a direct active fraction measurement.

Calcium plays a central role in a variety of essential physiological processes. Among the more important processes are muscle contraction, endocrine function, nerve conduction, enzyme conversions, blood coagulation, renal tubular function, bone formation, and membrane integrity.

The total calcium content of the adult human body is approximately one kilogram, most of which is in the skeleton. Extracellular fluid contains only about one gram of calcium divided into ionized (0.4 gram), complexed (0.2 gram), and protein bound (0.4 gram) fractions.

McLean and *Hastings* (1934), in a series of landmark experiments on the contractility of isolated frog hearts exposed to varying levels of ionized calcium, first showed that calcium physiological activity is solely a function of the ionized fraction. The development of the liquid membrane calcium electrode by *Ross* (1967) and subsequent research by *Moore* (1969) have led to the first practical procedures for the direct determination of ionized calcium in biological fluids.

II. Clinical experience with ionized calcium

The liquid membrane calcium electrode as developed by *Ross* has been used by many investigators during the past five years for the study of a variety of disease states associated with imbalances in calcium homeostasis. (*Moore* 1969; *Ladenson* and *Bowers* 1973; *Low, Schaaf, Earll, Piechocki* and *Li* 1973; *Muldowney, Freany, Spillane* and *O'Donohoe* 1973; *Fuchs, Paschen, Spieckermann* and *Westberg* 1972).

These investigators and many others have demonstrated the clinical utility of ionized calcium measurements for the differential diagnosis of such diseases as primary hyperparathyroidism, 'normocalcemic' hyperparathyroidism, multiple myeloma and renal failure. It has been shown not only that ionized calcium levels cannot be predicted from total serum calcium concentration, but more importantly, that in cases where free and total calcium measurements produce conflicting data, the ionized calcium value is more directly related to patient status.

An important recent development in the intensive care arena has been the use of the calcium electrode to monitor plasma ionized calcium levels in patients receiving transfusions of citrated blood (*Drop* 1974). Results from work in this area have shown that potentially dangerous alterations in ionized calcium homeostasis do occur in critically ill patients. It is expected that ionized calcium assays will play an increasingly important role in maintaining optimal hemodynamic performance in critical care situations.

III. Instrumentation

Figure 1 shows the Orion Model 99–20 Ionized Calcium System. The calcium and reference electrodes employed in this system evolved as simple flow-through adaptations of standard Orion dip-type sensors as shown in Figure 2. The 99–20 system was introduced more than five years ago and has, until now, provided the only available means for the direct measurement of ionized calcium in biological fluids. All the previously cited patient data were obtained by researchers employing this system.

However, it is only fair to say that full-scale clinical application of ionized calcium measurements has been hampered by certain inadequacies and inconveniences associated with this equipment. The quality of the results obtained with the 99–20 system was affected by such factors as calcium electrode mechanical tolerances, variations in membrane characteristics and operator expertise.

Recent work in our laboratories has resulted in new and markedly improved instrumental approaches for the measurement of electrolytes and blood gases with chemical sensing electrodes. Perhaps the most significant of these developments was the Space-Stat System, a fully automated seven-parameter blood gas and electrolyte system, developed for the National Aeronautics and Space Administration. A photograph of the fluid transport and

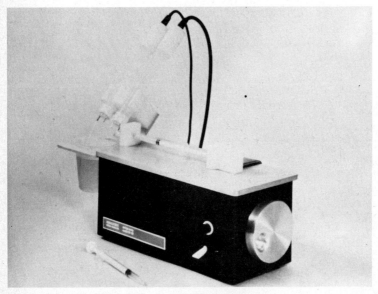

Fig. 1. Orion Model 99–20 Ionized Calcium System.

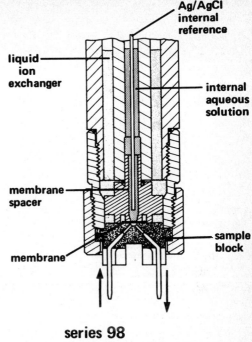

Fig. 2. Series 98 Flow-Thru Electrode.

electronics sections is shown in Figure 3. These units occupy only 0.8 and 0.35 cubic feet, respectively. A Digital Equipment Corporation PDP-8E mini-computer programmed by Orion controls system operation and data processing.

Fig. 3. Orion/NASA Space-Stat System.

Fig. 4. Space-Stat Na^+/K^+ Electrode Module.

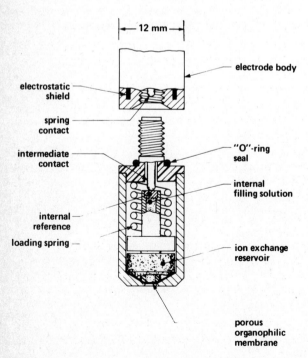

Fig. 5. Series 93 Liquid Membrane Electrode.

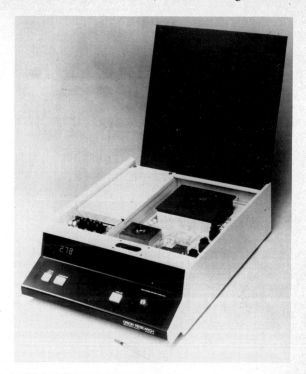

Fig. 6. Orion Model SS-20 Ionized Calcium System.

The system was designed to demonstrate the feasibility of performing simultaneous analyses for pH, carbon dioxide, sodium, potassium, chloride, total calcium and ionized calcium in serum or urine under weightless conditions. The only operations required for an analysis sequence are injecting the sample (approximately 1 ml) and pressing a single button. Each ten-minute analysis sequence includes two-point standardization for each of the seven parameters, data printout and system clean-out.

The sensing section consists of four combination electrode modules similar to the one shown in Figure 4. A typical flow-through electrode module contains a liquid membrane electrode, a glass or solid-state electrode and a solid-state reference electrode. The liquid membrane electrode internal configuration is shown in Figure 5.

Table I shows results obtained with the system on replicate serum samples. The data shows excellent agreement with that obtained by reference procedures for each of the test channels. The Space-Stat System has been in operation for almost one year at NASA with very few problems. Their current plans are to employ a system of this type to perform in-flight analyses for the upcoming Space-Shuttle program.

A second recent instrumental development, the Orion Model SS-20 Ionized Calcium System, is shown in Figure 6. This system incorporates the basic electrode designs developed for the NASA Space-Stat System for the automated assay of ionized calcium. Figure 7 shows a diagram of the system. The elements of the fluids section are the electrode module, a four-channel peristaltic pump, a septum sample inlet, an air-liquid switching valve, and a cartridge container for reagents and waste.

To initiate an analysis, the operator injects approximately 0.5 ml of sample into a holding line, and presses a single button. The sample is then automatically pumped through the

Table 1. Space Stat Serum Results (6 runs)

Species	Expected	Found	Std. Dev.
Na^+	136 meg/l	131.2 meg/l	2.4 meg/l
K^+	4.4 meg/l	4.2 meg/l	0.1 meg/l
Cl^-	103 meg/l	104.2 meg/l	2.2 meg/l
Ca^+	4.2 meg/l	4.1 meg/l	0.2 meg/l
Ca^{++} T	8.9 meg/l	8.7 meg/l	0.2 meg/l
pH	7.88	7.37	0.01
pCO_2	40 mm	40.3 mm	1.0 mm

electrode module and a potential reading is taken on the flowing sample and stored. A programmed series of alternating segments of air and standard solutions follow the sample through the holding line and electrode module, thereby rapidly and completely cleaning the line. A continuous segment of the internal standard then flows through the electrode module, a second reading is taken and the ionized calcium level is computed and displayed digitally in mEq/l. Meanwhile the sample holding line has automatically been filled with air to ready the system for the next sample injection.

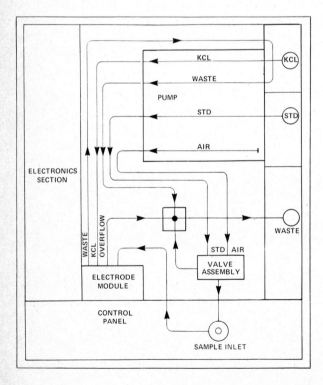

Fig. 7. SS-20 Fluid Manifold Diagram.

IV. Ionized calcium measurement in whole blood

The direct measurement of ionized calcium in whole blood offers several advantages over the plasma and serum samples used in all of the previously cited clinical studies. Anaerobic preparation of serum and plasma samples is difficult and time consuming and can lead to significant error in the final results.

Previous attempts to measure ionized calcium in whole blood have been unsuccessful. One report (*Sachs*, *Bordeau* and *Presle* 1971) gave electrode measurements of calcium in whole blood, but indicated problems which they attributed to such factors as the effect of erythrocytes on liquid junction potential and complexation of calcium by heparin. However, recent work shows that the heparin binding of calcium was due to impurities in the heparin (*Drop* 1974). Also new electrode designs may eliminate some of the response problems with blood. It was thus decided to repeat some of the earlier work with the new system.

V. Calcium system analytical performance

After three months of operating the SS-20 system in our laboratory we find that the standard deviation of assays on aqueous standards covering the clinical range for ionized calcium (1 to 4 mEq/l) is approximately 0.03 mEq/l. This performance compares very favorably with other analytical procedures for the determination of calcium at millimolar levels.

Table 2. Ionized Ca^{++} Values — Serum vs. Whole Blood

Donor	Serum mEq/l	Whole Blod mEq/l
F.G	1.99	2.08
	2.08	2.15
		2.16
		2.08
		2.23
R.L.C.	2.15	2.13
	2.24	21.6
	2.17	2.21
	2.16	2.14
		2.15
		2.13
L.F.	2.13	2.15
	2.05	2.12
		2.14
		2.15
		2.12
		2.13
L.W.	2.01	2.13
	2.01	2.15
		2.17
		2.12

Preliminary results comparing the performance of the new calcium system on serum and heparinized whole blood (10 units/ml, U.S.P. Sodium Heparin, Elkins-Sinn, Inc., Cherry Hill, N.J.) collected from four normal individuals are shown in Table II. The close agreement between the serum and whole blood values indicate that the determination of ionized calcium directly in whole blood is possible. The low value for the samples denoted L. W. may be due to loss of carbon dioxide from the sample prior to analysis. The normal values shown are in good agreement with those reported in previous studies. Also, the reproducibility of results on repeated samples is good.

VI. Summary and future

A number of investigators have established a strong case for using ionized calcium measurements in differential diagnosis. This, coupled with the improved instrumentation and the strong possibility of whole-blood analysis should greatly increase the clinical attractiveness of the procedure.

More important, as a result of work on the NASA Space-Stat System, the necessary elements of multi-parameter electrode-based clinical systems are now in hand. The real future of these devices does not lie in simple duplication of the capabilities of currently available instrumentation. Rather, it lies in the development of portable, compact, and reliable analytical systems which effectively utilize the unique properties of electrodes to provide rapid feedback of clinical data in the intensive care and emergency arenas.

References

Drop, L. J.: Interdependence Between Plasma Ionized Calcium and Hemodynamic Performance. Thesis, Katholieke Universiteit, Nijmegen 1974

Fuchs, C., Paschen, K., Spieckermann, P. G., Westberg, C. V.: Bestimmung des ionisierten Calciums im Serum mit einer ionenselektiven Durchflußelektrode: Methodik und Normalwerte. Klin. Wschr. 50 (1972) 824–832

Ladenson, J. H., Bowers, G. N.: Free Calcium in Serum. II Rigor of Homeostatic Control, Correlations with Total Serum Calcium, and Review of Data on Patients with Disturbed Calcium Metabolism. Clin. Chem. 19 (1973) 575–582

Low, J. C., Schaaf, M., Earll, J. M., Piechocki, J. T., Li, T. K.: Ionic Calcium Determination in Primary Hyperparathyroidism. J. Amer. med. Ass. 223 (1973) 152–155

McLean, F. C., Hastings, A. B.: A Biological Method for the Estimation of Calcium Ion Concentration. J. Biol. Chem. 107 (1934) 337

Moore, E. W.: Ion-selective Electrodes. In: Studies with Ion-exchange Electrodes: Some Applications in Biomedical Research and Clinical Medicine, pp. 215–285, ed. by *R. A. Durst.* National Bureau of Standards Special Publication 314, Washington 1969

Muldowney, F. P., Freany, R., Spillane, E. A., O'Donohoe, P.: Ionized Calcium Levels in „Normocalcemic" Hyperparathyroidism. Irish J. med. Sci. 142 (1973) 223–229

Ross, J. W.: Calcium Selective Electrode with Liquid Ion Exchanger. Science 156 (1967) 1378–1379

Sachs, C., Bordeau, A. M., Presle, V.: Ionized Calcium Measurement in Whole Blood: A New Application of the Calcium Selective Liquid Membrane Electrode. Europ. Etudies Clin. et Biol. XVI (1971) 1

C. Fuchs

Ion-selective Electrodes in Clinical Medicine*)

(8 Figures)

Introduction

In view of the large number of papers on the theory of ion-selective electrodes, this paper is intended to indicate what special significance potentiometric methods have in the sphere of clinical medicine.

Advantages of Ion-selective Electrodes Compared with other Methods of Measurement

Ion-selective electrodes offer fundamental characteristics and advantages compared with other analytical procedures which make their introduction into the field of medicine conceivable. They are economical to produce, compact and easily transportable measuring units, for which very little power is necessary. Electrode operation is without danger (no combustible gases) and usually simple. Within the wide range of measurement of the electrodes there is a rapid read out. Discrimination capacity and detection limit of these measuring systems are usually sufficient for medical problems. For the scientifically involved medical doctor, the possibility of measuring ion activities instead of ion concentration offers great advantages. It is especially important in the clinical laboratory that only small sample volumes are necessary, that the biological fluids can be measured without preparation of the sample, and that there is no loss of sample during the measurement. The use of ion-selective electrodes in the medical field is consequently thoroughly justified and of decisive significance in many problems.

Before discussing such problems in detail, a few of the elements which are determined potentiometrically by medical doctors at the present time, ought to be briefly mentioned: In the body fluids, whole blood, serum or plasma, urine, cerebrospinal fluid, sweat and gastric juice there are 7 different ions, the determination of which is at present of particular interest to the medical doctor: i.e., Na^+, K^+, NH_4^+, Ca^{++}, F^-, Br^-, Cl^-.

Comparative Serum Analyses

The following investigations show that it is possible to determine electrolytes in the serum potentiometrically with reliability: In 20 normal and pathological sera Na^+, K^+ and Cl^- were determined both potentiometrically and with the usual clinical chemical methods. The correlation of the results obtained by different methods of measurement was investigated. In

---

*) Supported by the Deutsche Forschungsgemeinschaft, SFB 89 – Kardiologie Göttingen.

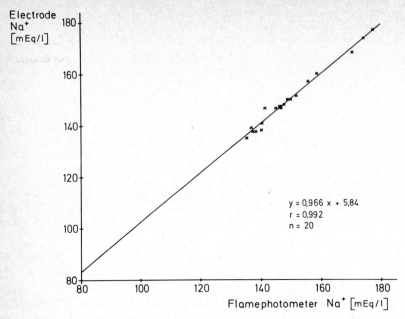

Fig. 1. Na⁺ determination in serum: comparative studies by flame photometry and potentiometry.

Figure 1, the results of flame photometric determinations of Na⁺ are compared with the results obtained with ORION flow-thru glass membrane electrode (Model 98–11). The two-point calibration of the electrode was performed with normal sera of known Na⁺

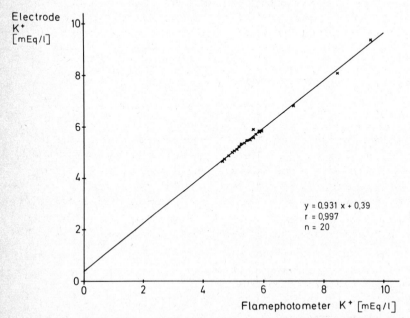

Fig. 2. K⁺ determination in serum: comparative studies by flame photometry and potentiometry.

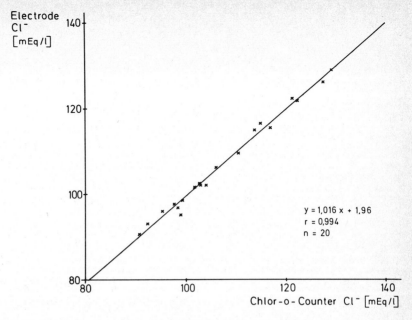

Fig. 3. Cl⁻ determination in serum: comparative studies by titrimetry and potentiometry.

concentration. Good agreement between the results can be seen. A similar level of correlation is seen in Figure 2: in this case the K^+ concentrations were also determined with a flame photometer and with the ORION liquid ion solvent flow-thru electrode (Model 98-19). Finally, in Figure 3 the Cl⁻ determinations were made titrimetrically with the MARIUS CHLOR-O-COUNTER and with the ORION liquid ion exchange elctrode (Model 98-17). Additional investigations on the reproducibility and recovery with the three electrodes showed a similar level of agreement with current reference methods.

The Importance of Potentiometric Determination of Ca^{++}:

It can be seen from Figure 4 just why the determination of Ca^{++} gave a boost to potentiometry in medicine. The total calcium in serum is subdivided into protein bound calcium and ultrafiltrable calcium. The latter is sub-divided into complex-bound and ionized calcium (Ca^{++}). Only Ca^{++} is biologically active. The proportion of this species in the serum depends on the serum protein, on the acid-base status and on the amount of complex-formation in the serum.

As the method of Ca^{++} determination in serum has already been reported in detail (*Fuchs, Paschen, Spieckermann* and *Westberg* 1972) only a few physiological and pathophysiological problems will be selectively discussed to underline the significance of this analytical procedure.

Examples of Application for Ca^{++} Determination:

Apart from a short communication by *Moore* and *Blum* (1968) which relates to only 5 hospital patients, there are no reports on the question of Ca^{++} in the cerebrospinal fluid

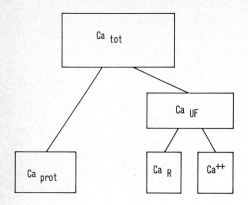

Fig. 4. Serum – Ca – Fractions.

(CSF). A study of normal levels was therefore undertaken in a larger number of subjects. The essential results of this study are summarized in Figure 5. In 24 persons, a CSF Ca^{++} of 1.62 meq/l was found, and this accounted for 73% of the CSF total calcium. In contrast to this, only 50% of the calcium in the serum occurs in the ionized form. This is explained by the low proportion of protein in the CSF; in the subjects investigated, the CSF protein was <45 mg% and in the serum 6.5 g%.

In this connection it is of interest to know how far shifts in the acid-base balance, which lower Ca^{++} in the serum as a result of hyperventilation and cause tetany, also lower Ca^{++} in the CSF. Since a lowering of the calcium in the CSF raises cerebral spasmophilia (*Prill* 1969) and leads to disturbances of consciousness, it was important to discover to what degree the organism is protected against a reduction of Ca^{++} in the CSF.

The results of this investigation are set out in Figure 6. Four patients had been hyperventilated for about 10 minutes. During this time distinct signs of tetany developed. At the beginning

N = 24	CSF *			Serum			$\frac{CSF}{Serum}$
	$\bar{x}$	s_x	V	$\bar{x}$	s_x	V	$\bar{x}$
Ca^{++} mEq/l	1,67	0,103	6,2 %	2,42	0,167	6,9 %	0,69
Ca_{tot} mEq/l	2,28	0,137	6,0 %	4,72	0,306	6,5 %	0,49
$\frac{Ca^{++} \cdot 100}{Ca_{tot}} = \%$	73,3	4,58	–	51,3	2,82	–	–
* CSF = Cerebrospinal Fluid							

Fig. 5. Ionised and total calcium in serum and cerebrospinal fluid.

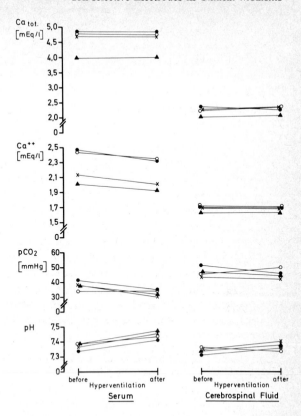

Fig. 6. Total calcium, ionized calcium, pCO₂, and pH in serum and cerebrospinal fluid before and after hyperventilation.

and the end of the investigation, total calcium (Ca_{tot}), Ca^{++} and the acid-base status of the serum and CSF were determined.

The shifts in the acid-base equilibrium in the serum correspond to those of a respiratory alkalosis with a fall in Ca^{++} as expected. The important point is that this change in Ca^{++} is not observed in the CSF, although slight acid-base shifts do take place. Ca_{tot} remains constant in both body fluids. This means, therefore, that the subarachnoid space of the central nervous system is protected against short-term changes of Ca^{++} caused by hyperventilation. Consequently the anticonvulsive function of the CSF calcium is retained.

A further example comes from the field of thoracic surgery. In operations on the arrested heart, the circulatory function must be taken over by a heart-lung machine. Relatively large amounts of bank blood or blood substitute solutions are required to fill the heart-lung machine before the operation. When the patient's circulation is connected to the heart-lung machine, extreme electrolyte shifts may occur. For the normal course of stimulation and contraction of the beating heart, however, a certain distribution of Ca^{++} in the intra- and extracellular spaces of the myocardium is necessary. In operative procedures with extracorporeal circulation, shifts within the calcium fractions lead to both arrhythmias and to impaired contraction of the usually damaged myocardium. It was the task of the following investigation to examine the Ca^{++} level during the operation, using various Ca contents in the prime solution of the heart-lung machine.

The plan of the investigation and the results of this study can be seen in Figures 7a and 7b. In the tests, total calcium, ionized calcium, and total magnesium were analyzed. It is clear from the Figures that Ca_{tot} does not reflect the behavior of Ca^{++}. Acid-base shifts and hemodilution are chiefly reponsible for this. It can be recognized that by the use of high Ca concentrations in the machine prime solution, the Ca^{++} level remains at the upper limit of normal during the bypass. In the other group, they are distinctly lowered at the same stage of the operation. On theoretical grounds and from earlier experimental findings (*Kübler*, *Hähn*, *Hellberg*, *Orellano*, *Reidemeister* and *Spieckermann* 1965) the low Ca concentration is to be preferred. Since energy requirements of the arrested heart and thus the velocity of ATP-breakdown during ischemia are closely related to the Ca^{++} concentration of the extracellular space, low plasma calcium levels are considered to be advantageous during cardiopulmonary bypass. Only at the end of partial bypass, before the heart fully takes over circulatory work, is calcium substitution recommended.

A last example comes from nephrology: Disturbances of Ca phosphate metabolism are among the limiting factors of prolonged hemodialysis treatment. Already in the early stages of renal failure, the patients develop a hypocalcemia which is due to increased Ca excretion

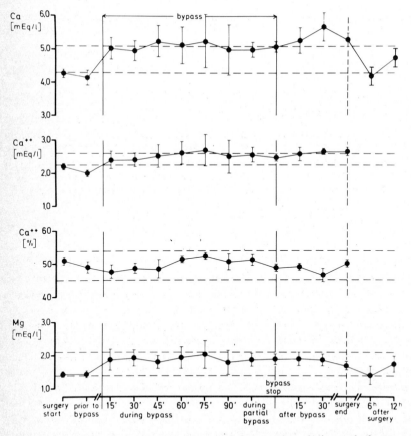

Fig. 7. Total calcium, ionized calcium, and total magnesium during and after open heart surgery with extracorporal bypass using a high calcium contents in the prime solution of the heart-lung machine Prime solution: $Ca_{tot} = 9.67$ mEq/l; $Ca^{++} = 6.88$ mEq/l; $Mg_{tot} = 1.85$ mEq/l.

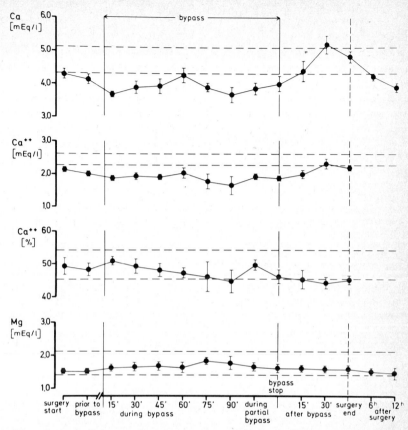

Fig. 7b. Total calcium, ionized calcium, and total magnesium during and after open heart surgery with extracorporal bypass using a low calcium contents in the prime solution of the heart-lung machine.
Prime solution: Ca_{tot} = 2.83 mEq/l; Ca^{++} = 1.98 mEq/l; Mg = 1.88 mEq/l.

and disturbances of vitamin D metabolism. This hypocalcemia leads to a stimulation of the parathyroid glands and to "hyperparathyroidism" which, especially if it has changed into an autonomous, unsuppressible form, can be a fatal complication under certain circumstances. The therapeutic efforts are concentrated particularly on a compensated Ca balance. With regard to hemodialysis therapy, a long dispute is now in progress as to the concentration of Ca in the dialysate which will achieve this compensated equilibrium. That is, what concentration of dialysate calcium is most likely to lower the risk of intermittent stimulation of the parathyroid glands?

The above-mentioned interrelationships were investigated by testing the effect of 3 different dialysate Ca concentrations on the serum Ca fractions during dialysis treatment and, in part, in dialysis-free intervals. Figure 8 shows the results of this investigation. In the first group of 10 patients with a dialysate Ca of 3.0 mEq/l a distinct and significant rise in total calcium can be seen. Other authors have discussed this finding as being due to a positive Ca balance, but this is not quite correct. The rise in Ca_{tot} can be explained by hemoconcentration resulting from ultrafiltration during dialysis treatment. The rise results from the increase of Ca_{prot}. Of central importance in this group of patients is the significant fall in Ca^{++} by 0.11 mEq/l. This means that, in spite of a clear rise in Ca_{tot}, the parathyroid glands are stimulated. It can be seen from Figure 8 that such a situation ought largely to be excluded with a dialysate Ca of 3.5 mEq/l in group II and that during dialysate treatment, the Ca balance becomes positive only with a Ca content of distinctly more than 3.5 mEq/l (in group III, 4.5 mEq/l). On the basis of these findings and from the long-term studies of other authors on the problem of parathyroid function with various dialysate Ca concentrations, the necessity to raise the Ca content from 3.0 to at least 3.5 mEq/l seems justified. (*Fuchs, Brasche, Donath-Wolfram, Kubosch, Quellhorst, Scheler* 1974).

Summary and Outlook

The examples described show the advantages of the ion-selective macro-electrodes in the medical field. These electrodes can be applied to measurements in almost all body fluids. They are without rival for the determination of electrolytes which are present in the body only partly in ionized form.

The future of ion-selective macro-electrodes in medicine will depend on their wider application to other important body elements; of the elements which are of interest to the medical doctor, surely magnesium must be added to those mentioned above, since magnesium is known to be distributed in serum fractions in a similar manner to calcium. Electrodes for the determination

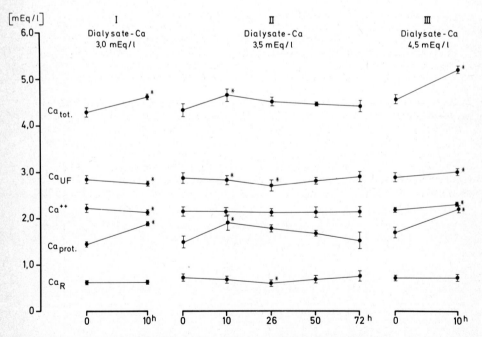

Fig. 8. Mean values and standard deviations of the mean of changes of total calcium (Ca_{tot}), protein-bound calcium (Ca_{prot}), ultrafiltrable calcium (Ca_{UF}), ionised calcium (Ca^{++}) and complex-bound calcium (Ca_R) using three different dialysate Ca-concentrations (* = $p < 0.05$).

of trace elements such as lithium, manganese, iron, cobalt, nickel, copper, zinc, cadmium, iodine, mercury and lead would also be desirable. That ion-selective electrodes are also suitable for the determination of trace elements has been shown by some investigations on the determination of F⁻ in the serum. These investigations presume optimal selectivity, detection sensitivity and stability of the electrode. The problem of providing suitable standards must also not be underestimated. In this, the appropriate efforts of the National Bureau of Standards may lead to more widespread use of electrodes in the field of medicine.

The future of ion-selective electrodes must be assessed differently for the individual medical areas. While electrodes may be introduced into scientific laboratories, even though there are technical and practical difficulties, appliances will only gain access to hospital routine laboratories if their handling requires a minimum of experience with ion-selective electrodes, i.e., they should be as fully automatic as possible. In this respect, the ability to analyze whole blood is of special interest. This unique possibility makes ion-selective electrodes markedly superior compared to other measuring methods. In all surgical procedures, and in intensive medical situations where larger electrolyte shifts may occur, it is of great importance to be able to carry out electrolyte checks directly at the bedside with small quantities of material without time-consuming centrifugation. In emergency situations such as diabetic coma, hypovolemic shock, burns or poisoning, much of the anxiety would be lessened if it were possible to construct suitable, reliable apparatus for analysis. Efforts in this field seem to be of the greatest importance for clinical medicine.

References

Fuchs, C., Brasche, M., Donath-Wolfram, U., Kubosch, J., Quellhorst, E., Scheler, F.: Dialysate Calcium and Plasma Calcium Fractions During and After Haemodialysis. Klin. Wschr. 53 (1975) 39–42

Fuchs, C., Dorn, D., Fuchs, C. A., Henning, H. V., McIntosh, C., Scheler, F.: Fluoride Determination in Plasma by Ion-selective Electrodes: A Simplified Method for the Clinical Laboratory Clin. Chim. Acta 60 (1975) 157–167

Fuchs, C., Paschen, K., Spieckermann, P. G., Westberg, C.,: Bestimmung des ionisierten Calciums im Serum mit einer ionenselektiven Durchflußelektrode: Methodik und Normalwerte. Klin. Wschr. 50 (1972) 824–832

Kübler, W., Hähn, N., Hellberg, K., Orellano, L. E., Reidemeister, C. J., Spieckermann, P. G.: Beziehungen zwischen aerobem und anaerobem Energieumsatz des Herzens unter verschiedenen funktionellen Bedingungen. Verhandl. Dtsch. Ges. Kreisl.-Forsch. 31 (1965) 86–92

Moore, E. W., Blum, A. L.: Studies with Ion Exchange Calcium Electrodes: The distribution of Ionized Calcium between Blood and Cerebrospinal Fluid (CSF) in Man and Dog. J. clin. Invest. 47 (1968) 70a

Prill, A.: Die neurologische Symptomatik der Niereninsuffisienz. In: Schriftenreihe der Neurologie, Bd. 2. Springer, Berlin 1969

E. Sinagowitz, R. Hagist, H. Sommerkamp

Measurements with Surface Electrodes of Interstitial Ion Activities and Local Tissue Po$_2$ on the Human Kidney (Preliminary Report)

(1 Figure)

Microelectrodes for measurements of local tissue Po$_2$ and for determining the activity of different ions in the interstitial space were until now used only in animal experiments.
Considering the objection of many clinicians that the conditions in the human body are not in all cases comparable with those in animals, it is important to make electrodes applicable for measurements in human tissue. On the other hand clinicians are searching for techniques which could give them valuable information about the actual state of organs of interest which normally can only be obtained indirectly by measuring such parameters as electrolytes or acid-base balance in the body fluids. Especially with respect to the kidney it is of great value to develop a dynamic test for the function of the kidney preserved for transplantation and to have a means of determining the borderline between normal and pathological tissue in order to perform subtle operations with the aim of preserving as much of normal parenchyma as possible.

The only electrodes suitable for these purposes are surface electrodes, since they can be used without causing additional tissue lesions. Such electrodes are now available as multiwire surface electrodes for measurements of tissue Po$_2$ (*Kessler* et al 1969) and as ion-selective surface electrodes as developed by *Kessler*, *Höper* and *Simon* (1974) and described in detail at this meeting.

For the first time, with such electrodes we were able to measure interstitial ion activities and local tissue Po$_2$ on the surface of human kidneys.

Two main problems are of great importance before performing intraoperational measurements on patients to test the applicability of these electrodes. The first is that all instruments which are to be used in the human body must be sterilized. This was achieved by gas sterilization with a mixture of ethyleneoxide (15%) and carbondioxide (85%) at a moderate temperature (50—55 °C) under high pressure (2.2 atm) and a pre- and postvacuum of 8 atm. We found that, after the electrodes had undergone the sterilization procedure several times, the slope of the sodium electrode used decreased from 57.5 over 55 to 52 mV at room temperature and that of the calcium electrode from 29 over 27 to 25 mV.

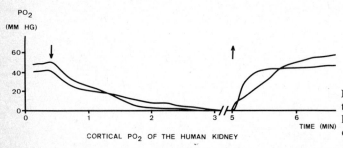

Fig. 1. Local tissue Po$_2$ on the surface of the human kidney during an ischemic cycle.

The other main problem is the selection of suitable patients. In order not to harm the patient during the experimental stage of testing, the applicability of the electrode measurements on the human kidney are permissible only on organs which are to be removed, i.e., mainly on kidneys with a tumor such as a hypernephroma, but with enough normal parenchyma left. The first measurements of sodium-ion activities on the surface of the human kidney showed values which were in close proximity to the serum levels.

The results of continuous measurements of local oxygen tension performed in the outer cortex of the kidney with the multi-wire surface electrode (*Kessler* and *Grunewald* 1969) are shown in Figure 1. After interruption of the renal blood flow with a tourniquet around the renal pedicle the Po_2 on the renal cortex decreased slowly from levels between 40 and 60 mm Hg and after 5 minutes of ischemia the Po_2 returned instantaneously to its original level. The same type of Po_2 fluctuation under ischemia has been observed in various animals.

These very first observations show that measurements in man with surface electrodes for ion activities and for local tissue Po_2 are possible in principle.

References

Kessler, M., Grunewald, W.: Possibilities of Measuring Oxygen Pressure Fields in Tissue by Multi-wire Platinum Electrodes. In: Oxygen Pressure Recording in Gases, Fluids and Tissues, ed. by *F. Kreuzer.* Progr. Resp. Res. 3 (1969) 147

Kessler, M., Höper, J., Simon, W.: Methodology and Application of the Multiple Ion-selective Surface Electrode (pH, pK, pNa, pCl) for Tissue Measurements. Fed. Proc. 33 (1969) 279

Discussion of Session IV B

Chairman: *D. W. Lübbers*

(3 Figures)

Discussion of Paper by R. C. Thomas: The Effects of CO_2 and Bicarbonate on the Intracellular pH of Snail Neurons

Heisler: You are confirming some results that I obtained some time ago in 1972. We found with the DMO method, a method quite different from that used in your investigations, no changes of intracellular pH within the range of extracellular pH between 7.4 and 7.15 both in metabolic and respiratory changes. This was very much doubted at that time and I am pleased very much that you now show a similar phenomenon in your experiments.
Secondly you told us that the beta value of bicarbonate buffers is much greater than the beta value of non bicarbonate buffers in your snail neuron. This cannot be extended to other animals, for example rats. This fits for some poikilothermic but not for warm blooded animals. In warm blooded animals the beta value of rat diaphragm for example is about 70 meq/pH/l cell water, the beta value of bicarbonate is much lower, about one third of that. Another fact I would like to know is what temperature you used.

Thomas: To answer your last question first, the temperature was room temperature, about 20–21 °C. I must admit I am not too familiar with the DMO work, although it is perhaps the best way of measuring intracellular pH indirectly.

Sies: As I have mentioned before, in the liver we have been able to show the intracellular acidification upon increasing total carbon dioxide at constant extracellular pH, but we have not studied that for a prolonged period of time. It is very interesting that intracellular pH returns to the original level in snail neurons. Have you tried to modify this process?

Thomas: Yes I have tried a number of ways of stopping the cell pumping out hydrogen ions, if that is what it is doing, I have not succeeded. You can either half kill the cell, in which case it goes acid inside anyhow, or you cannot stop the proton pump. Diamox slows down the pH changes, but does not seem to block the pump. The intracellular pH is also well controlled in a CO_2 free system, so I doubt whether bicarbonate is terribly important.

Discussion of Paper by A. M. Brown: Resting Ionic Permeabilities in a Neuron

Discussion of Paper by H. D. Lux: Simultaneous Measurements of Extracellular K^+ activity and Membrane Currents in Snail Neurons

Brown: I have three questions. First, are the currents with the point clamp similar qualitatively to the total currents? Second, are the localized measurements of potassium similar over the surface of the membrane? Third, as I recall the records show that even when there is considerable extracellular potassium there is an outward tail current; does this represent some persistent K^+ activation? Do the tail currents reverse at very depolarized potentials?

Lux: For clamp potentials of longer durations, the enveloping slope of the total current is similar to the current measured from a somatic area under 'patch' clamp control. However, the early inward current usually shows gross deviations due to activity of unclamped axonal regions which affects the total current recordings. The conclusions drawn from comparing K^+ measurements and membrane currents apply similarly for total currents if the pulse durations are sufficiently long.

The tail currents are usually outward since the holding potentials of -45 to -50 mV are considerably less negative than the inversion potential for the delayed current which is about -65 to -75 mV. The conditioned (second) current sometimes has a slightly less negative inversion potential than that of the conditioning (first) pulse. This may result from the inward current component. The inversion potential is also lower after higher and longer currents. This is obviously brought about by extracellular K^+ accumulation which is stronger during the larger currents of the first pulses.

Thomas: Does Dr. *Lux* think that the new enzyme techniques of dissociating nerve cells, so that you can get them without their axons, make the patch clamp technique obsolete? It would be nice if I did not have to make such an elegant technique work in my lab.

Lux: We have tried isolated snail nerve cells which were dissociated by the enzyme technique. For many purposes these are useful preparations which could possibly make the patch clamp method unnecessary. This should be checked. The leakage conductance of the cells in these experiments was considerably higher than in cells in the intact preparation. Reasonable equipotentiality, which is required when patch current control is omitted is warranted only by a high ratio of plasma conductance to membrane conductance.

Thomas: The cloride is notoriously unselective and I wonder what evidence you have that your intracellular values of chloride are in fact chloride and you are not having contributions from other ions.

Brown: We have checked this in a number of ways. One way is that acetylcholine applied to the cell makes it behave like a chloride electrode with regard to its voltage response to changes in external chloride.
Under these circumstances one gets an equilibrium potential for ACh that is identical to the measured chloride equilibrium potential. We have also recently made direct measurements of chloride in these cells with a chloridometer which indicates good agreement with the ion electrode values. We also find that our measured chloride activities in the resting cell in the steady state are much lower than can be predicted on the basis of passive distribution so that if there were interfering anions one would expect these values to underestimate the true values.

Sies: I have a question for Dr. *Brown*. It was interesting to hear that the tris and the methane sulphonate would penetrate into the cells. Has this been measured directly or is it inferred just from the measurements? It think it would be of interest to measure this with labelled compounds.

Brown: The evidence is that when you make a chemical measurement for net chloride flux using an ion electrode you get a value that is higher than the measurement you get from electrical data where the substituting anion is methane sulphonate. On the other hand if you substitute sulphate, despite the fact that you have problems with ionic strength and possibly calcium binding, the electrical and chemical values are similar. There was additional strong evidence that methane sulphonate is permeable, which is important since other ions of the same size such as propionate or acetate, which are frequently substituted for chloride, are also likely to be permeable. So that leaves you with sulphate and/or the possibility of paratoluensulfonate or salicylates as substituting anions. The problem with the later two is that they have direct effects on the surface charge of the membrane. In the case of tris the evidence is again that the chemical measurements of sodium flux are higher than the electrical measurements where you use tris as a substituting ion. But in those experiments where we measured the voltage changes when we substituted potassium for sodium and there was no tris in the solution, we obtained permeability values for sodium that agreed with the chemical data. I think that although we have not made direct measurement of tris or methane sulphonate inside the cells, there is no question from these data that they are permeable in this particular neuron.

Lux: Dr. *Brown*, have you performed the measurements of membrane conductance at a constant potential or at varied potentials? There is apparent rectification at usual resting potentials which may result from persistent voltage-dependent calcium and sodium currents in molluscan neurons.

Brown: We are worried about any voltage-dependent conductaneous being turned on in these experiments so we were very careful to make all our measurements at or close to the resting potential. That still leaves us with the problem of those conductances that are voltage-dependent that are present in the resting state. In the case of calcium we removed it from the solution in some of the experiments without affecting the results.

Zeuthen: I want to comment on the DC potential of the brain during various kinds of stresses, e.g., hypoxia and anoxia. The brain first responds with a small negative change in the potential; if the stress is prolonged, an anoxic depolarization develops and there is a big negative change in the potential. These two changes, the small and the large, in potential arise from different phenomena: when you have spreading depression or anoxic depolarization, the potential is very much linked to the influx and eflux of cellular ions. However, the small or pre-depolarization changes in potentials seem to depend on other things, for example the type of anaesthetic. The change in the chemical potential of the potassium ion looks like the small shift in potential but of opposite sign, when the animals are anaesthetized with Nembutal and immobilized with Flaxedil®, but not when we use Urethane. Flaxedil®, e.g., causes the baseline of the DC potential to move $+ 5$ mV. Furthermore, in measurements with two electrodes at two different places in the brain, we get exactly the same electrical potential shift on the two electrodes whereas one of them measured more potassium. Furthermore, in the ventricle the potential changed exactly like the potential in the cortical tissue whereas the potassium first decreases during hypoxia. So I do not think there is a direct connection between the small changes in electrical potential and the small potassium changes.

Lehmenkühler: We measured the DC potential near the potassium-ion sensitive tip of the electrode and not at a distance.

Grote: Dr. *Thomas*, during your experiments you found an increase of the buffer capacity of the nerve cells and you explained this as a result of a bicarbonate influx into the cells. I agree.

Aside from this, I think it should be taken into account that during your measurements the hydrogen-ion content in the cells increased and that the pH value came nearer to the pK values of the different intracellular buffer systems.

Thomas: Yes, but as intracellular pH moved in an acid direction, in fact the intracellular buffering power decreased, as you would expect from a loss of bicarbonate, and when there was no bicarbonate present, the intracellular buffering power seemed to be constant certainly over a range of about 7.6. It certainly is possible that there are other intracellular buffers coming into play and I have only really begun these experiments, so it is a little premature for me to speculate any further.

Discussion of Paper by Crowe, Mayevsky, Mela, Silver: Measurements of Extracellular Potassium, DC Potential and ECoG in the Cortex of the Conscious Rat During Cortical Spreading Depression

Brown: The DC potential that you are dealing with comes from a heterogenous population of cells some being glial and some being neurons. The glial cells show quite different electrical properties from cortical cells. Thus, the glial cells have a linear current voltage relationship over the tested range, i.e., they are nonexcitable. My question is: When you record DC potentials and correlate them to extracellular potassium changes, what is the contribution of these different cell populations to these potentials?

Crowe: Yes, we are measuring in a mixture of glia and nerve cells in an artificial extracellular space produced by the electrode tip. The DC and the potassium responses are very likely a composite of glial and neural responses. The contributions from different cell types at this point can not be separated.

Thomas: I wonder, I am unfamiliar with this field but are you able to produce the same effects with an electrical stimulus at the KCl application point or alternatively perhaps by cooling that point?

Crowe: Yes, mechanical or electrical stimulation or cooling can be used to initiate the response in place of an application of KCl.

Discussion of Paper by Lehmenkühler, Speckmann, Caspers: Cortical Spreading Depression in Relation to Potassium Activity, Oxygen Tension, Local Flow and Carbon Dioxide Tension

Betz: In acidosis one usually observes linear correlation between the increase of potassium concentration and hydrogen-ion concentration and the CO_2 in each instance. Did you observe changes in the systemic blood pressure or can you explain the changes in blood flow only by decreases of the local vascular resistance?

Lehmenkühler: There is no change of the systemic arterial pressure so that one can assume that the information to the local vessels is a direct one.

Kessler: Is there any theory about the oscillator which causes oscillations during spreading depression?

Lehmenkühler: That is an important question, but we have not investigated this problem.

Discussion of Session IV B

Discussion of Paper by D. Heuser und E. Betz: Measurements of Potassium-Ion and Hydrogen-Ion Activities by Means of Microelectrodes in Brain Vascular Smooth Muscle

Stockem: Dr. *Betz*, you changed the hydrogen- and potassium-ion concentrations in the perivascular space and you also altered the extracellular calcium-ion concentration by using calcium-ion antagonists like EGTA or EDTA. We know, however, that mainly intracellular calcium ions play an important rôle in the activation of the contractile system within the cytoplasma. How can the changes in concentration of extracellular calcium on the activation of the intracellular actomyosin-system be explained?

Betz: For this purpose we used calcium-binding agents which are assumed to remain outside the cell; indeed, we see that with these agents the pH-dependent reactions are partly or completely abolished, whereas the others – the electrically induced reactions – are still there in a very normal way. One cannot really discriminate completely between the intracellular pool and an extracellular one. With EDTA or with EGTA in the extracellular space, I think some of the calcium may diffuse out of upper layers of the membranes because of the concentration gradients. The idea that the calcium ion-activated ATPase plays a role in this context is an assumption made by *Fleckenstein* and *Grün*. Indeed it is difficult to understand how the intracellular compartment of calcium is influenced by the variation of the extravascular calcium ions. The only way seems to be the assumption that some of the calcium ions diffuse out of the membrane when the extracellular calcium-ion concentration reaches zero. The effect on the sodium-ion channel is still unclear in the smooth muscle cell.

Scarpa: Dr. *Betz*, as you probably know, a similar pH effect is present in perfused working hearts, where, if you decrease the pH of the perfusate to arround 6.5, you have a dramatic decrease in the contractility. Since we were interested in this pH effect, wo have isolated cardiac sarcolemma in a reasonably pure form for calcium binding studies. In isolated cardiac sarcolemma we found two classes of binding sites, one with an apparent K_m of 2 mM and another with an apparent K_m of 10 μM. At pH 6.5 the calcium ions bound were displaced from the high-affinity and not from the low-affinity binding sites. Similary, Verapamil at very low concentrations inhibits calcium ion binding to the high-affinity sites of sarcolemma. At this point, it is difficult to discriminate whether all the pH effect is due to a competition between calcium and hydrogen ions at some binding site of the sarcolemma or whether some other steps of the contractile mechanism are involved.

Betz: The tone of the vascular smooth muscle cannot only be varied by changes of the action potential but also by processes which are not dependent on action potentials. It is assumed that there are two basic mechanisms for smooth muscle concentration, one of which is triggered by action potentials and the other not.

Gebert: In the last slide Dr. *Betz* showed that during application of 100 mM potassium solution, electrical stimulation was effective. Normally stimulation is not effective in depolarized vessels and 100 mM K^+ solution is quite a depolarizing solution. I think that the true extracellular potassium concentration must be somewhere in between the plasma concentration and the rinsing concentration. That might be true for the 250 mM concentration too.

Betz: If I understand correctly, you think that the concentration of a substance in between the smooth muscle cells is the result of different interluminal and perivascular concentrations.

Discussion of Session IV B

This may be the reason why the observed reaction to K^+ do not correspond in their absolute values with those in which the intra- and extravascular concentration were the same, e.g., during artificial perfusion.

Clausen: I think *Reuter* in Berne has demonstrated that the tension in vasular smooth muscle contraction and calcium content are determined by sodium gradient and wouldn't it be possible that the effect you have of sodium/potassium ratio could be related to a decrease in the sodium gradient and the ensuing increase in the cytoplasmic level of calcium ions.

Betz: As far as the sodium concentration in the perivascular space is concerned, we kept that constant, but the transport of sodium into the cells of course may change despite the constant perivascular Na^+.

Lehmenkühler: It is well established that application of even isotonic KCl solutions elicits changes in the functional state of the cerebral cortex such as spreading depression. These changes in cortical activity are accompanied by high voltage negative DC shifts, lowerings of the local pO_2 and by considerable increases in extracellular potassium-ion activity. How can you exclude that these processes interfere with the mechanisms you discussed?

Betz: This was very simple when we had different set-ups. One of these questions was solved by pushing a little strip of plastic below the vessel- we isolated the vessel. Then we did experiments in exactly the same way or we rinsed the perivascular space constantly with mock CSF during the whole procedure so that no ions from the tissue could accumulate in the perivascular space. We again had the same effect.
Do you mean that there are nervous connections from the tissue through the perivascular space to the vascular wall?

Lübbers: It certainly would be nice if you could have a look inside the smooth muscle cell. We have these electrodes now, and, in the near future, we'll have beautiful tailored molecules.

Discussion of Paper by Höper, Kessler, Simon*): Measurement with Ion-selective Surface Electrodes (pK, pNa, pCa, pH) during No-flow Anoxia

Fuchs: Do you have any experience with the surface electrodes on human organs in vitro?

Kessler: No, not yet.

Fuchs: How can you be sure that this rise in ion calcium is not a pH or a magnesium ion effect?

Kessler: The great advantage of this ion exchanger is that the interference by hydrogen ion is so small that it doesn't play a significant rôle in the reaction. Our magnesium levels are very low.

Mela: I would like to ask you about the release of calcium from mitochondria. Some time ago it was shown that there is release of accumulated calcium during complete ischemia in tissues such as the kidney; would you care to comment on the correlation of your experiments to this?

---

*) and: Kessler, Hajek, Simon: Four-Barrelled Microelectrode for the Measurement of Potassium-, Sodium- and Calcium-ion Activity.

Discussion of Session IV B

Kessler: I know this reference, but this is still a hypothetical scheme and we have not yet investigated this question.

Lux: How high is the intracellular calcium content in the normal state? The general assumption is that potassium activity is in the range of 10^{-8} or 10^{-6}.

Kessler: That's certainly the situation also for calcium in the liver. Calcium must come from subcellular particles and there are three possibilities 1) the mitochondria, 2) the endoplasmic reticulum and 3) the membrane. But we haven't yet data confirming this.

Heisler: You reported a constant pH of 7.1; this is contrary to several determinations with DMO where it is shown that intracellular pH is changed with a delta pH i over delta pHe ratio about 0.5 to 0.7 which agree with our own measurements.

Kessler: Under which conditions, please?

Heisler: Under conditions in vivo. I don't think that the intracellular pH is so stable in your conditions yet you have really no change with the pH. Related to this I have a second question what was your equilibrium time for DMO?

Kessler: About 30 minutes.

Heisler: It is known that this is a quite insufficient time for DMO equilibrium in rat liver, you need at least 2 hours, so I would say this is an artefact.

Kessler: These values were measured not only with the DMO method, some of them also with the pH microelectrode. The equilibration time of DMO in the perfused rat liver is shown in Figure 1.

Gebert: Yesterday we saw that in the liver even a small microelectrode may do damage. What do you think about measurements with a *Thomas* type electrode and their relation to these morphological findings?

Kessler: There is no question that we cause damage. I have mentioned this in the discussion to the paper of Dr. *Thomas*. In a few cases, however, we think that we were able to measure the intracellular pH.

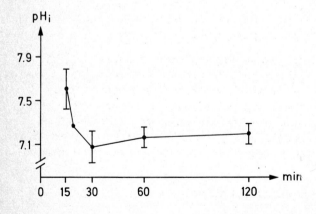

Fig. 1

Brown: Have you tried something like ouabain which is a reasonably specific sodium-potassium ATPase inhibitor and its effects on intracellular ATP concentration? Under your circumstances I can imagine that pratically every energy-requiring enzyme system is possibly involved in this mechanism, let's say calcium ATPase. The second question I don't know if I understood you correctly. Did you say that depolarization was due to the outward movement of hydrogen ions across the membrane? Formation of hydrogen ions is one important factor because we found that when we add more buffer to our system then these changes caused by ischemia are changed very much and so we think that suddenly the hydrogen ion is one of the very important factors.

Kessler: Concerning your first question we used ouabain, unfortunately the rat liver isn't sensitive to ouabain. At present we are analyzing the effects of protons which are formed during ischemia. There is evidence that they induce a protonation of the plasma membrane which subsequently induces a depolarization.

Discussion of Paper by L. Mela, C. W. Goodwin and L. D. Miller: Mitochondrial Calcium Transport Activity Following Acute and Chronic Changes of Tissue Oxygenation

Dora: I would like to ask if you have any data about potassium activity of the mitochondrial membranes in newborn animals.

Mela: No, unfortunately not yet. The mitochondrial potassium transport activity is very slow in mitochondria from adult animals. Although it can be enhanced by inducers it is not physiologically important. I do not know whether the newborn mitochondria behave differently from adult in this respect.

Lübbers: Have you also tried other organs.

Mela: Yes, the data presented were all from the heart, but we have also studied brain and liver mitochondria. These preparations in the newborn are much more difficult than the heart, and thus the data are less consistent. The brain in particular is a very difficult organ to deal with when comparing data from one animal to another. The trends, however, are identical in all the organs that we have looked at.

Lübbers: I do not know how this development of the arterial Po_2 is in other animals, but in humans we know that Po_2 rises very rapidly after birth and goes up to almost normal values. If the newborn respires oxygen you can immediately get a very high Po_2.

Mela: In our experiments the newborns were not allowed to breath oxygen, only air! We have also worked on other animals – the piglet and the lamb. In both of these animals the Po_2 rises faster than in the puppy. Still we obtain linear relationship between the arterial Po_2 and the Po_2 rather than anything else. There is a very interesting point in this respect. Some of the puppies show a Po_2 change which is much slower than the one I showed here. Mitochondrial studies in these animals also showed a slow change in respiratory capacity. Thus there is always a linear relationship between the arterial Po_2 and the mitochondrial respiratory capacity.

Pressman: Did I understand your data correctly that dog hearts totally lose their ability to transport calcium after several days?

Discussion of Session IV B

Mela: In the puppies at the age of seven days, yes. This is during the phase when the respiratory activity drops as well. However, these changes are transient. During the next two weeks of life, the mitochondrial calcium transport capacity recovers and reaches adult levels.

Discussion of Paper by Zeuthen and Monge: Microelectrode Recording of Gradients of Electrical Potential, Potassium and Chloride Activity Across the Epithelial Cell Layer of the Rat Ileum in vivo.

White: I have had experience in measuring membrane potentials in the small intestine of two amphibian preparations, bull frog and amphiuma. Most recently I have done in vitro measurements on the amphiuma small intestine which has mucosal cells that are about 20 μm in diameter. I have used a Kopf microelectrode advance in which you know the distance which your microelectrode is moving. I have not made deep penetrations into the cells but I have just looked for the mucosal membrane potential. In this tissue I find data similar to bull frog in that the mucosal membrane potentials are around 40 mV, maybe higher. I do not know whether there is a gradient of potential but there is definitely more than 5 mV in these amphibian preparations. Also I would like to ask if you feel there is a 5 mV potential difference in the mucosal membrane and something of the order of 40 mV at the serosal membrane. Don't you need to suppose that the resistance between the cells is rather high and doesn't that differ from data that have been recently accumulated on these extracellular shunts?

Zeuthen: Firstly I have focused on the longish cells of the warm-blooded animals because I think it is partly the geometry of the cells that causes the gradients. However, in the case of amphibians there have also been reports of small mucosal membrane potentials (*Wright*) as well as high (*White* and *Armstrong*). The most reasonable explanation of the gradients at the moment is that of a diffusion potential; where you have to assume only 5–10 times less mobility of the sodium ion compared to the potassium ion and partly fixed anions. The extracellular shunt in the rabbit has only been demonstrated *in vitro* where the adsorbtion is four times less than *in vivo*, so it might arise from experimental damage.

Thomas: Did you repeat your potential measurements with the sharpest possible single-barrel microelectrode? With the double-barrelled electrode you may be causing more damage than you realize.

Zeuthen: The double-barrelled electrodes I used had tip diameters of roughly 0.3–1.0 μm. In assessing the damage caused by the electrode, it is important to note that the same gradients were recorded whether the electrode was advanced from the luminal or the serosal side.

Walker: Just a point of information. You probably mentioned it but I overheard it: What did you have perfusing the mucosal side?

Zeuthen: I tried various solutions in the lumen since one never finds *Krebs* solutions in normal gut. Gradients in the cells are present if for instance one uses 75 mM $NaHCO_3$ and 75 mM NaCl in the lumen.

Gebert: You have given an explanation for your sodium transient which I think is not quite correct, because though the response of the sodium itself is slow, if you have a DC-potential, it will respond fast, provided that your electronic equipment is all right. But there might be another thing you have overlooked. With the sodium electrode you have the difficulty that it responds transiently to hydrogen ions and if you have a decrease in pH you will find a transient response of the sodium electrode. About 4 years ago we observed transient potential changes of sodium-ion electrodes at the surface of pial arteries during respiratory acidosis and we almost thought that it was a sodium change, but that was just a pH effect, and that might be the same in your experiments.

Zeuthen: Due to the noise problem, the time constant of my sodium electrode plus the electronics was 5 second higher than that of the DC-electrode (0.1 seconds) which makes the subtraction in the fast-changing phases incorrect for the sodium-electrode.
Your second point: the pH-changes during anoxia do not exhibit a jump the way potassium and chloride ions do. So I don't think the jump on the sodium-ion reading is due to a hydrogen-ion change.

Eisenman: A comment to this same point. Since this sodium transient occurred simultaneously with your sharp change in the potassium concentration, it is reminiscent of Sidney Friedman's observations of a transient potassium response of the sodium electrode.

Zeuthen: I don't believe either that the sodium is recorded correctly during the onset of the depolarization, but I do believe in the sodium readings in phases where the DC and the ions change only little.

Kessler: You mentioned that either ions can move or water can move. I have some difficulties in understanding how water movement can take place without ions or other osmotically effective substances. You mentioned also that you think there is some swelling in anoxia. Did you make any electron microscopic studies or other studies to find out how big the swelling may be?

Zeuthen: Water movements are caused by changes in intra- and extracellular osmolarity caused by changes in metabolic end-products. Van Harreveld demonstrated that there is cell swelling in the brain during anoxia, but the question is also when it happens. The swelling might happen even before the depolarization, as was indicated by Reneau's model.

Lübbers: Certainly the blood flow changes also affect the impedance measurements and I think under these conditions of anoxia there are also considerable blood flow changes. It is difficult to differentiate between these effects.

Zeuthen: Impedance measurements are suspect but seen together with the ion movements, they seem to fit.

Discussion of Paper by J. L. Walker and R. O. Ladle: Intracellular Chloride Activity in Heart Muscle

Thomas: Did you try raising the external potassium at all? In snail neurons for example you can have all kinds of funny effects on the chloride when you change the potassium?

Walker: No, we run these at a potassium concentration ot 2.5 mM, which is the standard recipe for frog Ringer's. Replacement with glucuronate does not appear to change activity coefficients appreciably.

Discussion of Session IV B

Hinke: What is the actual magnitude of potential change in the chloride electrode when you go from outside to inside? Before you start making any gradient calculations – just the straight experimental change from inside to out.

Walker: It is about -50 mV, in these experiments, until you start the chloride replacement, and then the absolute value of the change becomes larger.

Discussion of Paper by J. F. White and J. A. M. Hinke: Use of the Sodium Microelectrode to Define Sodium Efflux and the Behavior of the Sodium Pump in the Frog Sartorius

Eisenman: A very minor point: Since, thanks to *Willi Simon* and his colleagues, we no longer are restricted to only one kind of sodium-selective electrode it would be a good idea to identify sodium electrodes as being of the "glass" type or of the neutral ligand type. They are not totally identical and they should not be confused in the literature. This is not totally trivial because some of each type of electrode can serve as a control on the others, particularly when you encounter transient responses. I think for some period of time one will have to perform controls on the liquid ligand sodium electrode with the conventional glass sodium type electrodes.

White: I use the Thomas-type electrode and I thought I had made that clear. Corning NAS 11-18 glass was used inside pyrex insulating glass.

Thomas: I would like to congratulate you on your beautiful results.
A graduate student of mine has managed to get two or three results on frog muscle which look similar to yours though not nearly as good, and many more results on crab muscle which is extremely easy to work with. With the crab muscle the situation seems to be that the ouabain-sensitive fraction of the sodium efflux when you go to low sodium is almost invisible.
Do you have any view on what the ouabain-insensitive sodium efflux was? My guess is that it is a sodium-calcium exchange.

White: The best that we can do at the moment is to suggest that it is a passive loss of sodium. We do not know what alterations in bath calcium would do to the sodium flux. As I mentioned, our calculations show the loss to be of the same order of magnitude as sodium influx so we have chosen to define the loss as passive.

Thomas: We get the same effect even if you have sodium reduction to only 10%, so it could not be a net passive efflux.

White: I did a few experiments in low sodium and got a decline but I am afraid I did not carry those out long enough to see whether there was a slow component.

Discussion of Paper by R. N. Khuri: Intracellular Potassium in Single Cells of Renal Tubules

Hinke: If I understood you correctly, you said that most cells, including muscle, have no binding of potassium. There is already some evidence for potassium binding inside cells. May I also suggest that if you find potassium binding in your proximal tubule you will probably also find significant sodium binding as well.

Discussion of Session IV B

Khuri: I have done some work on the intracellular potassium of rat muscle under a wide variety of states. The entire chemical potassium content of about 160 mM can be sensed electrometrically.

Hinke: I am not just referring to my work. Have you seen, for example, the work by *Lee* and *Armstrong* (J. Membrane Biol. 15, 331–362 [1974]; Science 171, 415 [1971]).

Khuri: No I have not.

Discussion of the Paper by Hirche, Steinhagen, Hosselmann, Manthey and Bovenkamp: The Interstitial pH of Isolated Skeletal Muscle of the Dog at Rest and During Exercise.

Thomas: I would call these mini rather than macro electrodes. How did you measure the lactic acid? Did you take samples?

Hirche: Lactic acid release from the working gastrocnemii was determined according to the Fick principle. Lactic acid concentration in the muscle tissue was measured by taking biopsies using a special method (*Hirche* et al.: Pflügers Arch. 343, 89, 1973).

Betz: I saw in your first slide that blood flow and interstitial pH did not correspond, at least not in the initial phase. Did you never see initial decrease in flow but always a sudden increase, despite the change in pH?

Hirche: We could never observe a relationship between the interstitial pH and muscle blood flow, neither in the initial phase nor under steady state conditions; this is also true for the experiments with severe metabolic acidosis and alkalosis.

Discussion of Paper by G. Gebert: Tissue Electrolytes and Blood Flow in Skeletal Muscle

Hirche: You observed a dependence of the acidosis in the interstitial fluid on your stimulation frequency. I think the explanation is that if you use low stimulation frequency only a very small amount of lactic acid is released and vice versa.

Gebert: That is quite right because if you use a stimulation frequency of 0.5 Hz or below you normally don't see any acidosis at all. With 0.5 Hz you can obtain almost the maximum blood flow if the load is big enough; but then of course you might get a slight acidosis too.

Discussion of Paper by R. L. Coleman: Recent Advances in Areas of Clinical Utility and Instrumentation for the Determination of Free Calcium in Body Fluids

Baum: Two questions. One, I do not understand the flowing junction of your external reference electrode. Secondly, how do you calibrate? What do you use for standard in flowing blood?

Coleman: For the blood measurement I use 15 mM sodium chloride with 2 mM calcium chloride and a little bit of tris to keep the pH from changing.

Baum: But you need a blood reference.

Discussion of Session IV B

Coleman: No, I do not I need a reference. The values I obtain from blood are in good agreement with determinations by other complicated but reliable procedures. We are getting very close to the real values here.

Baum: What about pH? How sensitive is your system to pH changes and how important is it to run anaerobic samples?

Coleman: Do you mean the calcium electrode itself?

Baum: Well, when you want to do free calcium.

Coleman: The normal pH for blood serum is about pH 7.4. We have found, as I pointed out the other day, that the selectivity of our exchanger can be made somewhat better than we thought it was by employing better membrane materials. We have been measuring for two years now, total calcium by proton displacement in the system. That is the approach that we use and we can operate at constant pH down at pH 3.5.

Baum: What about the flowing junction?

Coleman: I wish I knew the reasons why it works so well. I think I have said about as much as I know about it.

Fuchs: I refer to Dr. *Baum*'s question. When there is a question about pH I would say one has to consider two things. One is that you have to measure anaerobically and that means that this must be a very suitable machine to make these measurements. The other thing is that there may be some pH interference, but this is quite independent of the need for anaerobic measurement. Another thing I want to mention is that I think these whole blood measurements are really very important. We did some preliminary experiments with commercially available sodium, potassium chloride and calcium ion electrodes and these gave quite correlation coefficients. This is the big advantage of the ion-selective electrode measuring whole blood. You do not need to centrifuge the blood and I think in emergency situations this can be a great help.

Simon: From your sketch on the board I must say that I still do not understand how the liquid junction really is made. Would you please comment on that?

Coleman: Essentially I have a flow of KCl into the sample stream, radially at 360°, so there is a very instantaneous mixing of it and I pump it in under some pressure.

Baucke: Is there turbulent flow in your system?

Coleman: There is very turbulent flow, that was the idea. It turned out that under these conditions I can start and stop the pump during measurements and I get minimal charge on this junction. It has real advantages for biological application that it cannot clog, it cannot be affected by protein.

Lübbers: Such measurements have been tried before with a stream of KCl and blood.

Coleman: Yes, it turns out that just putting in a stream of KCl does not do.

Hebert: I spoke with Dr. *Drott* about two weeks ago and he had done a very extensive study of calcium electrodes and their applications in serum and whole blood and he insisted that these could not be used in whole blood. What is the trick that allows you to make these measurements in whole blood?

Coleman: The electrodes that he was using were the old variety of the 1920's. He does not have this system as yet. There is a lot of dispute in the literature about the whole blood response characteristics system. I am not sure if all the things that were attributed to whole blood were really the effect of the whole blood. After all if you measure pH in whole blood on blood gas basis then calcium is just another ion. I think we have cleaned up the reference anomalies in terms of the red cells which have a respectable charge on them; there may have been some quivocating response on first exposure of the red cells to establish that junction. They are very large masses and they do not move easily and they are charged.

Discussion of Paper by C. Fuchs: Ion-selective Electrodes in Medicine.

Betz: I find it difficult to understand that during hyperventilation you found a change in the calcium activity in the serum but not in the CSF. The association should change and therefore there must be an additional process if calcium remains completely constant.

Fuchs: Well that is what we found. It may be explained by the low proportion of protein-bound calcium in the cerebral spinal fluid, because it is a very complicated thing in the serum. There you have the three calcium fractions and shifts may be seen in the serum. Even slight shifts of pH changes may result in a big difference of ionic calcium in the serum. This must not take place in the cerebral spinal fluid when there is a low proportion of protein bound calcium. This might be an explanation. I just wanted to present this result and I think as it is very easy to measure the cerebral spinal fluid ionic calcium I do not have any doubts.

Clark: I agree that measurements in whole blood are better, but there is one word of caution and that is illustrated by the case with glucose, where the human has about 70% as much glucose in his red cells as in his whole blood. That is also true for other primates. But all non-primates have no measureable glucose in their red blood cells, so that if you get a reading of 50 mg/% in a dog for example, that is more like 100 mg/% in the plasma.

Fuchs: You are quite right. I postulated this whole-blood measurement only for electrolytes, not for glucose and the other enzymes. This is quite a different thing.

Clark: But in the case of calcium one assumes that the concentration in the red cell is the same as in the plasma.

Fuchs: No, not at all. It is much lower.

Clark: Well, then I do not see how you could have the agreement that you had, but it looked to me as if whole blood and plasma agreed as to the calcium.

Fuchs: The big advantage is that when you make a potentiometric analysis of whole blood, you do not destroy the erythrocyte membrane. You really measure the activity of plasma and this is the real theoretical advantage, and practical advantage, too as you can see in the next slide (Figure 2) the correlation between whole blood and serum is quite good.

Clark: I understand what you are saying, I agree with that. This has to do more with sample size, because if you take a sample of dog blood the red cells are in effect glass beads so therefore you have half as much substance there. It has more to do with content than it does with activity.

Discussion of Session IV B

Sample	Na⁺ [mEq/l] Plasma	WB *	K⁺ [mEq/l] Plasma	WB *	Ca⁺⁺ [mEq/l] Plasma	WB *	Cl⁻ [mEq/l] Plasma	WB *
1	143,6	143,6	5,79	5,85	1,93	1,94	110,9	110,1
2	151,5	149,2	6,30	6,11	2,29	2,20	119,2	118,8
3	150,3	149,8	6,01	5,83	2,24	2,33	113,0	111,7
4	159,2	161,7	6,01	5,94	2,47	2,41	123,0	123,9
5	163,6	160,5	5,42	5,50	1,62	1,70	135,9	132,8
	r = 0,961 y = 0,958x + 5,788		r = 0,962 y = 0,656x + 1,971		r = 0,976 Y = 0,851x + 0,320		r = 0,991 Y = 0,929x + 7,567	

Fig. 2

* WB = Whole Blood

Fuchs: I think if we compromised and talked about plasma activity, you would probably agree with me. You must not forget that even if the sample is whole blood, the measurement result refers to the ionic activity in plasma.

Coleman: Two comments. First of all, Dr. *Fuchs'* explanation about protein is probably reasonable since protein provides about 90% of calcium buffering capability in the serum. Secondly, I do not think you can draw parallels between the glucose situation in red cells in whole blood versus the other measurements and in the electrolytes, because we know for example that in the preparation of serum for potassium assays there is an artefactual loss of potassium out of the cells into the serum once the red cells get below about 30 °C. So this is an error as a result of preparing the serum, and if you avoid stressing the cells you can get some very valid measurements on whole blood provided you don't go through the storage gains. The idea is to measure it at the site as opposed to carrying them.

General Discussion

Dora: I would like to ask Dr. *Kessler* whether he measured creatin phosphate in the liver when he had either kind of anoxia, a non-flow or the normal anoxia.

Kessler: No, we have not done that.

Dora: According to Siesjö you have an early creatin phosphate decrease in brain and your ATP changes after about 1½ minutes.

Kessler: Yes, of course, in the brain the fraction of creatin phosphate is extremely important, whereas in the liver it does not have this significance.

Simon: Did you use a solid internal reference with the electrode you just described for sodium?

Sinagowitz: Yes, I got it from Dr. *Kessler*.

Kessler: It was the electrode shown in Jens Höper's paper, it has a diameter of 5 mm.

Discussion of Session IV B

Fuchs: Why do you think you can only use the electrodes in organs which have to be removed?

Sinagowitz: We have to remove the renal capsule and that would be an additional trauma if we perfomed an operation on a normal kidney.

Lübbers: We have certainly seen that the use of ion-selective electrodes, enzyme electrodes and also photometric assay methods have considerable advantages in application, but we were also aware that there are shortcomings and possible artefacts in these methods.

Eisenman: First I would like to congratulate both *Coleman* and *Fuchs* because their most impressive studies clearly show that one can use existing kinds of electrodes in exactly those conditions necessary for making clinical measurements, which is not the conclusion that most people would reach from reading the existing literature based on many people's poor experience with the infererior quality electrodes often available commercially.

Stockem: I would like to make one comment. It is true that you should try to make the electrodes as small as possible. However, one can also suggest that this will not solve all the problems which have been reported in this workshop. It is much easier for the cell to defend itself against a small electrode than a large one. In addition to this the possibility of meeting the nucleus or a big vacuole with a very small electrode is rather high. As these compartments may differ in ion concentration from the remaining cytoplasm, the exact localization of the tip of the electrode seems to be very important. I think the main problem will be to develop electrodes with special surfaces, which can prevent the cell from building new membranes at the tip region but do not influence normal cell activity.

Lübbers: But is it proven that the mammalian cell reacts in the way you have shown for the slime molds and amoebae?

Stockem: If you introduce an electrode into a big cell, the ratio of the cell surface or the cell volume to the hole which is made by the electrode may be 1 : 100 or even more. Hence the amount of membrane material stored within the cell in order to repair such damage may be large enough to close the hole very quickly. But if we have a small cell, this ratio normally is in the range of 1 : 1 or 1 : 2, so that a small cell is not capable of closing such a hole at all. This may explain the fact that, in contrast to many other smaller cell types, up to now it has not been possible to get any normal values in electrophysiological experiments with slime molds. On the other hand, if a small cell cannot repair the damage caused by the electrode, this may mean that it cannot survive the experiment, or in other words, that it is going to die very soon. From this point of view I would like to ask you if it is possible to exclude that some of the values which have been reported here have been taken from dying or even dead cells?

Lübbers: From electrophysiological measurements people know very well that, when they put a micropipette through a cell membrane, they can measure the resistance of the membrane. I think the problem is different for different cells and we have to look at this in the future carefully for each special cell. I think it is good that we have shown so clearly what can happen and that it is not safe just to assume that nothing happens.

Eisenman: I would like to make a specific suggestion to Dr. *Stockem* that if he uses a non-polar microelectrode (e.g. polyéthylene) the membrane might be penetrable since the hydrocarbon interior of the membrane would be compatible with the polyethylene and the slime mould would never know that it had been penetrated. This goes back to the Langmuir-Blodgett way or orienting lipid tails versus polar head groups when penetrating with hydrophobic versus hydrophilic surfaces.

Clark: Perhaps you could put an enzyme on the end of the micro electrode and this would be a biochemical cutting edge.

Baum: I was just going to make a suggestion along those lines that the enzyme collagenase could be used to strip away connective tissue. Perhaps those people who are working on penetrating cells and the problem of whether one is measuring cellular electrolytes or interstitial electrolytes might use collagenase to strip away connective tissue. A naive suggestion from an organic chemist!

Buck: Several years ago I learned that it was necessary to cover an ion selective electrode with some kind of dialysis membrane or cellophane in order to make measurements on whole blood. Do I understand from today's comments that we were being over-cautious in those days, in that protection is no longer necessary?

Coleman: Yes, but it depends on the ion-selective membrane and the analytical cycle. The disadvantages of the cellophane approach are longer electrode response times and the fact that the protective membrane can also be adversely affected by long-term exposure to blood samples. For the calcium electrode, I have found it was more effective to perform direct measurements on a moving blood sample with a programmed cleaning and standardization cycle after each measurement.

Simon: My comment is just the opposite. I do not think that you can generalize. I know that the Radiometer Group in Copenhagen is trying to determine potassium in whole blood and they are now highly successful. However, they have to use a membrane in between the ion-selective liquid membrane electrode and the whole blood. I think someone is here from Radiometer and he might comment on that.

Jensen: We have done some studies on calcium electrodes and potassium electrodes and we have found protein effects on both, so I am not confident that you will have the same calibration, when you standardize with a pure solution and use it on plasma afterwards. I can only confirm that we prefer membranes on the electrodes.

Fuchs: This is an old problem. We discuss this at every ion-selective electrode conference. The problem is how can we prove that ionic calcium is measured accurately when we use a protein-free standardizing solution. This is extremely difficult. You cannot make any recovery experiments, you do not have any reliable reference methods, and you cannot make any dilution experiments. This basic problem is just a problem of the standardizing solution. You will not solve that problem by using another membrane.

Coleman: I would like to add that a review of the literature over the past twenty years indicates that protein effects on electrode membranes have been a "catchall" explanation for poor results in clinical samples. Other factors, such as the reference electrode and the plumbing leading to the measuring chamber must also be considered and eliminated as possible sources of error.

Jensen: The indication for seeing this protein shift is a long-term drift of the electrode, if you do not have it covered by a dialysis membrane. So when we change from a protein solution to a pure solution, we would see a long-term drift of the calibration. Have you similar indications, Dr. *Fuchs*?

Eisenman: Directly to this point. I would like to reiterate the plea that since we now have several classes of elctrodes selective for a given ion, it becomes important to identify the particular class of electrode used – even designating the solvent and the carrier and the batch number and the glass composition. By comparing data from one electrode type to another one can often estimate such a drift or "poisoning" problem. This is because what leads to trouble with a valinomycin K^+ electrode in a particular solvent will not be what leads to troubles with a liquid ion exchanger K^+ electrode or with a glass K^+ electrode. Also K^+ electrodes of two different glass types will show different transient responses, which will also allow you to sort this out. One of the things the manufacturers and retailers should do from now on is to identify the compositions of the electrodes they are selling. The researchers in turn should state not only the type of electrode (e.g. "NAS 11-18 from Corning" or "Beckman 78178") but should also indicate the date of which it was received. This is because as in wines one can find that 1964 was a good "vintage" year whereas 1966 was ghastly!

Lübbers: The only thing is as Dr. *Fuchs*' mentioned, the position with calcium is much more difficult than sodium and potassium because the question of ionisation in calcium is much more difficult to solve.

Fuchs: I was asked if we observed an electrode drift. I will explainit in Figure 3. This is an original recording of a reproducibility experiment measuring ionized calcium in a pooled serum. The procedure was as follows: we calibrated the recorder read-out. In the middle you see the zero potential point of the electrode corresponding to a 2 mEq/l aqueous standard. We measured a pooled serum. The observed overshoot is what has just been referred to as a drift. I would not say it is a drift because we did have a very stable potential after about 1 minute and this is not only a stable potential but also a very reproducible one. What you observe also is that you only find this overshoot when you come from an aqueous solution to a serum. It is reversible when you return to an aqueous solution. You do not find this overshoot if you come from 2 mEq/l aqueous solution to 4mEq/aqueous solution (upper part of the Figure).

We dicussed this overshoot often and as far as I remember *Rechnitz* has also discussed the overshoots using pH glass membrane electrodes and he explained this as an ion interference on the membrane. But we are not quite sure because this ion interference should be there in aqueous solution, too. My question is, could it possibly be a problem of the reference electrode?

Clark: I repeat that I do not believe you can have a glass pH electrode or an ion-selective electrode in contact with blood for a long period of time without it going down hill and in need of calibration. It has been routine practice for twenty years to alternate sample and standard for blood gases and pH.

Eisenman: With glass electrodes for example one can dip them through a monolayer of barium stearate which I have done, and after you run them through 8 or 20 times you can wreck the response by the multilayers you have plated on. The conditions for plating out of protein are very variable and it sounds to me quite reasonable that if you take precautions to

Discussion of Session IV B

avoid layering – on surface-active material or things that will build up that you can in fact get away with using these as *Coleman* and *Fuchs* have suggested and I do not think that is at all in conflict with what Dr. *Buck* or the Radiometer people have said. You have to avoid the buildup of things that can absorb to or can plug the holes in your electrodes; and it may very well be that the act of continually calibrating really reflected this by keeping the electrodes clean.

Coleman: I agree with Dr. *Clark's* statement that continous exposure to blood will adversely affect the analytical performance of any known electrode. One major difference in the way we have been doing things is the fact that this is not a static measurement. Surface nucleation sites do not have quite the chance to form with a continuously moving sample. This, combined with attention to such details as electrode to sample exposure time, a programmed scrubbing cycle for recovery of the original baseline, and automatic restandardization on every cycle has led to substantial improvements in long-term electrode performance with both blood and serum.

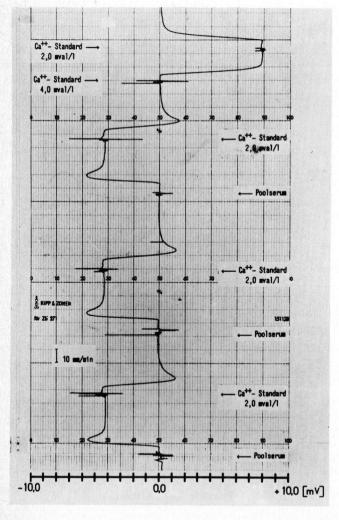

Fig. 3

Discussion of Session IV B

Lübbers: I think the discussion has made it clear how important it is to have different methods for the same ion or for the same experimental question. If you do not have the different tools for the same ion or for the same experimental question you can sometimes run into difficulties. I think to produce new tools for important scientific work is very worthwhile.
I thank everybody who contributed this afternoon to the scientific results of our conference and also to the discussion.

Kessler: Ladies and gentlemen, before we come to the end of this meeting in the name of the organising committee, I would like to express our thanks to all participants who made this Workshop really successful by their investigations, their presentations and their discussions. I hope there will be an opportunity for us to meet in about 2 or 3 years again and therefore I would like to say "auf Wiedersehen" to all of you.

Clark: On behalf of the participants and the guests I have been selected to sincerely thank all those who have worked so hard to make this conference the success I really think it is. I have attended a few conferences and workshops and this one is near the top for exchange of ideas and information. And if we add the environment to it, I believe it is at the top.
There has been good contact between physiologists, neuro-physiologists physical chemists, enzymologists and engineers, between academy and industry.

Index

acid-base balance 368, 370
acid diffusion 269
acidosis 338
aldesterone 370
amebae 205 f
ammonia uptake 264
ammonium ions 264
c-AMP 247
anaerobic glycolysis 332, 336
anoxia 264, 335
anoxic depression 295
Aplysia 283
asolectin 4
ATP-ase 317, 338
ATP-depletion 335

Ba^{2+}-ligand 31
bicarbonate accumulation 288
– injection 291
brain 191
– cortex 294, 299, 311
– metabolism 294
bulk activity 51
t-butylhydroperoxide 262

calcium activity, cytoplasmic 238
– current 309
– determination 391
– ions 237
– transport 239, 341
carrier 182 f
catecholamines 251
cation exchange 266
– migration 79
cerebrospinal fluid 316, 322
chaos chaos 205 f
chloride 73 f
– activity 351
– permeability 353
cholesterol ester hydrolase 163
– oxidase 163
choline ester 194
CO_2 290 coenzyme, immobilized 187
complex-formation-constant 24
connective tissue 191
constant field equation 284
colometric titration technique 59 ff
cyanide 65
cytochalasin B 244
cytochrome 341

DC-potential 294, 299, 311, 322
dehydrogenase 175
derivate technique 179
dielectric constant 26
dielectrics, thin layers of 110 f
differentiator 195
diffusion 51 f
dinactin 4
2,4-dinitrophenol 240
DMO 336

EDTA 317
EGTA 317
electrocorticogram 294, 299
electrode acetylcholine 193
– calcium 381
– cholesterol 161 f
– clinical use 389
– enzyme 187
– fluoride 61
– glassy carbon 174 f
– glucose 198
– internal reference 106
– liquid membrane 22 f
– miniature glass 131
– miniature reference 132
– multilayer 113
– pH 322, 336
– pO_2 336
– reference 200 f
– resistance 110
– „second kind" 38 f
– surface 331, 398
electron-exchange 41
electron microscopy 205, 217
enzymes immobilized 182
– oligomeric 182
equilibrium 39 f, 51
– domain 4 f
Eriochrom Blue 252
exocystosis 251
extracellular space 333, 357

field strength 18
flame photometry 390
formation constant 42
frog ventricle 351

glucose 189
– diffusion 269
– oxidase, immobilized 198
– transport 237

glutamate 187
glyceryl dioleate 5
– oleate 5
gramicidin A 16

heart muscle 351
Helix aspera 288
hormonal control 237
hydrogen peroxide 161
hypoxia 313, 341

ileum 345
injection technique 59, 63 f
insulin 243
integration method 176
interstitial ion activities 138
– space 374
iodide 65
ion-exchange (r) 41
– – liquid 116 f, 119 f
ion-selective glass 123
ion sputtering 77 f
– transport 270
ionic contacts 42 f
ionophores 251

K^+ excretory rate 368
kidney cortex 366
– human 398

layer 80
ligand Ca^{2+} 26
– cation selective 52
– ion selective 23 ff
– Li^+ 32
– Na^+ 30
lipid bilayer 3
Li^+ 77

macrotetralide actins 7
magnesium 252
membrane regeneration 205 f
– potential 283, 289, 319, 332, 352
– solid state 38 f
– stabilizers 244
– thickness 82
metabolism, intermediary 237
metallochromic indicator 252
microelectrode 133
– bulb-type 376
– Ca^{++} 117
– Cl-sensitive 143
– double-barrelled 127, 294, 299, 302, 345, 365
– flow-through 123
– four-barrelled 136
– glucose 189

– liquid-ion-exchanger 126
– multi-parameter 119
– Na^+-glass 141
– pH 77, 103 f, 110 f, 124, 143, 372, 375
– recessed tip 141 f
– resistance 144
– storage of 82, 146
micropipette 116
– sealed type 123
micro-spectro-photometry 252
mitochondria 256, 341
monactin 4, 27, 262
murexide 341
myofibrils 241

NADH 173, 294, 336
– amperometric measurement 173
Nernst-Planck equation 348
neuron 283, 288, 302
no-flow-anoxia 331, 335
nonactin 4, 27, 262
norepinephrine 323
normal-flow-anoxia 335

ouabain 359
oxidative phosphorylation 342
oxygen consumption 313
– diffusion 268

papaverine 326
permeability coefficient 333
– ratio 3, 284
pH interstitial 372
– intracellular 288, 336
Physarum polycephalum 205 f
plasma membrane 205 f, 336
P_{O_2} 313, 398
potassium balance 368, 369
– current 307
– intracellular 364
– interstitial 378
potential, interfacial 39 f
– liquid junction 200
– standard 43
– streaming 139
puller, electromagnetic 137

rat, isolated perfused liver 217, 261, 331
– kidney 366
– liver cells, alternations in 217
regulatory ion 237
renal tubules 364
response dynamic 51 f
– time 54, 114, 146

salt bridge 201
sartorius muscle 355

421

Index

selectivity 3f, 23, 194
shock 335
silane 117
sinusoids 219
skeletal muscle 372, 376
slime molds 205f
sodium efflux 358
spreading depression 299, 311
sputtering technique, RF- 103, 110
squid axons 256
stoichiometry 39f
streaming solutions 56f

tetraphenylborate 34, 331
Thalamid® 203

tissue alteration 219
trinactin 4
tyrosinase 179

vacuoles 212, 220
valinomycin 9, 25, 251, 262
– hexa deca 12
– HyIv 10
– lac 10
vascular construction 317
– resistance 320
– smooth muscle 316
vesiculation 212
voltage-clamp 283, 306